ADVANCES IN THE BIOSCIENCES

Volume 46

NEUROBEHAVIORAL METHODS IN OCCUPATIONAL HEALTH

ORGANIZING INSTITUTIONS

Institute of Occupational Health "Clinica del Lavoro L. Devoto" of the University of Milan (Clinic of Occupational Diseases and Department of Preventive Medicine), Milan, Italy.

Center for Occupational and Environmental Health, the Johns Hopkins University School of Hygiene and Public Health, Baltimore, U.S.A.

Institute of Occupational Health, University of Bari, Bari, Italy

NEUROBEHAVIORAL METHODS IN OCCUPATIONAL HEALTH

Proceedings of the International Symposium on Neurobehavioral Methods in Occupational Health (State of the Art and Emerging Trends) — Como and Milan, Italy, June 23–26, 1982

Editors:

RENATO GILIOLI
MARIA GRAZIA CASSITTO
VITO FOÀ
University of Milan, Italy

PERGAMON PRESS

OXFORD · NEW YORK · TORONTO · SYDNEY · PARIS · FRANKFURT

U.K.	Pergamon Press Ltd., Headington Hill Hall, Oxford OX3 0BW, England
U.S.A.	Pergamon Press Inc., Maxwell House, Fairview Park, Elmsford, New York 10523, U.S.A.
CANADA	Pergamon Press Canada Ltd., Suite 104, 150 Consumers Road, Willowdale, Ontario M2J 1P9, Canada
AUSTRALIA	Pergamon Press (Aust.) Pty. Ltd., P.O. Box 544, Potts Point, N.S.W. 2011, Australia
FRANCE	Pergamon Press SARL, 24 rue des Ecoles, 75240 Paris, Cedex 05, France
FEDERAL REPUBLIC OF GERMANY	Pergamon Press GmbH, Hammerweg 6, D-6242 Kronberg-Taunus, Federal Republic of Germany

First edition 1983

Library of Congress Cataloging in Publication Data

International Symposium on Neurobehavioral Methods
in Occupational Health (State of the Art and
Emerging Trends) (1982: Como and Milan, Italy)
Neurobehavioral methods in occupational health.
(Advances in the biosciences; v. 46)
Includes index.
1. Central nervous system—Diseases—Diagnosis—
Congresses. 2. Occupational diseases—Diagnosis—
Congresses. 3. Neurotoxic agents—Congresses.
4. Neuropsychology—Congresses. 5. Neurologic
examination—Congresses. I. Gilioli, Renato. II. Cassitto,
Maria Grazia. III. Foà, Vito. IV. Title. V. Series.
[DNLM: 1. Nervous system—Drug effects—Congresses.
2. Behavior—Drug effects—Congresses. 3. Environmental
exposure—Congresses. W3 AD244 v. 46 / WL 100 I609n 1982]
RC365.I6 1982 616.8 83-13466

British Library Cataloguing in Publication Data

International Symposium on Neurobehavioral Methods
in Occupational Health (State of the Art and
Emerging Trends) *(1982: Como)*
Neurobehavioral methods in occupational health.—
(Advances in the biosciences, ISSN 0065-3446;
v. 46)
1. Nervous system—Diseases 2. Neurotoxic agents
I. Title II. Gilioli, Renato
III. Cassitto, Maria Grazia IV. Foa, Vito V. Series
615.9 RC365
ISBN 0-08-029802-8

In order to make this volume available as economically and as rapidly as possible the authors' typescripts have been reproduced in their original forms. This method unfortunately has its typographical limitations but it is hoped that they in no way distract the reader.

Printed in Great Britain

Foreword

Enrico C. Vigliani

Past President, Permanent Commission and International Association on Occupational Health
Past Director of the Institute of Occupational Health "Clinica del Lavoro L. Devoto", University of Milan

I am very pleased indeed to introduce the Proceedings of the First Symposium on Neurobehavioral Methods in Occupational Health held in Como and Milan, which was organized by the Institutes of Occupational Health of Milan and Bari Universities and by the Center of Occupational Diseases of the Johns Hopkins University, Baltimore.

I think the initiative taken by the University of Milan to host this Symposium was most appropriate since work now demands a much higher activity of the psychological and nervous systems than the manual work of several decades ago.

Neuropsychological toxics may have an effect long before any clinical symptoms appear and the early detection of these effects, both on the nervous system and on the psychological state, is one of the most important tasks of modern occupational health practice.

Many studies are likely to show that occupational exposure to neurotoxics contributes significantly to the severity of several pathological states, such as early ageing and organic brain syndrome, which pose important public health problems and result in considerable expense to society and much personal suffering due to the decreased quality of life of the individual. In occupational health, there is therefore a growing awareness of the problems of deterioration of cognitive functions that may occur in subjects occupationally exposed to a variety of workplace chemicals. The core of the problem is the possible increase in neurological disorders among individuals still at work who, in most cases, due to improved hygienic standards at the workplace in industrialized countries with consequent achievement of low levels of airborne neurotoxic agents, may show clinical manifestations only at a later age or even after retirement.

I feel sure that the results of this Symposium will be highly rewarding and a useful aid to a better understanding of the subtle alterations which the type of work can produce in the neuropsychological system of the worker, and that they will, in addition, be a stimulus for the activity of all those involved in occupational health practice in particular and in public health in general.

Neurology and psychology have established a number of tests that can be used to assess the neuropsychological condition of each individual and the proper choice of these tests and their exact interpretation could lead to conclusions that could induce an improvement of the work methods and thus

to better health and welfare of the workers.
The results of this Symposium will, I am sure, have a great impact on occupational health all over the world.

Preface

In the first meeting on Behavioral Toxicology held in Cincinnati in 1973 by Charles Xintaras, Barry Johnson and Ido de Groot, the need was felt to periodically review the state of the art of the discipline and to make every effort to develop a common approach to the study of neurotoxicity effects in the working environment.
The proceedings of the Symposium of Como and Milan of June 1982 aim to meet some of the needs which arose from the discussion. The Como and Milan Symposium will be followed by a WHO-NIOSH workshop on the neurobehavioral effects of work-place chemicals to be held in Cincinnati in 1983.
We hope that the contents of the present volume will provide imput for that meeting and that, in the future, exchange of information in this rapidly developing discipline will take place more frequently.
The Editors wish to acknowledge the Patronage of the Office of Occupational Health of the World Health Organization, the International Labour Office, the Health and Safety Directorate of the Commission of the European Communities, the Scientific Committee on Occupational Psychophysiology, and Behavioral Toxicology of the Permanent Commission and International Association on Occupational Health, the Ministries of Education and Health of the Republic of Italy and the Italian Society of Occupational Health and Hygiene.
The Editors are particularly grateful to the Fidia Research Laboratories for supporting the Symposium and to the Personnel of the Research Organization for the assistance given in organizing the Workshop.
The Co-sponsorship of the Department of Health of the Regione Lombardia, the University of Milan, the Province of Milan, the Municipality of Milan, the Municipality of Como, and of the Istituti Clinici di Perfezionamento is also most gratefully acknowledged.
The substantial help given by Ms. Giovanna Radaelli, organizing secretary, and by Ms. Kathleen White for the English language revision of the manuscript is greatly appreciated.

THE EDITORS

Contents

RECENT BEHAVIORAL DEVELOPMENTS

CONCLUSIONS

PART TWO - APPLICATIONS OF METHODOLOGIES

APPLICATION OF SINGLE TECHNIQUES TO THE STUDY OF EXPOSED POPULATIONS

Welcome Address

G. Chiappino

Institute of Occupational Health "Clinica L. Devoto",
University of Milan, via S. Barnaba 8, 20122 Milan, Italy

The first occupational disease producing neurological and behavioral symptoms was described by Bernardino Ramazzini more than two hundred and fifty years ago.

The disease was observed very frequently in workers manufacturing mirrors with mercury in the island of Murano, near Venice.

Ramazzini wrote: "In Venice, on a island called Murano where large mirrors are manufactured,workers who use mercury show motor impairment . . . and with grim expression contemplate their sorrowful state in their mirrors. . . and detest the work that they have chosen".

After so many years, today we know that toxic damage of the nervous system may be produced by a great number of chemicals and we have various clinical and instrumental methods for detecting nervous damage.

However, our methods are sometimes inadequate for a number of reasons:

- First of all, alterations in the emotional state and other behavioral variations may represent the earliest detectable sign of the toxic effect on sensory motor and integrative functions, without modification of other parameters.
- On the other hand, emotional instability and more severe phychic and behavioral alterations may also be produced by completely different factors, that can act in a much more complex way than chemicals: such factors, that are linked to general aspects of work (as, for instance: work rhythms, shiftwork and consequent deprivation of sleep, problems in interhuman relations and many others) can exist at the workplace with or without chemical agents and can exist, of course, also outside work, in social life.
- We know also that the nervous system possesses a great functional reserve against potentially damaging situations (both toxic and psychological): the functional effects may then be masked, until certain individual limits are exceeded.

- Experimentally, in fact, cytologic or neurochemical modifications can be observed when functional damage is not yet detectable by neurophysiological methods or by studies of performance.
- Moreover, the nervous system can develop tolerance, adapting itself to a certain degree of negative situations: hence, for instance, function can return to normal during continuous exposure to a toxic.
- When functional alterations appear in the central or peripheral nervous system, they are always aspecific and qualitatively highly variable.
- In individual cases, however, aetiological interpretation of signs and symptoms can be particularly difficult.
- Finally, when working populations are considered, many selective situations may significantly influence the responses to the neurotoxic agents in both directions.

So in Occupational Medicine, the neurologist and the psychologist have to evaluate effects that are often the only visible last part of a chain of unknown biochemical and functional reactions.

The complexity of the problem is similar to what we are confronted with in the field of carcinogenesis.

As I said before, the development of neurophysiological and psychological techniques has provided us with various methods: some of these are now showing their limitations, others are quite new and have raised our hopes considerably.

I think that today we have to keep some considerations in mind:

- The detection of neurolesive effects of occupational factors always requires the study of multiple parameters.
- The probability of a satisfactory evaluation of the effects of factors that can be multiple on a very complex system characterized by a great functional reserve and tolerance capacity, increases with the increase in sophistication of the methods employed.
- We also need methods for controlling great numbers of persons, which, of course, have to be very simple, but without significant loss of sensitivity compared to the much more sophisticated programs followed in highly specialized centers (this is the most difficult practical aspect of the present problem in my opinion).

I am sure that in the latter years, considering the variety of the present needs, during your daily activity aimed at the study of neurological, behavioral and neuro-behavioral occupational diseases, you more and more acutely feel the desire of a mutual evaluation of experiences, of comparison of results and of a pause for meditation.
This is, in fact, the feeling that we experienced.
Much more in general, this is the type of human behaviour, when problems have grown to an excessive degree of complexity.

This Symposium is meant to be a pause for meditation, and their Organizers, the Institute of Occupational Health of the University of Milan, the Institute of Occupational Health of the University of Bari, and Center of Occupational and Environmental Health of the Johns Hopkins University of Baltimore have created a good occasion for laying the basis for future work.

The prompt and enthusiastic participation of the W.H.O., the International Labour Office, the Commission of the European Communities and the Permanent Commission and International Association on Occupational Health, and the most important Italian Authorities and Organizations in the field of Occupational Health and of Public Education strengthens our conviction that the time has really come for an exhaustive methodological discussion.

Our Institute wants to continue in the role of a pole of attraction as regards the problems and, at the same time, in that of a center of propagation of experience and knowledge; together with all other Institutions that have a similar approach, we can do something immediately useful, for example for the developing countries.
We are ready to accept requests and discuss practical programs and future contacts.

Now, ladies and gentlemen, let me welcome you on behalf of all the staff of the Institute of Occupational Health of Milan and also on behalf of Prof. Schiavinato, Chancellor of our University, who will be the host for the second part of the Symposium,and of Prof. Mantegazza, the Dean of our Medical Faculty.

I should also like to sincerely thank our host of today, the Mayor of Como, Dr. Spallino and, everyone who has made this meeting possible, overcoming unexpected difficulties.

I should like to thank you all who, by participating, have enhanced the significance of our initiative.

I am sure that the next few days will see important scientific developments and also that there will be precious contacts between scientists working towards the common objective of ever greater effectiveness in the solution of Occupational Health problems.

The Symposium is now open.

PART ONE
METHODOLOGY

The State of the Art

Chairmen: R. Gilioli (*Italy*)
B. Johnson (*USA*)

Neurological Examination and Assessment of the Syndromes caused by Exposure to Neurotoxic Agents

J. Juntunen

Neuroscience Group, Institute of Occupational Health, Helsinki, Finland

ABSTRACT

One inherent feature of nervous system dysfunction in toxic states is the great variability of symptoms and signs. This variability and the nonspecificity of the clinical picture make it extremely difficult to draw etiological conclusions on the basis of any single test or measure of nervous system function. This is particularly true in cases of chronic intoxication, which are the greatest challenge for the clinician in the field of occupational neurology. This review attempts to elucidate the advantages and difficulties arising in the clinical examination and assessment of neurotoxic syndromes on the individual level.

KEYWORDS

Neurology; neurotoxic syndrome; intoxications; nervous system; occupational neurology; occupational health.

INTRODUCTION

The neurological examination comprises detailed social and medical histories, clinical tests of the nervous system functions, and the critical evaluation of the results of all laboratory tests and ancillary examinations. In each individual case a comprehensive neurological evaluation of the symptoms and signs of nervous system dysfunction is necessary to assure that the correct diagnosis is made and that the proper treatment is given. There are several substrate-specific pathways by which neurotoxic agents exert their effects on the nervous system. Therefore, to a certain extent, there are specific clinical syndromes for some chemicals. Some of these chemicals mainly affect the central nervous system, whereas the effects of others are restricted to the peripheral nerves. In all instances, the early symptoms and signs are nonspecific; and because of high interindividual variation, there is no uniform clinical picture.

The role of clinical experience cannot be overemphasized in the evaluation of individual cases. It has also been stated that the clinician is the key to neuroepidemiologic investigations (Schoenberg, 1982). This article briefly reviews the problems arising in the clinical examination of neurotoxic syndromes. The main emphasis is placed on the clinical characteristics of chronic intoxications of the nervous system.

EXAMINATIONS OF THE CENTRAL NERVOUS SYSTEM IN TOXIC STATES

The clinical examination of the central nervous system in toxic states essentially comprises the detection of diffuse cerebral dysfunction, which is referred to as organic brain syndrome. From the diagnostic point of view, acute intoxications provide no problems because of the obvious temporal association between the (usually accidental) exposure and the resulting symptoms and signs. Chronic exposure to neurotoxic agents, which may cause slight organic brain damage is a very difficult occupational health problem.

Organic brain syndromes constitute a group of disorders in mental functions caused by permanent or temporary brain damage (Lipowski, 1980). According to the American Psychiatric Association's (1980) Diagnostic and Statistical Manual of Mental Disorders (DSM-III), organic brain syndromes can be divided into seven descriptive subclasses: delirium; dementia; amnestic syndrome; organic delusional syndrome; organic hallucinosis; organic affective syndrome; and organic personality syndrome. Damage to the central nervous system following exposure to toxic agents can lead to any of these syndromes. Therefore, a brief summary of their main characteristics is given here.

Delirium is a relatively acute organic brain syndrome characterized by the following: in milder forms a reduced capacity to shift, focus, and sustain attention on environmental stimuli (i.e., disturbances in cognitive functions), anxiety and increased motor activity, and restlessess. The full-blown forms also involve clouded consciousness and hallucinosis. Dementia is characterized by the loss of intellectual abilities and deteriorated cognitive performance (which may range from mild memory disturbances to total dullnes, without clouded consciousness). Amnestic syndrome refers to a focal affection or dysfunction of the neurons of the central nervous system responsible for memory functions, particularly the recent memory. This often leads to confabulation, which is usually accompanied by the individual's lack of awareness of the defect. Consciousness is not clouded. Organic delusional syndrome is characterized by the presence of delusions, but consciousness is not clouded. These symptoms are typically manifested in withdrawal syndromes (e.g., alcohol withdrawal). In organic hallucinosis auditive hallucinations are usually present, with no clouding of consciousness. Organic affective syndrome comprises disturbances of mood (i.e., manic or depressive episodes with no clouding of consciousness and no loss of intellectual abilities). Organic personality syndrome often involves emotional lability, impaired control of impulses, apathy, indolence and a paranoid imagination.

The clinical diagnostics of the above syndromes requires considerable experience, and psychiatric consultation is often necessary. The examination and scoring of mental status can be performed in many ways. The most important end point should be the correct impression of whether or not the patient has organic brain syndrome. In practice, the following items should be examined (Strub and Black, 1981):

Level of consciousness
-attention
-observation
-history
-digit repetition
-vigilance

Behavior
-physical appearance
-emotional status

Language
-handedness
-spontaneous speech
-comprehension
-object naming
-repetition
-reading and writing

Memory
-orientation
-short-term memory
-remote memory
-new learning

Constructional ability
-reproduction drawing
-drawing according to instructions

Higher cognitive processes
-proverb interpretation
-similarities
-calculation

The comprehensive assessment of the individual's mental state is laborious and time-consuming. Nevertheless it should be performed for each patient with neurotoxic syndrome as the first part of the clinical examination. This is because the interpretation of symptoms and signs in subsequent examinations is highly dependent on the results of this assessment. Particularly mild syndromes are problematic, and the diagnostic procedure often requires neuropsychological examinations.

After the assessment of the patient's mental state, the detailed examination of the other parts of the central nervous system is performed in the usual way. Toxic states may involve increased muscle tone together with tremor and exaggeration of the muscle tendon reflexes. Careful evaluation of both cerebellar functions and gait and station are particularly important - i.e., the careful evaluation of those central nervous system functions easily disturbed in diffuse toxic states (e.g., Juntunen et al, 1982). The clinical tests for cerebellar functions include tests for muscle tone, co-ordination, diadochokinesis, nystagmus, speech, ataxia, and balance. Gait and station should be examined in detail because the complexity of the neural elements regulating these functions makes gait and station the most susceptible targets for toxic agents.

One problem arising in the assessment of the central nervous system functions is the proper scoring of the findings. Since the experience of different examiners may vary considerably, a rather rough scoring method should be used (e.g., 0 = normal; 1 = slightly abnormal; 2 = clearly pathological). Severe chronic intoxications are rare in modern industrial communities with well-designed occupational safety programs, and the number of suspected slight cases is relative large. Thus the importance of clinical experience is strongly emphasized, as slight abnormalities are readily overlooked by the unexperienced examiner.

EXAMINATION OF THE PERIPHERAL NERVOUS SYSTEM

The selective vulnerability of the peripheral nervous system accounts for the varied clinical pattern in different toxic neuropathies (Thomas, 1980). The clinical examination of the peripheral nervous system in toxic states is mainly concerned with the detection of signs of polyneuropathy. In this context, it should be emphasized that isolated pareses and entrapment neuropathies are common in polyneuropathic patients. Although opinions differ as to how polyneuropathy should be defined, it is generally accepted that polyneuropathy is essentially a clinical diagnosis supported by neurophysiological, morphological, or other examinations in borderline cases.

From the clinical point of view, polyneuropathies fall into one of the four categories suggested by Thomas (1975): 1) acute severe polyneuropathies; 2) mononeuropathies; 3) subacute or chronic distal symmetrical sensorimotor polyneuropathy; 4) recurrent and hypertrophic polyneuropathy. These clinical types may be further subdivided on the basis of their rate of progress, their distribution, the type of fiber involved, their pattern, and their pathology. Most toxic and metabolic polyneuropathies fall into the category of subacute or chronic sensorimotor polyneuropathies (Thomas, 1980). This type of polyneuropathy has a characteristic pattern of development: sensory disturbances of stocking-glove distribution and diminished distal muscle tendon reflexes. Nerves lose their function according to their length, regardless of which of the peripheral nerves they traverse in their course to the limbs (Sabin et al, 1978). The typical distal proximal development of toxic polyneuropathies can be explained by the metabolic disturbances which occur in the neuron and can first be seen in the longest ones. Disturbances in the axoplasmic transport mechanism also lead to the disruption of the longest nerves (Spencer and Schaumburg, 1978). The random scattering of a myriad of lesions within a peripheral nerve leads to disturbances of the longest nerves first. In short, the probability of nervous dysfunction in toxic states increases with increasing nerve fiber length. The spatial distribution of typical sensory deficits and reflex abnormalities in various stages of polyneuropathy is seen in Fig. 1. Sensory deficit extends above the knee before the hand is affected (Fig. 1., A, B). When the deficit extends to the proximal parts of the extremities, a typical sensory defect, a "teardrop", is present over the abdomen (Fig. 1., C). The border between normally and pathologically sensitive skin is not sharp, in contrast to the border of hysteric sensory loss.

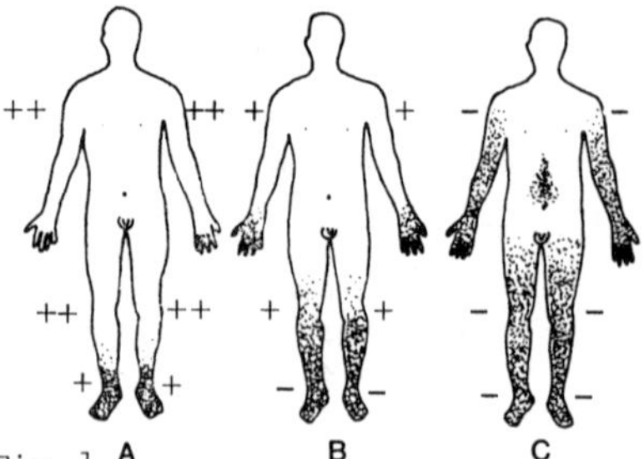

Fig. 1.

Three important prerequisites for the proper evaluation of the clinical findings in peripheral neuropathies are required: First, the conditions for examination should be optimal (a peaceful room, the patient's mental state reasonably clear etc.). For instance, if the patient is not co-operative, it is difficult to interpret the results of sensory tests, for example. Second, the clinical examination of all cases should include tests of the motor, the sensory, and the autonomic systems. Third, the clinical examination should provide data for the accurate quantitation of the severity of polyneuropathy.

The sensory tests at the very least should include tests for light touch, pain, vibration, and stereognosis. In skilled hands, these tests provide enough data to warrant the diagnosis of polyneuropathy. Sensory abnormality should be defined in proper terms, i.e., hypoesthesia, hyperesthesia, anesthesia, hyperalgesia, hyperpathia, paresthesia. Scoring these findings is difficult, and a rather crude scale is preferable (e.g., 0 = normal; 1 = slightly abnormal; 2 = clearly abnormal). During the past few years, many techniques and instruments for the quantitative analysis of sensory modalities have been developed, and they are very useful tools for population studies and follow-up studies. In practice, however, clinical examination reveals if the patient has sensory disturbances. The motor examination includes tests for muscle power, muscle bulk, and muscle tone as well as tests of muscle tendon reflexes. The interpretation of the muscle tendon reflexes requires experience, particularly if the patient cannot relax himself or if he has concomitant central nervous system disorders.

The muscle tendon reflexes are usually classified into five categories: absent; diminished; normal; slightly exaggerated; and grossly exaggerated. Autonomic neuropathy can best be assessed by a careful history of disturbances in cardiovascular, gastrointestinal, sudomotor, and genitourinal functions as well as by changes in orthostatic blood pressure and pulse. It should be emphasized that autonomic neuropathy not accompanied by concomitant sensorimotor neuropathy is rare. In practice, autonomic neuropathy should be sought only in patients with polyneuropathy and clear autonomic symptoms.

The most useful ancillary examination of patients with polyneuropathy is electroneuromyography (ENMG), which can usually detect processes and generic groups of disorders even though ENMG does not allow the identification of specific diseases (Asbury, 1980). Although sophisticated electrophysiological tests have been developed during the past few years and some of them are currently available even for routine study, conventional methods are sufficient for clinical practice on the individual level. Biopsies of nerves and muscles (though they may be very useful aids in problem cases) are seldom necessary in clinical routine (Ochoa, 1980).

In practice, the cases fulfilling at least two of the following criteria can be diagnosed as having polyneuropathy: 1) the patient has subjective symptoms compatible with polyneuropathy; 2) the clinical examination reveals diminished or absent muscle tendon reflexes, typical sensory disturbances or both; 3) electroneuromyographic examinations show at least one abnormally slow nerve conduction velocity in both the upper and the lower extremities, or two abnormally slow nerve conduction velocity in the lower extremity, and evidence of neurogenic muscle damage is found.

DIFFERENTIAL DIAGNOSTIC PROBLEMS IN OCCUPATIONAL NEUROLOGY

Wide interindividual variability, the selective vulnerability of the nervous system, and the great nonspecificity of the symptoms and signs in toxic states make differential diagnostics the most important step in the diagnostic procedure (Fig. 2.).

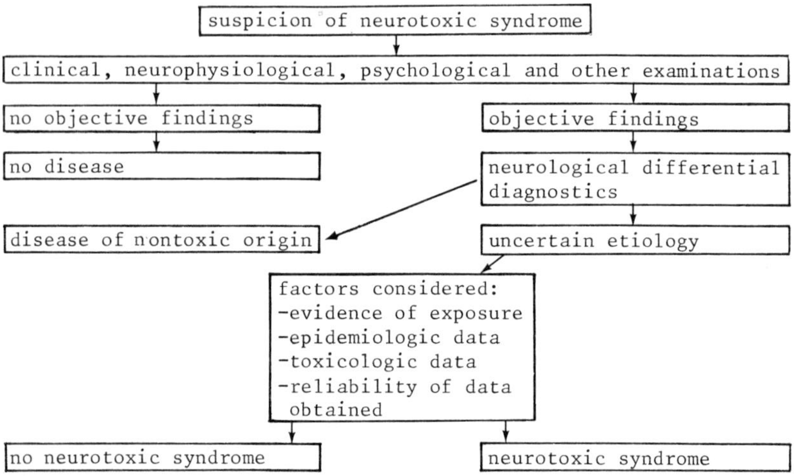

Fig. 2.

The diagnosis of either organic brain syndrome or polyneuropathy reveals nothing about the etiology. The differential diagnostic procedure often requires special examinations and laboratory assays (including sophisticated cerebrospinal fluid analyses, electrophysiological tests, neuroradiological, neuro-ophthalmological and other examinations). In some cases it is also advisable to follow the patients for some period of time, before the diagnosis is made.

As an example of the problems encountered in the individual diagnostics of neurotoxic syndromes, a list of reported possible etiological factors underlying polyneuropathy is given below:

Nutritional factors	Metabolic diseases	Drugs	Chemicals
Thiamine	Diabetes	Nitrofurantoin	Acrylamide
Pyridoxine	Cancer	Chloramphenicol	TOCP
Pantothenic acid	Amyloidosis	Chloroquine	Arsene
Folic acid	Rheumatic diseases	INH	Mercury
Niacin	Disturbances in lipid metabolism	Diphenylhydantoin	Lead
	Hypothyreosis	Indomethacin	Benzene
	Sarcoidosis	Hydralazine	Toluene
	Uremia	Gold	Xylene
	Porphyria	Levodopa	Styrene
	Hepatic disease	Metaqualone	n-hexane
	Tangier's disease	Vincristine	MBK
	Infections	Disulphiram	TCE
	Macroglobulinaemias	Ergotamine	CS_2
			Manganese
			Alcohol

These etiologic factors should be reasonably excluded for each patient with suspected toxic polyneuropathy. As far as chronic neurotoxic syndromes are concerned, ageing and the abuse of alcohol are perhaps the most common confounders in modern industrial societies. The coexistence of several etiological factors underlying polyneuropathy and organic brain syndromes is a well-known clinical problem still relatively poorly understood.

CLINICAL VERSUS LABORATORY EXAMINATIONS IN OCCUPATIONAL NEUROLOGY

There are inherent problems in judging the relative value of clinical and other examinations and test results. The dangers of excessive reliance on laboratory-generated data on the one hand, and the dangers of technologic nihilism on the other are well recognized in this era of rapid technological progress. It should be emphasized that neurotoxic syndromes often develop gradually via the progressive deterioration of the compensatory mechanisms of the nervous system. This is particularly true in chronic states. Therefore, the disturbances may be masked for a considerably long period of time. The stage at which the neurotoxic syndrome becomes clinically manifest depends on individual factors and on the examination techniques used. The examination techniques used highlight the core problem of neurobehavioral research in occupational health: the more sensitive the method, the more nonspecific the results. The situation is illustrated schematically in Fig. 3., where the maximal sensitivity (cutoff point A) and the maximal specificity (cutoff point B) of the test used affect the results. In the diagnostics of neurotoxic syndromes on the individual level, the comprehensive neurological examination and assessment of the test results, excluding other possible etiologies, yield the maximum specificity in its proper relationship to sensitivity. Unfortunately, laborious neurological examinations can rarely be included in large epidemiologic studies. Yet the two cornerstones of the inferences drawn from neuroepidemiologic studies are the representativeness of the population studied (which is an epidemiologic problem), and the accuracy of the neurological diagnoses within that population (which is a clinical problem) (Schoenberg, 1982). Because of the many aforesaid subtleties in making an early neurologic diagnosis, it is essential that researchers active in the field of occupational neurology be thoroughly familiar with clinical neurology.

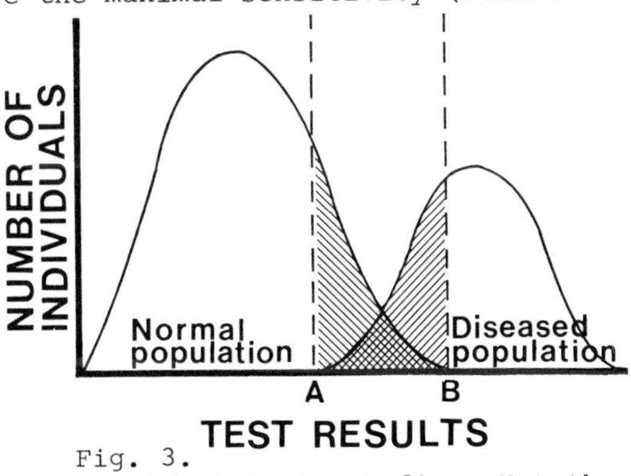

Fig. 3.

ACKNOWLEDGEMENTS

The author wishes to thank Ms. Sheryl Hinkkanen for revising the English manuscript and Ms. Tuula Moisio for expert secretarial assistance.
This study was in part supported by the Finnish Foundation for Alcohol Studies.

REFERENCES

Asbury, A. (1980). In: A. Sumner: The Physiology of Peripheral Nerve Disease, W.B. Saunders Company, Philadelphia, London, Toronto.

Juntunen, J., M. Antti-Poika, S. Tola and T. Partanen (1982). Acta Neurol. Scand., 65, 488-503.

Lipowski, Z.J. (1980). In: H.I. Kaplan, A.M. Freedman and B.J. Sadock: Comprehensive Textbook of Psychiatry/III, Williams & Wilkins, Baltimore, London.

Ochoa, J. (1980). In: P.S. Spencer and H.H. Schaumburg: Experimental and Clinical Neurotoxicology, Williams & Wilkins, Baltimore, London.

Sabin, T.D., N. Geschwind and S.G. Waxman (1978). In: S.G. Waxman: Physiology and Pathobiology of Axons, Raven Press, New York.

Schoenberg, B.S. (1982). Neuroepidemiology, 1, 1-16.

Spencer, P.S. and H.H. Schaumburg (1978). In: S.G. Waxman: Physiology and Pathobiology of Axons, Raven Press, New York.

Sterman, A.B. and H.H. Schaumburg (1980). In: P.S. Spencer and H.H. Schaumburg: Experimental and Clinical Neurotoxicology, Williams & Wilkins, Baltimore, London.

Strub, R.L. and F.N. Black (1981). Organic Brain Syndromes. An Introduction to Neurobehavioral Disorders, F.A. Davis Co, Philadelphia.

Thomas, P.K. (1975). In: R.J. Dyck, P.K. Thomas and E.H. Lambert: Peripheral Neuropathies, W.B. Saunders Co, Philadelphia.

Thomas, P.K. (1980). In: P.S. Spencer and H.H. Schaumburg: Experimental and Clinical Neurotoxicology, Williams & Wilkins, Baltimore, London.

Electromyography

A. M. Seppäläinen

Institute of Occupational Health, Haartmaninkatu 1,
SF-00290 Helsinki 29, Finland

ABSTRACT

Equipment used in electromyography is described. Spontaneous and voluntary electrical activity of the muscle is reviewed in normal and diseased state. Automatic analysis of the electromyographic signal and single fibre electromyography are briefly presented. Applications in occupational medicine are emphasized.

KEYWORDS

Electromyography; denervation activity; motor unit potentials; automatic analysis; single fibre electromyography.

INTRODUCTION

Electromyography (EMG) means recording of the electrical activity of a muscle and it has developed out of electrophysiological methods established during the past hundred years. EMG started to flourish after the Second World War and has since achieved widespread use in clinical diagnosis. A few authors applied EMG to occupational diseases in the 1960's, but largescale use in neurotoxicological studies and occupational health started in the early 1970's.

The development of electronics and microprocessors during the last decade has greatly changed the construction and appearance of electromyographs - devices specifically made for EMG - and also widened and improved the possibilities of studying various aspects of the peripheral nervous system. Computers with analog-digital converters have made an elaborate statistical analysis of the EMG signal possible.

EQUIPMENT

A basic EMG apparatus is composed of amplifiers, a cathode-ray oscilloscope and a loudspeaker, but it should always also contain an electrical stimulator for nerve stimulation. Several compact and relatively inexpensive commercial systems are available for this purpose. A more sophisticated apparatus is equipped with some kind of memory, preferably digital, an averager, and a photographic, electrical or UV recording system for permanent records. Some EMG devices are equipped with an X-Y writer, these are also suitable for evoked potential studies. Digital data acquisition and analysis possibilities are also offered in more expensive commercial electromyographs.

Diagnostic EMG is performed with a needle electrode. Concentric needle electrodes are the most widely used. They consist of a hypodermic type needle, in the core of which is cemented an insulated wire made of stainless steel or platinum. The wire is bare at the tip. The potential difference between the tip of the central wire and the outer cannula is recorded. Concentric needle electrodes are manufactured in various sizes, from small ones suitable for insertion in the facial muscles to larger ones needed for the study of deep paraspinal muscles. Multilead electrodes contain several insulated wires within a common steel cannula, and the recording surfaces of the wires are situated along the side of the cannula. Multilead electrodes are used for mapping of motor unit territories. A single fibre electrode has a small recording surface on the side of the cannula and thus records the activity of 1-2 muscle fibres only.

Needle electrodes have to be meticulously cleaned and sterilized, preferably with an autoclave, between patients to avoid the risk of transmitting viral hepatitis. Surface electrodes, small plate electrodes taped on the cleaned skin, can be used with an electrode paste for nerve conduction studies and to EMG in ergonomic and kinesiologic applications. The patient also has to be grounded with a plate or string electrode on the skin to prevent interference from outside currents.

STANDARD ELECTROMYOGRAPHY

The muscle should be studied in the area of its motor point and charts of motor points in various muscles are available (Licht, 1971). The patient should be reassured and calm at the beginning of the examination. The needle electrode is inserted through the cleaned skin into the muscle with the limb relaxed and relaxation is best achieved with the patient in a reclining or half reclining position. While inserting the needle the insertion activity may be recorded. The high frequency insertional activity is usually a very short burst only, in neurogenic lesions it may last for up to a number of seconds.

If the normal muscle is well relaxed there is no electrical activity recordable outside of the end-plate zone. Possible spontaneous activity is searched for in several needle positions. The muscle can be lightly tapped, which may slightly change the needle position and may activate some spontaneous activity, too. In the end-plate zone brief monophasic negative discharges of up to 2 ms duration and 100 uV amplitude can be recorded, and these are considered to represent miniature end-plate potentials extracellularly (Lenman and Ritchie, 1977). Diphasic potentials of 2-3 ms and of up to 200 uV may occasionally be recorded in the end-plate zone, too; they may arise from small bundles of intramuscular nerve fibres touched by the needle; this touching induces pain and prompts relocation of the needle.

Fibrillation potentials are recorded especially in denervated or partially denervated muscles, although they may be very occasionally noted in a healthy muscle or in certain forms of myopathy, especially in myositis. Fibrillations are small potentials with a duration of 0.5 to 2 ms and an amplitude of 30 to 150 uV. They may have an initial positive phase followed by a negative deflection, and may have up to three phases. They occur at a rate of 2 to 10 per second and probably arise from single spontaneously contracting muscle fibres.

Another variety of denervation activity is positive sharp waves. They consist of an initial abrupt positive spike followed by a slow change in the negative direction, possibly followed by a prolonged negative phase, so that the total duration may be up to 10 ms or more. They frequently accompany fibrillation potentials or occur alone repetitively and can be evoked by an electrode movement. They may occur also in polymyositis and in myotonia.

When small groups of muscle fibres contract spontaneously, fasciculations are noted as visible contractions without movement and as concomitant small bi- tri- or polyphasic potentials on the oscilloscope screen. Fasciculations occur at a rate of about once every three or four seconds in motor neuron disease and then usually a dull thump is heard over the loudspeakers. Fasciculations may be also completely benign (Trojabord and Buchthal, 1965), and as a solitary finding should not be considered of any pathological meaning.

In myotonia a high frequency myotonic discharge may be evoked by electrode movement or percussion of the muscle. This is characterized by a "dive-bomber" waxing and waning sound based on gradual increase and decrease in the frequency of short-duration action potentials.

In certain neurogenic atrophy or myopathic weakness pseudomyotonic discharges or bizarre high-frequency discharges may be noted. These start and stop abruptly and the individual potentials are generally of constant amplitude, frequency and shape. These discharges may last several seconds and up to several minutes.

Motor unit action potentials (MUP) are recorded during voluntary contraction of the muscle. MUPs are summated action potentials from groups of muscle fibres getting their innervation from one and the same motor neuron and contracting nearly synchronously. MUPs may be monophasic, diphasic or triphasic, occasionally polyphasic with five or more phases.
Their duration varies from 2 to 10 ms and the amplitude from 100 uV to 2-3 mV in normal muscles.

In a weak contraction a few motor units start firing at a low rate, but the rate increases to a steady state of around 15-20 Hz in increasing contraction force, when more motor units are added. During a strong voluntary contraction many motor units are recruited,and MUPs are superimposed one upon the other to form an interference pattern on the oscilloscope. The pattern during a weak contraction is scarce, called pattern of single motor units and so-called mixed pattern is a transition form between the former two. When the maximal muscular force is clinically graded as 5, grade 3 representing the force capable to counteract the gravity and grade 1 representing a tiny muscle twitch barely visible, in a normal situation we see interference at force 5, mixed pattern at force 4 and single MUPs are usually separable at force 3.

The pattern of maximal voluntary contraction can be greatly affected by neurogenic diseases; in denervation the number of recruitable MUPs decreases and only a mixed pattern or a pattern of single motor units can be achieved. In peripheral neurogenic lesions the muscle force usually seems greater than the amount of activated motor units. On the other hand, in myopathies where the number of motor units has not decreased, but diseased fibres within the unit have lost their contractile properties, many units may already be activated during a weak contraction resulting in an interference pattern. When the disease progresses, the differences in the pattern of clinically weak maximal muscle contraction are obscured, and greatly diminished activity is then a significant feature in both neurogenic and myogenic disorders.

The evaluation of the possible loss of motor units needs great skill and experience on the part of the electromyographer. The electrode has to be in a good position in the muscle, the muscle has to be activated in a correct manner and the patient should be cooperative. Slight changes in the position of the needle electrode may induce great changes in the pattern, the effort of the patient may be inadequate, etc. The electromyographer has to do the examination himself to be able to follow the recruitment on the oscilloscope. An evaluation of photographically recorded EMG activity afterwards cannot substitute an on-line visual and auditory analysis of the signal together with a clinical evaluation of the applied contraction force and effort. The muscle force can be measured with straingauges, but this is only supplementary to the clinical evaluation, since the maximal force varies depending on body build and sex.

The evaluation of the duration and amplitude of individual MUPs is very important, because neurogenic and myogenic disorders induce different changes in MUPs. In neurogenic disorders, especially in reinnervation, the duration of MUPs increases and the amplitude may be increased also. On the other hand, in myopathy the duration and amplitude of MUPs may both be decreased. In both types of diseases the proportion of polyphasic MUPs may increase, but they are of short duration in myopathy and of long duration in neuropathy or motor neuron disease.

The actual measurement of MUP parameters is time consuming when done manually from serial photographic recordings of single motor units (Buchthal *et al*, 1954). The parameters of various muscles are different also, consequently an accumulation of a large amount of reference material is a prerequisite; this type of material has been collected by Buchthal and his group.

AUTOMATIC EMG ANALYSIS

Automatic means of data collection and measurement have been designed, and individual potentials can be selected by an amplitude or rate of rise criterion and then arithmetically averaged (Lang and Tuomola, 1974). The averaged MUPs can easily be measured with computers and the EMG signal during a light to moderate muscle contraction can be analyzed. This method gives somewhat different values for motor unit parameters than manual measurement, probably because bigger motor units are recruited in a more powerful contraction.

The interference pattern can also be quantitated. The quantitation techniques are based on analysis of features related to the frequency of the waveforms or their components. They use turns or baseline crossings in some combination with amplitude (Willison, 1971; Hayward, 1977; Hayword and Willison, 1977). Spectral analysis has also been applied to frequency analysis of the EMG signal (Larsson, 1975).

SINGLE FIBRE EMG

Single fibre EMG is performed with a specific electrode with a miniature recording surface, which enhances recording from a small number of muscle fibres near the electrode, while suppressing the activity of more distant fibres. A single fibre electrode records from individual muscle fibres within 300 um of the electrode as biphasic or triphasic potentials with a rise time of less than 300 us and amplitude exceeding 200 uV up to 25 mV. The electrode has to be localized extremely precisely and maintained in this specific location while a series of measurements is made. Reliable single fibre EMG is therefore highly dependent on the electromyographer's skill in manipulating the needle electrode.

In single fibre EMG one must isolate and display a series of sweeps from single potentials. This can be made easier with a trigger and signal delay unit. Generally during mild voluntary activation single fibre EMG will record an action potential from only one muscle fibre in a motor unit, but pairs of potentials from the same unit in a time-locked form occur at approximately 35% of electrode placements (Stålberg and Trontelj, 1979), and very rarely three or more time-locked potentials are found. Fibre density is the mean number of single fibre potentials recorded from motor units in 20 or more recording sites. Normal fibre density ranges from 1.3 to 1.8 in normal persons less than 70 years old (Stålberg and Trontelj, 1979). It reflects the density of muscle fibres in one motor unit within the recording area and corresponds most directly to the number of notches seen on a standard MUP recorded with a concentric needle electrode. Fibre density increases with disorders that results in histological grouping of muscle fibres. Grouping is most common in neurogenic disorders with reinnervation, e.g., in peripheral neuropathy or anterior horn cell disease. Some grouping may be seen in myopathies as a result of fibre splitting or regeneration after necrosis.

When two or more single fibre potentials are recorded from a single motor unit, the interpotential interval between the two shows a small fluctuation referred to as jitter. This jitter is increased in myasthenia, i.e., disturbance in neuromuscular transmission. In myasthenia too, blocking of the second potential frequently occurs (Sanders et al, 1979). Single fibre EMG is the most sensitive way of detecting abnormal neuromuscular transmission.

APPLICATIONS IN OCCUPATIONAL MEDICINE

Standard EMG has been applied to detecting toxic neuropathy in several studies: see for example Buiatti et al, (1978), Gilioli et al, (1978), Giuliano et al, (1975), Perticoni and Cianchetti (1976), Seppäläinen et al, (1972). Often the preponderance of toxic neuropathic signs in the distal parts of the limbs are emphasized. We should remember, however, that, e.g., the extensor digitorum brevis muscle of the foot may be damaged by many etiological factors, for example sciatic syndrome and external pressure from tight protective shoes, etc., consequently findings limited to that muscle only should not be regarded as a sign of toxic polyneuropathy. Quantitative analysis of the EMG signal may offer valuable information in epidemiological studies if reliable and repeatable recordings can be secured by experience and skill of the electromyographer. Single fibre EMG might prove a sensitive way of detecting reinnervation phenomena in toxic neuropathy, but the method is time consuming and at times more painful than ordinary EMG. Controversial results have been reported on possible

disturbance in the myoneural junction caused by organophosphorous pesticides (Jusic and Jurenic, 1980; Roberts, 1976), using repetitive stimulation; consequently jitter studies could be applied to investigations of possible neuromuscular transmission disturbance in connection with pesticide exposure or carbon disulfide exposure (Seppäläinen *et al*, 1972).
Standard EMG seems quite reliable and also sensitive in combination with nerve conduction studies for the study and follow-up of toxic neuropathies and can be performed with relatively simple and inexpensive equipment, but necessitates an experienced electromyographer. The usefulness of more elaborate techniques like automatic analysis and single fibre EMG in occupational medicine has yet to be proved.

REFERENCES

Buchthal, F., P. Pinelli and P. Rosenfalck (1954). *Acta Physiol. Scand.*, *32*, 200-218.

Buiatti, E., S. Cecchini, O. Ronchi, P. Dolara and G. Bulgarelli (1978). *Br. J. Ind. Med.*, *35*, 168-173.

Gilioli, R., C. Bulgheroni, P.A. Bertazzi, A.M. Cirla, M. Tomasini, M.G. Cassitto and M.T. Jacovone (1978). *Med. Lav.*, *69*, 130-143.

Giuliano, G., O. Ronchi, L. Rossi and L. Focardi (1975). *Lav. Um.*, *27*, 33-41.

Hayward, M. (1977). *J. Neurol. Scienc.*, *33*, 397-413.

Hayward, M. and R.G. Willison (1977). *J. Neurol. Scienc.*, *33*, 415-423.

Jusic, A. and D. Jurenic (1980). *Arch. Environ. Health*, *35*, 168-175.

Lang, A.H. and G. Tuomola (1974). *Electromyogr. Clin. Neurophysiol.*, *14*, 513-525.

Larsson, L.E. (1975). *Electroenceph. Clin. Neurophysiol.*, *38*, 69-78.

Lenman, J.A.R. and A.E. Ritchie (1977). *Clinical electromyography*. Pitman Medical, Bath, 214 pp.

Licht, S. (1971). *Electrodiagnosis and Electromyography*. 3rd edn. Elizabeth Licht, New Haven, Conn.

Perticoni, G. and C. Cianchetti (1976). *Electromyograph. Clin. Neurophysiol.*, *16*, 493-506.

Roberts, D.V. (1976). *Int. Arch. Occup. Environ. Health*, *36*, 267-274.

Sanders, D.B., J.F. Jr. Howard and T.R. Johns (1979). *Neurol. (Minneapolis)* *29*, 68-76.

Seppäläinen, A.M., M. Tolonen, P. Karli, H. Hänninen and S. Hernberg (1972). *Work-Environ-Health*, *9*, 71-75.

Stålberg, E. and J.V. Trontelj (1979). *Single fibre Electromyography*. *Miravalle Press, Old Woking, Surrey*, UK.

Trojaborg, W. and F. Buchthal (1965). *Acta Neurol. Scand.*, *41*, *suppl. 13*, 251-254.

Willison, R.G. (1971). Quantitative Electromyography. In *Electrodiagnosis and Electromyography* (ed. S. Licht) 3rd edn. Elizabeth Licht, New Haven, Conn. 390-411.

Electroneurography

A review — with regard to personal experience with antidromic sensory conduction velocity of N. ulnaris and N. medianus and H-reflex velocity of N. tibialis

E. Lukàs

Institute of Hygiene and Epidemiology, Centre of Industrial Hygiene and Occupational Diseases, Prague, Czechoslovakia

ABSTRACT

Besides classical electroneurography, a method of investigation of sensory conduction velocity of N. ulnaris and N. medianus by means of antidromic stimulus in 115 healthy persons is described.
In a further 437 persons (130 controls and 307 persons with potential polyneuropathy due to lead, carbon disulphide and diabetes) a two-point estimation of conduction velocity of the H-reflex in the area of the N. tibialis is demonstrated.
The decreased H-reflex velocity showed a good correlation with the objective clinical findings(L5-S2)reflexes. Because of the fact that such a decrease is greater than in the simultaneously investigated motor fibres, the demonstrated method enables an earlier EMG diagnosis of the initial forms of polyneuropathies.

KEYWORDS

Antidromic sensory conduction velocity; N. ulnaris, N. medianus; H-reflex conduction velocity, N. tibialis; early EMG diagnosis; polyneuropathy (lead, carbon disulphide, diabetes mellitus).

Neurology of the peripheral nervous system is continually being enriched with new information on polyneuropathic complications in different chronic internal diseases, in traumatology, in pharmacology (use and abuse of drugs), in industrial and environmental toxicology and/or in problems of the influence of physical factors (e.g., vibrations). All this information is mainly based on electromyography (in all its variants) because these electrophysiological methods permit achievement of our primary goal, that is, to confirm our clinical findings in time.
These methods are, however, also suitable for long-term follow-up studies, because the development of impairment and/or improvement can be very well surveilled at any time.

The basic aim of electroneurography is to investigate the function of the myelin sheath, e.g., the conduction velocity of the nerve.
The nerve trunk is electrically stimulated at one or two points and from

the competent muscle an action potential is amplified and registered. From the onset of the stimulus to the point of first deflection from the base line the latency is measured and from the distances between stimulation and registration points the conduction velocity can be calculated, when conduction velocity (CV) in m/s = distance : latency time.
For stimulating the nerve a rectangular pulse of 0.2-0.5 ms and of supramaximal intensity (e.g., 30-50% over the maximal intensity which evokes the greatest constant amplitude of action potential) are generally used and thus the fastest fibres are investigated (maximal conduction velocity-MCV).
For registration a surface electrode (less frequently a needle electrode) is placed over the competent muscle and the skin is prepared with ether to reduce its impedance. For an exact investigation it is important to get the same shape (duration and amplitude) of the evoked potential from each stimulation spot.
The distances are measured from the middle of the stimulus point to the middle of the registration electrode.
Because of the fact that conduction velocity is dependent on body and skin temperature (a decrease of 1°C is followed by a decrease of approximately 1-2 m/s) it is recommanded to measure the local temperature and ensure a constant level of about 29-30°C by warming the region under examination.
The dependency of conduction velocity on age (decrease of approximately 1 m/s for each decade above 40 years) should be eliminated by use of correction tables.
The nerves of the upper extremities conduct the stimulus faster (55-65 m/s) than the nerves of the lower extremities (45-55 m/s) and the motor conduction velocity along the trunk of the nerve is twice as high as in the terminal part, where a delay in the neuromuscular junction decreases the speed of transmission.

Not only the fast fibres are investigated. The conduction velocity of slow fibres (SFCV) can be determined by the method of paired stimuli according to Hopf (1962), based on the antidromic block-technique and/or by the desynchronisation method of Pinelli (1961) used, for example, also in Gilioli's laboratory (1978).
The normal values in slow fibre velocities are approximately about 10 m/s slower than in the fast fibres.

Using a repetitive stimulation it is possible to investigate the situation in the neuromuscular junction. In each patient with polyneuropathy of various origin, a myasthenic syndrome (of the Lambert-Eaton type) could be anticipated. The muscle fatigability due to alteration of neuromuscular transmission is demonstrable in low frequency stimulation (4-5 c/s) by a decrease in the amplitude of the evoked potential over time. After stimulation with higher frequency (30-50 c/s) in patients with myasthenic syndrome a paradoxical elevation of the amplitude can be seen, while in myasthenia gravis the decrease in the amplitude continues further. Another difference between these two types of symptoms could be seen after Tensilon test (10 mg), which elevates the decreased amplitude, but only in the cases with classical myasthenia gravis. These two variants of responses are dependent on the quality of the postsynaptic membrane, which in myasthenia gravis is a result of destroyed autoaggressive processes and thus, by blockin the cholinesterase activity by Tensilon, an amelioration effect can be seen. In myasthenic syndrome, on the contrary, the postsynaptic membrane is unaffected; only the release of acetylcholine is disturbed, and for that reason the higher repetitive stimulation can evoke a greater amount of transmitters, the transmission is of higher quality, and, independently of Tensilon, an increase in amplitude can be demonstrated.

Besides the estimation of motor conduction velocity the usefulness of sensory conduction velocity has been reported in recent years (Kaeser, 1966; Thomas et al, 1967; Wiederholt et al, 1970; Duensing et al., 1974; Benecke et al, 1980; Ludin, 1980; Campbell et al, 1981, etc.). In the method described by Dawson (1956) and specified by Gilliatt et al (1961), the orthodromic stimulus is used. It is applied to fingers or toes, and the recording is made with surface or needle electrodes placed along the nerve stem on the wrist, elbow or shin, etc. A greater amplification must be used, due to the fact that the evoked neurogram is of lower amplitude (mean 20 μV), and for an exact registration the averaging method is used.
The relatively laborious technique of registration requiring expensive equipment influences the availability of this method, so that sensory conduction velocity is not examined in a standardized manner.
Other ways were studied to find an easier method for sensory conduction velocity estimation. An interesting method uses the antidromic stimulus. On the basis of data in the literature (Kaeser, 1966; Jebsen et al, 1967) 10-12 years ago we started to examine the sensory fibres by means of antidromic stimulus using the following procedure (Lukäs 1969, 1970, 1977):

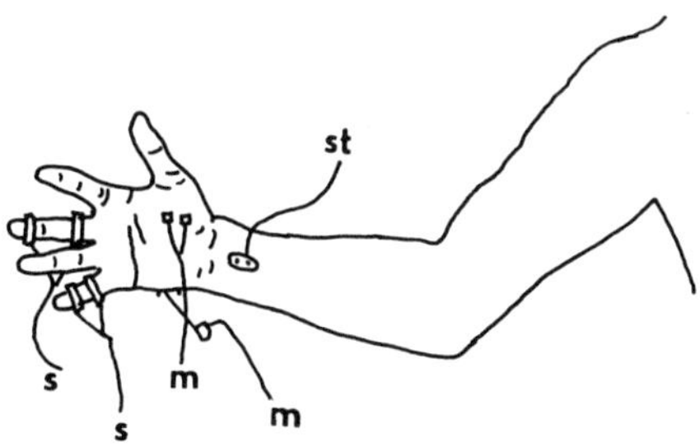

Fig. 1. The neurogram is recorded by a silver-foil ring electrode (S-electrode) which is connected to the original DISA connector. We applied the electrode to the little finger in the N. ulnaris area and to the middle finger in the N. medianus area. The distance between two parts of the electrode is 1.5-2.0 cm according to the length of the finger. The standard bipolar stimulation electrode (St-electrode) is localized in standard stimulation points used for examination of motor fibres and a stimulus lasting 0.3 ms and of maximal voltage is applied. We carried out photographic recording by superimposing 5-6 neurograms in order to demonstrate constancy and reproducibility. The registered potential is a slightly higher voltage than the "orthodromic" potential (50-70 μV) which corresponds to the findings of Kaeser (1966). The latency times and distances between stimulation and registration spots are measured by the same method as described in motor fibre investigations.
An advantage of this method is the fact that on the second channel of the monitor the motor conduction velocity can be simultaneously registered and measured by classical means.

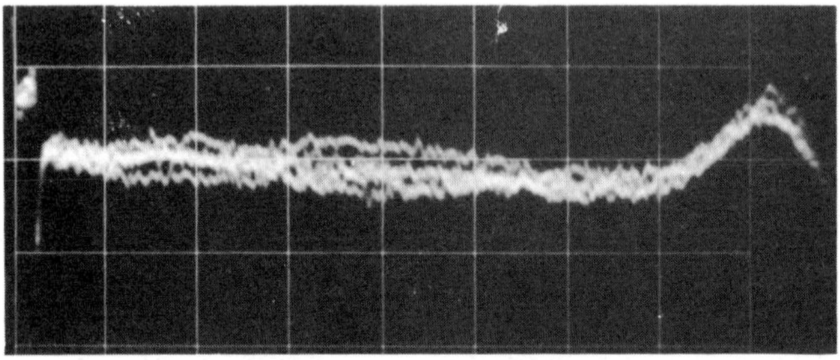

Fig. 2. Stimulation of N. ulnaris in the sulcus. Calibration: each square horizont. 1 ms; vertic. 60 μV.

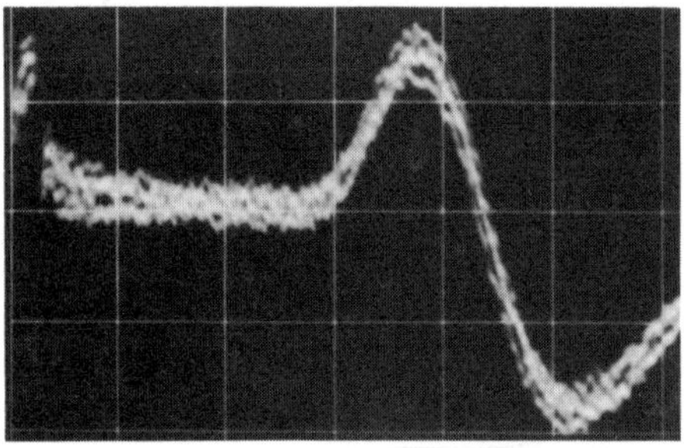

Fig. 3. Stimulation of N. ulnaris above the wrist. Calibration: each square horizont. 1 ms; vertically 30 μV.

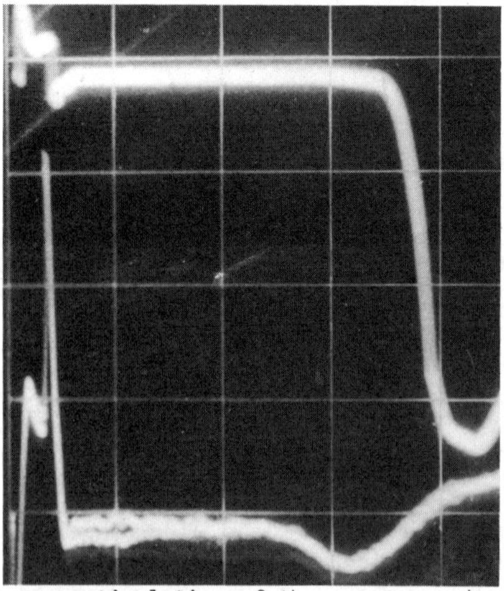

Fig. 4. Simultaneous stimulation of the sensory and motor fibres of the N. medianus. Upper trace - motor fibres, vert. calibration: each square 200 μV. Lower trace - sens. fibres, vert. calibration: each square 60 μV. Time calibration: 1 ms each square.

Table 1. Results of sensory CV of N. ulnaris and N. medianus of 115 controls (35 adult males, mean age 33.2 ± 5.8 years, and 80 healthy children, 11-12 years old).

	N. ulnaris		N. medianus	
	E - F	W - F	E - F	W - F
mean values of both groups	57.1±5.5	56.2±6.5	56.0±6.1	56.6±6.6
lower limits (mean - 2 SD)	46.1 m/s	43.2 m/s	43.8 m/s	43.4 m/s

E - F = velocity from elbow to the finger
W - F = velocity from wrist to the finger

The results demonstrate that the determined values of sensory conduction velocity using antidromic stimulus correspond to the values reported in the literature after orthodromic stimulation. Kaeser (1966) and others report values for the forearm area or distal parts of the hand within limits of 40-66 m/s with a mean value between 55-60 m/s. Agreement of the velocities we determined which also agree with the results of antidromic examination reported by Kaeser (1966), Jebsen (1967), Cruz-Martinez et al (1978 a-b), Mavor et al (1971), Murai et al (1975), Ludvingsson et al (1979), confirms that we are measuring really sensitive fibres. This is further confirmed by the fact that the measured distal velocities are twice as great as the values measured in motor fibres of acral parts of the hand, where the lower limits are approximately 20 m/s. In agreement with Kaeser we could also register some higher amplitudes of the neurogram when values between 20-90 μV were recorded.

The described method proved to be very useful in diagnosing the initial stages of, for example, compressive tunnel syndromes, when sensory-motor dissociations have been detected (i.e., pathological values in sensory conduction: velocity with normal values in the motor conduction times). This corresponds to the clinical experience that hypoesthesia precedes the motor function disturbances, thus demonstrating the reliability of this method.

Another method of investigation of afferent fibres of lower extremities used in our laboratory is by means of estimation of H-reflex velocity in N. tibialis of the shin.

Objective clinical neurological findings have confirmed that polyneuropathy starts in its initial phase in the afferent fibres of the lower extremities. This experience is entirely in agreement with the experience of older clinicians and also agrees with the dying-back theory of Cavanagh (1964) and with the experimental findings in distal axonopathy of Spencer and Schaumburg (1978).
This fact, and a desire to find a sensitive technique for early detection of subclinical forms of polyneuropathies led us to investigate the conduction velocity of the H-reflex which is a classical example of afferent fibres of the lower extremities.

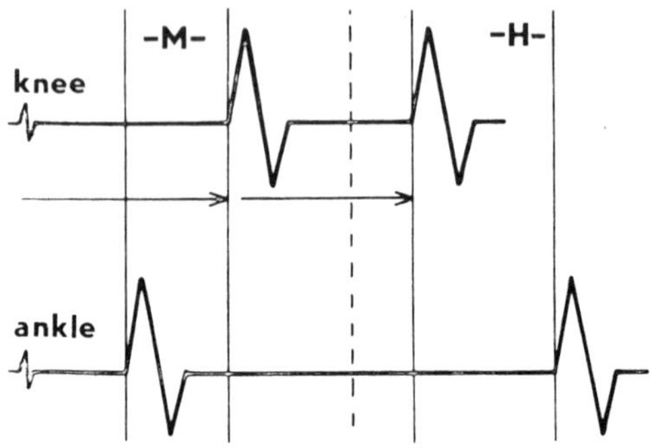

Fig. 5. The H-reflex, described by Hoffmann (1922) is the second response to the stimulation. It appears within 40-50 ms after stimulation in the popliteal pit, or a little later following stimulation behind the ankle. This longer latency than the response of motor fibres (-M-) is due to the fact that the stimulus also passes along the afferent pathway to the spine and returns by way of the spinal arch to the muscle. Since the mere measurement of H-reflex latency depends on an accurate estimation of the length of the lower limb, which is not easy, we use the two-point stimulation (knee and ankle) to obtain two latencies in the region of N. tibialis. A surface registration electrode is placed in the region of M. flexor pollicis brevis.

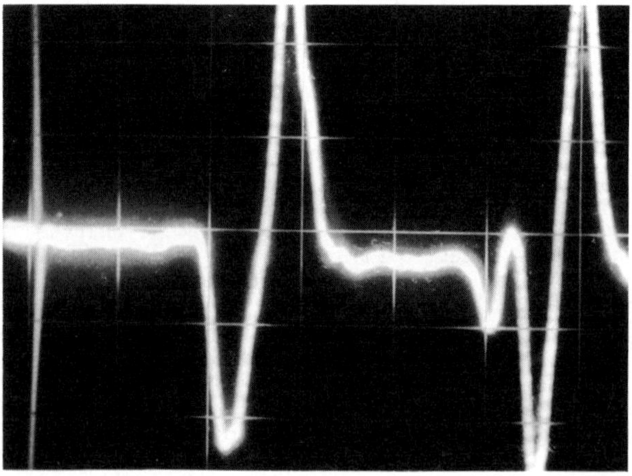

Fig. 6. The original registered -M- and -H- responses. Calibration: each square horizontally 10 ms, vertically 200 μV.

The velocity can be calculated from the differencies in latency times and from the distance between the two stimulation points.
The bottom limits for each type of conduction velocities are defined by subtracting 2 SD from the mean values of the group (calculated on a 1% significancy level).

The H-reflex response could be found in 2/3 of the investigated persons, which corresponds to the known fact that this reflex is absent in part of a normal healthy population; such a situation cannot, of course, be considered as pathological.

437 subjects were examined:

130 controls (50 adults aged 40.5 ± 7.1 years and 80 healthy children aged 11 - 12 years)

307 persons with potential polyneuropathy (80 children aged 11-12 years exposed to high doses of lead from motor-car exhaust; 51 workers aged 42.7 ± 8.0 years exposed to carbon disulphide at concentrations of 30-50 mg CS_2/m^3; 24 workers aged 21-51 years exposed to carbon disulphide at concentrations of 150-200 mg CS_2/m^3 and 152 diabetics aged 56.5 ± 7.5 years).

The simultaneous estimation of motor fibre conduction velocity (-M) and H-reflex conduction velocity (-H-) were compared with the clinical findings in L5-S2 reflexes, evaluated in 3 degrees: normal response (+ or ++), hyporeflexia (↓) and arreflexia (ϕ).

Results in controls:

	r.r. L5-S2			N. tiabialis	
	+ ++	hypo	ϕ	-M-CV	-H-CV
n = 130	89.2%	10.8%	ϕ	55.9±7.4	59.4±8.9
	lower limits (mean - 2 SD)			41.1 m/s	41.6 m/s

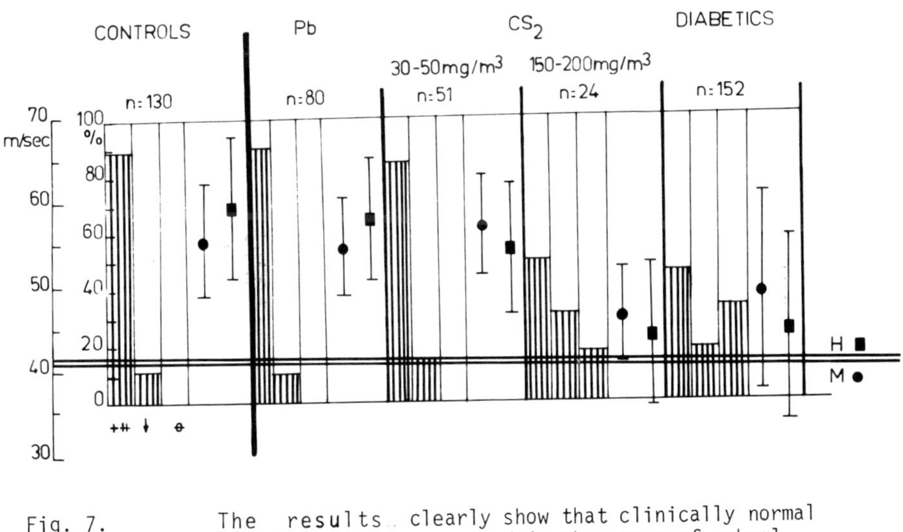

Fig. 7. The results clearly show that clinically normal findings in the control group and in the group of school children exposed to high doses of lead motor-car exhaust, also have normal values of EMG indicators.

In the group of patients exposed to low concentrations of carbon disulphide, the group mean values of conduction velocity in the afferent fibres (velocity of the H-reflex) was significantly lower (5% level of statistical significance) than in the controls.
There was a slightly higher number of hyporeflexia (14.9%), too.

In the group of patients exposed to higher concentrations of carbon disulphide, both the clinical findings and the EMG score were more seriously altered. In both the two indicators monitored (velocity of the -M- and -H- responses) the values were significantly lower than in the control group (1% level of statistical significance). The clinical findings agreed: frequency of hyporeflexia increased to 31.2%, arreflexia appeared in 18.8%.

Similarly, the EMG findings in the group of diabetics corresponded to their pathological clinical results (19.7% of hyporeflexia and 33.6% of arreflexia).
In all these last three groups, the mean value of the decreased H-reflex conduction velocity was greater than the decreased motor conduction velocity which is in good agreement with the opinion of those authors who described a greater decrease in conduction velocity in sensory than in motor fibres in patients with polyneuropathy (Stöhr et al, 1978; Vasilescu, 1976, Duensing et al, 1974).

In our opinion, the above results show that:

1. the simultaneous measurement of conduction velocity in the motor and afferent fibres of N. tibialis proved to be a suitable method for epidemiological studies and/or for individual diagnosis because of the short time needed to perform the measurement and the simple type of equipment used.

2. there is a correlation between the objective clinical findings (reflexes L5-S2) and the values of the chosen EMG indicators.

3. the diagnostic importance of hypo- and arreflexia can be potentiated by the estimation of H-reflex conductivity because of the fact that its decrease is earlier than the decrease in the conduction velocity in the motor fibres.

4. the proved relationship between H-reflex latencies and the sensory conduction velocities (found by Wager *et al*, 1974, by means of orthodromic stimulation) and the fact that our velocity values agree with the classical sensory conduction velocity data of Lovelace *et al*, 1973, demonstrate the same usefulness of evaluating conduction velocities both in proprioceptive and in sensory fibres.

Finally, one remark: the main problem in the estimation of conduction velocities is, however, the definition of normal values and of the lower limits in healthy persons.
According to our experience and in our opinion, the estimation of mean values of the monitored control groups should be ranged by ± 1 standard deviation (SD), but it should be emphasized that this SD *must be calculated* on the 1% significancy level and the determination *of the lower limit of* the normal conduction velocity values made by subtraction of 2 SD from the mean value.

This development should be used by all means for each statistical evaluation of electroneurographic parameters, be it in motor or sensory fibres or both, fast or slow fibre velocities, measurement of various latencies, of duration and amplitudes of evoked potentials, etc, mainly in epidemiological studies.

Only in this way will the findings be real and many discrepancies between various laboratories be eliminated.

REFERENCES

Benecke, R., B. Conrad (1980). J. Neurol., 223, 231-239.
Campbell, W.W., L. Ch. Ward, T.S. Swift (1981). Muscle and Nerve, 4, 520-523.
Cavanagh, J.B. (1964). Int. Rev. Exp. Path., 219-267.
Cruz-Martinez, A., M. Barrio, M.C. Perez-Conde, M.T. Ferrer (1978). a. J. Neurol. Neurosurg. Psych., 1097-1101; b. J. Neurol. Neurosurg. Psych., 1092-1096.
Dawson, G.D. (1956). J. Physiol., 436.
Duesing, F., K. Lowitzsch, V. Thorwirt and P. Vogel (1974). Ztsch. Neurol., 267-284.
Gilioli, R., C. Bulgheroni, P.A. Bertazzi, A.M. Cirla, M. Tomasini, M.G. Cassitto and M.T. Jacovone (1978). Med. Lavoro, 2, 130-143.
Gilliatt, R.W. and T.A. Sears (1961). J. Neurol. Neurosurg. Psych., 305.
Hopf, H.C. (1962). Dtsch. Z. Nervenheilk, 579-588.
Jebsen, R.H., H. Tenckhoff and J.C. Honet (1967). N.E.J. Med., 327-333.
Kaeser, H.E. (1966). Dtsch. Z. Nervenheilk, 289-299.
Lovelace, R.E., S.J. Myers and L. Zablow (1973). J. Neurol. Neurosurg. Psych., 942-950.
Lowitzsch, K., H.C. Hopf and J. Galland (1977). J. Neurol., 181-188.
Ludin, H.P. (1980). G. Thieme Verlag, Stuttgart, New York, 173 pages.
Ludvigsson, J., G. Johannesson, L. Heding, A. Hager and Y. Larsson (1979). A. Ped. Scand., 739-743.
Lukàs, E. (1969). Int. Z. Klin. Pharmak., Therap., Toxicol., 4, 354-358.
Lukàs, E. (1970). Work Environ. Health, 1, 67-79.
Lukàs, E. (1977). Int. A. Occ. Envir. Health, 283-294.
Mavor, H. and R. Shiozawa (1971). EEG, Clin. Neurophysiol., 210-221.
Murai, Y. and I. Sanderson (1975). J. Neurol. Neurosurg. Psych., 1187-1189.
Pinelli, P., G. Lanzi, F. Savoldi and F. Zebri (1961). see Gilioli, R.
Seppäläinen, A.M., M. Tolonen, P. Karli, H. Hanninen and S. Hernberg (1972). Work Environ. Health, 71-75.
Seppäläinen, A.M, and M. Tolonen (1974). Work Environ. Health, 145-153.
Seppäläinen, A.M. (1975). Scand. J. Work Environ. Health, 1-14.
Spencer, P. and H.H. Schaumburg (1978). Envir. Health Perspect., 97-105.
Stöhr, M., F. Schumm and R. Ballier (1978). EEG Clin. Neurophysiol., 44, 172-178.
Thomas, J.E., E.H. Lambert and K.A. Czeus (1967). A. Neurol., 635-641.
Vasilescu C. (1976). Eur. Neurol., 447-457.
Wager, E.W. and A.A. Buerger (1974). Neurol. (Minn.), 8, 711-714.
Wiederholt, W.C. (1970). J. Neurol. Neurosurg. Psych., 438-441.
Willer, J.C. and H. Dehen (1977). EEG Clin. Neurophysiol., 42, 205-212.

Current Behavioral Techniques

M. G. Cassitto

Institute of Occupational Health, Department of Preventive Medicine, University of Milan, Italy

ABSTRACT

Review of the current behavioral techniques used in neurobehavioral toxicology is introduced by a discussion on the definition of behavior in terms of an operational instrument which determines the choice of the techniques to be used. The danger of an oversimplification of the mechanisms controlling administration and output of tests is stressed.
An analysis of the most currently used behavioral batteries is made focusing attention on their different loadings and attempting to identify the behavior range they can explore.

KEYWORDS

Behavior; behavioral toxicology; behavioral instruments; higher nervous functions assessment; occupational health.

Behavior, which is the main subject of our symposium, is also an intriguing problem for many of those who deal with the problems related to its measurement and evaluation. Just two months ago, at the Congress on Neurobehavioral Pharmacology in Capri, we heard Prof. Zbinden ask: what do we mean by behavior? This should be the right place to try and find an answer because since our purpose is to revise well established methodologies and discuss the validity of recently emerging ones, we certainly are expected to know what we are looking for and why, in order to adopt the instruments that are most suitable for our purposes. We must however, admit that behavior is sometimes difficult to define in terms of operationally useful constructs. The first definition that comes to the mind is that behavior is in fact any answer given by the living organism as a whole in response to any stimulus coming from the internal and/or external environment of the subject.
But, however true this definition might be, it is not operationally useful because any answer given by the organism is never a simple reaction to a stimulating situation but a highly complex pattern of integrated moments. Does that mean that behavior is not measurable? It is certainly difficult with the instruments available today to measure the single components which constitute final behavior; what we measure is in fact the final and visible

result. Let us take a clarifying example. We all know the Wechsler Digit Span Test, which is a mechanical memory test requiring the subject to repeat forward and backward series of casual numbers. We all have a certain number of certainties about that test: 1. it is one of the simplest and most direct measures of mechanical memory, 2. it is free of cultural influence, 3. any subject with a sufficient capability of retention should be able to cope with its requirements, 4. the result is certainly a direct measure of short-term memory efficiency in its simplest form. In fact, the process is not so simple. I have tried to illustrate the functions activated and the other variables influencing the subject's production in Table 1.

Table 1. Functions activated and abilities requested to cope with the Digit Span Test requirements

TEST PARADIGM	FUNCTIONS AND ABILITIES
- Test requirements made by the operator	sufficient level of vigilance
	orienting reflex
- Interval between instructions and stimulus administration	maintenance of the orienting reflex
- Stimulus administration	ability to resist to previously established number connections
	ability to establish new connections
	resistance to number reciprocal influence both positive and negative

Operator related variables which may affect the elaboration process: voice intensity, interstimuli interval, spontaneous operator grouping.

We can see from the table that the process is all but a simple one, requiring first of all a satisfactory level of vigilance, ability to isolate scattered numbers from previously established number connections, to concentrate on a meaningless task, to quickly establish connections which must be utilized for reproduction, to cancel the mutual conditioning influence of each number on the others. Conditions contingent to the situation can also influence the subject's performance: variables induced by the operator, namely his voice intensity, the interstimuli interval, any spontaneous operator grouping as well as variables arising from the test situation, namely the interval between operator requirements and stimulus administration, interfering factors originating from the environment and/or interfering factors originating from the subject, namely, his emotional and physical state.

This scheme is valid both for the three-number span, which is the minimum to be reproduced as well as for the seven-unit span. What do we measure then when we give a simple test like Digit Span? Certainly we measure what we call behavior in its final result. But when such behavior is inadequate, do we have the possibility of knowing which ring of the response chain is too strong or too loose? In general we do not have this possibility and even less so when the instruments are more complex. What we need is an operationally useful construct more useful than the oversimplification given at the beginning. We could tentatively propose a definition and leave it to the rapporteurs who will speak about behavioral methodologies in these four days, to define, modify or completely change it to a better one. We may consider *behavior as the final product of a series of complex units of responses through which the subject progressively selects the appropriate pattern leading to the attainment of a specific goal. Of course, the seletion of the pattern to be followed is a result of the subjective evaluation of the significance of the stimulus in the individual motivational context.* If this has a meaning for all of us, the problem now is to know whether our instruments are such as to allow measurement of the units of behavior which takes place at levels below those usually explored by our tests. Moreover most of these instruments have been developed with purposes totally different from ours. As shown in Table 2, they had been conceived and developed with the primary goal of measuring the intellectual level of the individuals (IQ) in school or occupational counselling. Later on they were also applied to verify the presence or absence of specific attitudes and/or abilities.

Table 2. Main objectives underlying the construction and development of the psychological testing batteries

1 - Measure the intellectual level (IQ) of individual and/or groups for scholastic and/or occupational counselling

2 - Check the presence or absence of aptitudes and/or abilities

3 - Identify central nervous function deficits referable to neurological syndromes

Only later were they used for verifying and measuring intellectual deficit referable to neurological syndromes. The first and the second of these goals are certainly not our special concern. High or poor as the intellectual level of our subjects may be, what we want to know is whether or not it can be preserved in spite of occupational neurotoxic exposure, and if not, what the real damages are. Secondly, we are not interested in knowing whether or not they possess certain abilities but to verify whether the occupational history has somehow accelerated the aging processes.

The same is true for the last aspect though the goals are in this case much more similar. When we started to use these tests for neurotoxic effects, the levels of exposure were much higher and the involvement of higher nervous functions was dramatically near those found in severe and irreversible neurological syndromes. At that time, these instruments proved to be extremely valuable since they supplied the bulk of knowledge which today justifies our presence at this Symposium in search of more sophisticated instruments. Today, when the impairments we are looking for are slight and subtle, and we are more interested in knowing the underlying mechanisms of such impairments, are these instruments still so valuable? We shall now examine the state of the art with a review of the tests currently used in epidemio-

logical and clinical investigations.
The test batteries used in behavioral toxicology are basically of two kinds: in the first the batteries consist almost exclusively of performance tests while the second also employs personality questionnaires and projective techniques. The two groups are certainly not neatly divided. We should rather speak of a continuum where the range of the psychological functions explored becomes more extensive. The information yielded by the two extremes of this continuum are of course substantially different, on the one side little more than information on psychomotor effectiveness is obtained,on the other an integrated portrait of the subject tested is given. An overall analysis of these batteries led to the construction of Table 3 which shows the main function and personality traits explored by the tests used.

Table 3. Psychological functions and traits measured by the behavioral techniques most frequently used in behavioral toxicology investigations

FUNCTIONS	TRAITS
- Vigilance	Intro-extroversion
- Focused attention	Neurotic traits
- Visual acuity	Depression
- Auditory acuity	Anxiety
- Time orientation	Disphoria
- Space orientation	Bodily preoccupations
- Motor abilities	Hostility
- Eye-hand coordination	Self-perception
- Mnestic abilities	
- Verbal abilities	
- Perceptive abilities	
- Abstract perceptive abilities	
- Abstract reasoning	

It is immediately evident that the functions listed in the table can be referred to progressively higher levels of mental functioning and of functional integrity. The table might therefore be transformed into a new one (Table 4) based on hierarchies of functions from those which supply the necessary neurophysiological and psychophysiological substratum of any intentional activity to the most elaborate and complex. I mentioned that these functions have been derived from the test batteries referred to in the literature regarding behavioral toxicology of the last twelve years. Some batteries seem to have been chosen specifically for specific neurotoxic agents; others, usually more extensive, seem to have been applied in any situation in which neurotoxics were used. In the end, only those of the last five to six years appear to be representative of the state of the art, since almost all groups working in this field seem to have progressively structured their approach, passing from the use of single testing instruments to wider batteries supplying either a more extensive range of data on mental functioning or a deeper knowledge of specific functions.

Table 4. Psychological functions and traits, measured by the test batteries considered, grouped in terms of progressively more complex levels of psychic functioning

FUNCTIONS	LEVELS
- Vigilance Focused attention	Orienting reflex Maintenance of a sufficient level of arousal
- Visual acuity Auditory acuity	"Information carriers" reflecting the integrity of the sensory pathways
- Time orientation Space orientation	Information location in time and space
- Mnestic abilities	Integration of the new information into the past experience
- Motor abilities Eye-hand coordination	When necessary, production of an elementary output
- Verbal abilities	Possibility of deeper and wider levels of comprehension and acquisitions
- Abstract perceptive abilities Abstract reasoning	Higher mental functioning
- Personality traits Personality alterations	Mental functioning conditioners and goal motivators

We shall now examine some of the more representative test batteries used at present in the various laboratories, derived from the available literature. This review does not pretend to be exhaustive as regards all the methodologies used. Most of those which have not been mentioned are reduced copies of the one examined here. The purpose of the analysis is also to show that, more or less we all use the same instruments and what changes is mainly the emphasis placed on the various functions measured. I apologize in advance to those of you who might find something missing in the tables presenting their batteries; the articles I went through were not always totally exhaustive and I certainly did not have all the existing literature. Equally, I apologize for possible wrong presentations of the single tests.For some of the instruments, I know little more than the name and I may well have overlooked some important function measured.
The main purpose of this analysis is not only to present the state of the art in terms of currently used methodologies but to supply enough information in order that we might verify together whether our methods are penetrating and specific so as to give a stratified analysis of behavior which makes it possible to also measure those complex units of behavior referred to in my tentative definition of behavior. Fig. 1 illustrates the first battery of tests analyzed. One word on this series of tables:

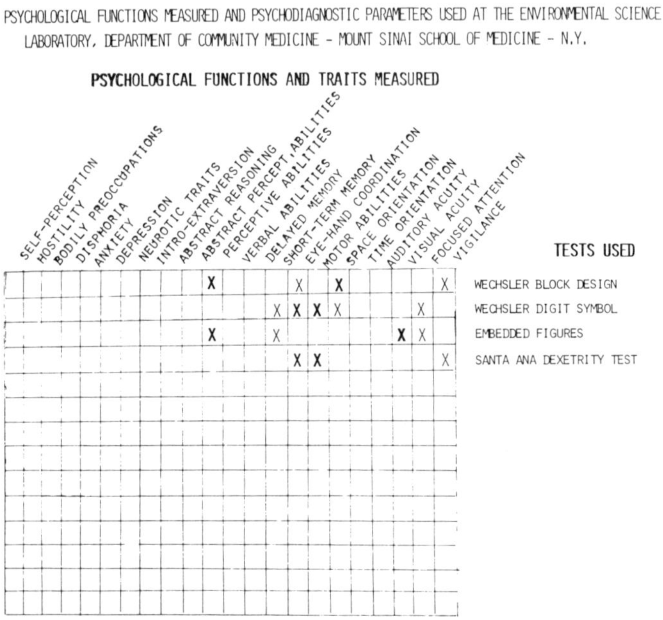

Fig. 1.

the specific loading of each test in given in black while hatching indicates partial loadings. The first two colums are always filled up since each test situation requires an optimum level of vigilance; for some it is the main requirement, for others a real focused attention capacity is requested. Let us go back to this battery; it is a short battery, with a prevalent loading on visuomotor and spatial abilities. Referring to table 4 of function hierarchies, this battery seems to be sufficient to give exhaustive information on the level of vigilance of the subject and of his possibility of supplying simple motor outputs integrating perception and space elements. It has been used mainly for lead-exposed populations.

The next battery (Fig. 2), though having more or less the same structure and length, supplies information regarding essentially the possibility of prompt answers to both sample and complex environmental stimulations.

Moreover, the scale of subjective symptoms add further information on the possible relationship existing between presence of altered psychic functionality and subjective consciousness of deficit. This battery is usually reported in the literature in association with solvent exposure investigations. The same observations can be made for the next battery, reported for use in lead exposure (Fig. 3). However, we can observe here that the range of functions measured progressively increases with the inclusion of measurement of a higher level of mental functioning, like abstract reasoning. Here, too, the subjectivity of the person has been considered through a self-applied questionnaire. The next battery (Fig. 4) has been found in investigations on solvents and especially styrene. Here, too, memory and visuomotor functions are the main sphere of activity measured by these tests. It should be observed that, in this case, other function loadings have been neglected due

PSYCHOLOGICAL FUNCTIONS MEASURED AND PSYCHODIAGNOSTIC PARAMETERS USED AT THE DEPARTMENT OF OCCUPATIONAL HEALTH - NATIONAL BOARD OF OCCUPATIONAL SAFETY AND HEALTH - STOCKHOLM (SWEDEN)

PSYCHOLOGICAL FUNCTIONS AND TRAITS MEASURED

SELF-PERCEPTION	HOSTILITY	BODILY PREOCCUPATIONS	DISPHORIA	ANXIETY	DEPRESSION	NEUROTIC TRAITS	INTRO-EXTRAVERSION	ABSTRACT REASONING	ABSTRACT PERCEPT, ABILITIES	PERCEPTIVE ABILITIES	VERBAL ABILITIES	DELAYED MEMORY	SHORT-TERM MEMORY	EYE-HAND COORDINATION	MOTOR ABILITIES	SPACE ORIENTATION	TIME ORIENTATION	AUDITORY ACUITY	VISUAL ACUITY	FOCUSED ATTENTION	VIGILANCE	TESTS USED
																			X		X	CRITICAL FLICKER FUSION
								X							X				X	X		REACTION TIMES ADDITION
															X				X		X	SIMPLE REACTION TIMES
													X						X		X	SHORT TERM MEMORY TEST
													X	X	X					X		CHOICE REACTION TIMES
		X	X	X	X																	SUBJECTIVE SYMPTOMS SCALES

Fig. 2.

PSYCHOLOGICAL FUNCTIONS MEASURED AND PSYCHODIAGNOSTIC PARAMETERS USED AT THE PERFORMANCE RESEARCH LABORATORY - UNIVERSITY OF LOUISVILLE (KENTUCKY)

PSYCHOLOGICAL FUNCTIONS AND TRAITS MEASURED

SELF-PERCEPTION	HOSTILITY	BODILY PREOCCUPATIONS	DISPHORIA	ANXIETY	DEPRESSION	NEUROTIC TRAITS	INTRO-EXTRAVERSION	ABSTRACT REASONING	ABSTRACT PERCEPT, ABILITIES	PERCEPTIVE ABILITIES	VERBAL ABILITIES	DELAYED MEMORY	SHORT-TERM MEMORY	EYE-HAND COORDINATION	MOTOR ABILITIES	SPACE ORIENTATION	TIME ORIENTATION	AUDITORY ACUITY	VISUAL ACUITY	FOCUSED ATTENTION	VIGILANCE	TESTS USED
															X				X	X	X	SIGNAL DETECTION
										X					X				X	X		PATTERN DISCRIMINATION
								X				X	X							X		ARITHMETIC COMPUTATION
										X									X	X		ORTHORATER
																		X			X	AUDIOMETRY
															X					X		TREMOMETRY
														X							X	EYE-HAND COORDINATION
													X							X		DIGIT SPAN
X			X	X	X																	MULTIPLE AFFECT ADJECTIVE CHECK LIST

Fig. 3.

PSYCHOLOGICAL FUNCTIONS MEASURED AND PSYCHODIAGNOSTIC PARAMETERS USED AT THE INSTITUTE OF OCCUPATIONAL HEALTH, LONDON SCHOOL OF HYGIENE AND TROPICAL MEDICINE - LONDON (ENGLAND)

PSYCHOLOGICAL FUNCTIONS AND TRAITS MEASURED

TESTS USED	SELF-PERCEPTION	HOSTILITY	BODILY PREOCCUPATIONS	DISPHORIA	ANXIETY	DEPRESSION	NEUROTIC TRAITS	INTRO-EXTRAVERSION	ABSTRACT REASONING	ABSTRACT PERCEPT. ABILITIES	PERCEPTIVE ABILITIES	VERBAL ABILITIES	DELAYED MEMORY	SHORT-TERM MEMORY	EYE-HAND COORDINATION	MOTOR ABILITIES	SPACE ORIENTATION	TIME ORIENTATION	AUDITORY ACUITY	VISUAL ACUITY	FOCUSED ATTENTION	VIGILANCE
SIMPLE REACTION TIMES																X				X		**X**
VIGILANCE TEST																X						**X**
WECHSLER DIGIT SYMBOL														X	**X**	X	X				X	
WECHSLER DIGIT SPAN														**X**							X	
WECHSLER BLOCK DESIGN									**X**						X		X					X
TAPPING																**X**						X
TRAIL MAKING																**X**						X
MOTOR TASK																**X**						X
NELSON ADULT READING TEST												X	**X**									X
BUSCHKE MEMORY TEST									X				X	X							X	
VISUAL SEARCH TASK											**X**									**X**	X	
SUBJECTIVE SYMPTOMS SCALE	X		X	X	X	X																

Fig. 4.

to my imperfect knowledge of tests like the Trail-making test or the Buschke memory test.

The last three examples belong to the more extensive batteries. They cover most of the measurable psychological functions. Projective techniques are included as well as personality questionnaires. In the first battery of this group (Fig. 5), the total Wais is performed with its eleven subtests. We must observe that this battery also places special emphasis on mnestic function assessment. If we go back to our function hierarchy, we realize that with this battery we enter in the group of assessment instruments giving an integrated portrait of the individual tested. The last two batteries (Fig. 6 and 7) have been used in their actual form for more than ten years by the respective groups.

They have been used in all the studies on behavioral toxicology carried out by the two groups. Of course they are more time-consuming that the other shorter batteries and sometimes also not completely accepted by the subjects, especially patients, who find the many hours necessary to go through the whole paradigm tiring and not always comprehensible. As far as our group is concerned, the whole battery is used both in epidemiological and clinical situations. From a comparison of the results obtained in field studies and the daily experience gained with the patients, and not only those hospitalized for suspected intoxication from neurotoxics, a certain number of observations and questions arises. Independently of their structure and the mode of application, do these batteries as such supply the kind of information we need in this stage of knowledge about the action of neurotoxic agents? I do not think it is too optimistic to say that at present we have the basic data concerning their action at least for the most

PSYCHOLOGICAL FUNCTIONS MEASURED AND PSYCHODIAGNOSTIC PARAMETERS USED AT THE DANISH NATIONAL INSTITUTE OF OCCUPATIONAL HEALTH - HELLERUP (DENMARK)

PSYCHOLOGICAL FUNCTIONS AND TRAITS MEASURED

TESTS USED	SELF-PERCEPTION	HOSTILITY	BODILY PREOCCUPATIONS	DISPHORIA	ANXIETY	DEPRESSION	NEUROTIC TRAITS	INTRO-EXTRAVERSION	ABSTRACT REASONING	ABSTRACT PERCEPT. ABILITIES	PERCEPTIVE ABILITIES	VERBAL ABILITIES	DELAYED MEMORY	SHORT-TERM MEMORY	EYE-HAND COORDINATION	MOTOR ABILITIES	SPACE ORIENTATION	TIME ORIENTATION	AUDITORY ACUITY	VISUAL ACUITY	FOCUSED ATTENTION	VIGILANCE
VISUAL GESTALTS														X						X		X
WORD PAIRS												X		X					X	X		X
GRAPHIS CONTINUOUS PERF. TEST																	X				X	
FINGER TAPPING																X						X
REPETITION OF A SENTENCE												X		X					X			X
STORY RECALL									X			X		X					X			X
DIGIT LEARNING														X							X	X
TOTAL WAIS									X	X	X	X		X	X	X	X	X	X	X	X	X
RORSCHACH		X	X				X				X											X

Fig. 5.

PSYCHOLOGICAL FUNCTIONS MEASURED AND PSYCHODIAGNOSTIC PARAMETERS USED AT THE INSTITUTE OF OCCUPATIONAL HEALTH - HELSINKI (FINLAND)

PSYCHOLOGICAL FUNCTIONS AND TRAITS MEASURED

TESTS USED	SELF-PERCEPTION	HOSTILITY	BODILY PREOCCUPATIONS	DISPHORIA	ANXIETY	DEPRESSION	NEUROTIC TRAITS	INTRO-EXTRAVERSION	ABSTRACT REASONING	ABSTRACT PERCEPT. ABILITIES	PERCEPTIVE ABILITIES	VERBAL ABILITIES	DELAYED MEMORY	SHORT-TERM MEMORY	EYE-HAND COORDINATION	MOTOR ABILITIES	SPACE ORIENTATION	TIME ORIENTATION	AUDITORY ACUITY	VISUAL ACUITY	FOCUSED ATTENTION	VIGILANCE
WECHSLER DIGIT SPAN														X							X	
WECHSLER SIMILARITIES									X			X										X
WECHSLER BLOCK DESIGN										X					X		X					X
WECHSLER PICTURE COMPLETION											X									X		X
WECHSLER DIGIT SYMBOL														X	X	X				X		
BOURDON-VIERSMA														.			X			X	X	
SANTA ANA															X	X						X
BENTON VRT									X					X			X					X
SYMMETRY DRAWING															X	X	X					X
MIRA																X	X					X
FINGER TAPPING																X						X
SIMPLE REACTION TIMES																X				X		X
CHOICE REACTION TIMES														X	X	X						X
RORSCHACH		X	X				X				X											X

Fig. 6.

PSYCHOLOGICAL FUNCTIONS MEASURED AND PSYCHODIAGNOSTIC PARAMETERS USED AT THE NEUROTOXICOLOGICAL UNIT OF THE INSTITUTE OF OCCUPATIONAL HEALTH OF THE UNIVERSITY OF MILAN - MILAN (ITALY)

PSYCHOLOGICAL FUNCTIONS AND TRAITS MEASURED

TESTS USED	SELF-PERCEPTION	HOSTILITY	BODILY PREOCCUPATIONS	DISPHORIA	ANXIETY	DEPRESSION	NEUROTIC TRAITS	INTRO-EXTRAVERSION	ABSTRACT REASONING	ABSTRACT PERCEPT. ABILITIES	PERCEPTIVE ABILITIES	VERBAL ABILITIES	DELAYED MEMORY	SHORT-TERM MEMORY	EYE-HAND COORDINATION	MOTOR ABILITIES	SPACE ORIENTATION	TIME ORIENTATION	AUDITORY ACUITY	VISUAL ACUITY	FOCUSED ATTENTION	VIGILANCE
RAVEN PM38									X	X							X			X		X
WECHSLER BLOCK DESIGN										X					X		X					X
WECHSLER DIGIT SPAN														X							X	
BENTON VRT										X				X		X	X					X
REY MNESTIC PROFILE											X	X	X	X			X					X
DIGIT SYMBOL														X	X	X					X	
VISUAL/AUDITORY SIMPLE R.T.																X			X	X		X
CHOICE REACTION TIMES														X	X	X						X
TREMOMETRY					X										X	X						X
EYSENCK M.P.I. OR E.P.I.				X			X	X														
IPAT ANXIETY SCALE					X		X															
C.D.Q. IPAT SCALE					X	X																
WARTEGG DRAWING COMPLETION TEST	X	X	X		X	X	X	X														

Fig. 7.

well known and most commonly used substances. The same kind of information is now currently repeated in the studies reported almost monthly in the literature by different authors in different countries. We can therefore be quite confident that these findings are objective and transferable from one country to another. But I do not believe that we can stop at this level. Already obsolete literature on this subject reports many tables of which table 5 in an example. It lists, like all the others of the same kind, the main signs and symptoms consequent to occupational exposure to neurotoxic agents. Do we know more today? Of course, on a theoretical basis,yes. These data have been quantitatively confirmed; we no longer have heavy exposure levels and serious neurological and psychological syndromes to deal with. We are now looking for earlier, slighter and more subtle signs and of course we do find them. That table can be trasformed into a more exhaustive and sophisticated one. We can supplement the correlations with indexes of exposure and effect, we can distinguish between signs and symptoms indicating early stages of involvement or the presence of real deteriorating processes. But it remains basically the same table, listing the same signs and symptoms. All neurotoxic substances induce meomory deficits, visuomotor impairments, reduced vigilance, etc., and the same is true for subjective symptoms, like depression, erethism, insomnia, etc. Clinical experience adds other problems to these questions. Not only the patient's subjectivity but also these same batteries indicate the presence of mechanisms that only seldom fit the scheme we have developed in field studies. One year ago our group carried out a research to verify which were the most frequently occurring signs and symptoms of individuals sent to our Laboratory of Industrial Hygiene for assessment of specific job fitness. Only a few of these subjects were occupationally expo-

Table 5. Main neurotoxic agents investigated and most frequently associated behavioral disfunctions

	Solvents	CS_2	Mercury	Lead	Arsenic	Co
Tremor			x			
Asthenia	x	x	x	x	x	x
Headache	x	x	x	x	x	x
Dizziness		x	x		x	
Sensory loss	x	x			x	
Memory deficit	x	x	x			x
Sleep disturbances	x	x	x	x	x	x
Anxiety	x	x				
Mental impairment	x	x		x		
Irritability	x	x	x	x		
Uncontrolled anger		x	x	x		
Reduced rational control	x		x	x		
Sexual impotence	x	x				
Depression	x	x				x
Confusion						x
Disorientation						x
Changes in attention	x	x	x	x	x	x
Diminished concentration	x	x				x
Reduced motivation		x		x		
Paranoid ideas	x	x				
Motor disorders			x	x		

sed to neurotoxics. Their problems were rather noise, shiftwork and work rhythms. The effects of these factors on their mental and psychological wellbeing were more or less the same as listed in the table mentioned above. Why? There are two possible reasons. Either the underlying mechanisms of functional and organic alterations are the same or our instruments are not sufficiently discriminating. I am certainly for this second explanation. It is perhaps now clear why at the beginning I spoke about a definition of behavior. If we go back to that definition we see that we do not have the possibility of detecting those complex units of responses leading to the final output, which is the only aspect of behavior we are able to measure. We know almost nothing about all the compensatory mechanisms that each of us employs in case of difficulty, and certainly our sophisticated batteries tell nothing about them. We should work in this direction to obtain deeper knowledge both of the mechanisms underlying behavior and of the characteristics of our measuring instruments.
I really hope that these four days of work will see the beginning of a new fruitful period of reflection and constructive novelties, at least like the one we have experienced in the past years and which has for the first time brought behavioral toxicologists together at this Meeting.

ACKNOWLEDGEMENTS

Thanks are extended to Mrs Giovanna Radaelli for expert secretarial assistance and Mr. Gianni Ciboldi for technical assistance in making the tables suitable for printing.

REFERENCES

Benton, A. (1972). The Visual Retention Test: Clinical and experimental applications. The Psychological Corporation, New York, N.Y., 92 p.

Cassitto, M.G., V. Foà (1982). Assessment of Behavioral Toxicity in Occupational Health.Raven Press, in print (International Workshop on Application of Behavioral Pharmacology in Toxicology, Capri, 1982, April 26-28).

Cattell, R.B.,G.F. Stice (1959). The 16 Personality Factors Questionnaire IPAT, 1602, Coronado Dr.; Champaign, Ill.

Cherry, N., B. Rodgers, H. Venables, H.A. Waldron, G.G. Wells (1981). Brit. J. Ind. Med., 38, 346-350.

Eysenck, H.J. (1959). Manual of the Maudsley Personality Inventory. London: University of London press.

Feldam, R.G., N.L. Rick, E.L. Baker (1980). Am. J. Ind. Med., 1, 211-227.

Foà, V., M.G. Cassitto, M. Forzi, C. Bulgheroni (1976). Adverse effects of Environmental Chemicals and Psychotropic Drugs, 2, Elsevier Scientific Publishing Company, Amsterdam.

Foà, V., M.G. Cassitto, R. Gilioli (1980). Advances in Neurotoxicology.Pergamon Press,Oxford and New York.

Gamberale, F., G. Annwall, M. Hultengren (1978). Scand. J. Work Environ Hlth 4, 204-211.

Grandjean, P., E. Arnvig, J. Beckmann (1978). Scand. J. Work Environ Hlth, 4 295-303.

Hänninen, H., K. Lindström (1976). Behavioral Test Battery for toxicopsychological studies Used at the Institute of Occupational Health in Helsinki. Institute of Occupational Health, Helsinki, 51 p.

Kinget, G.M. (1952). The Drawing-Completion-Test. A Projective technique for the Investigation of Personality, based on the Wartegg Test Blank, New York.

Krug, S.E., J.E. Laughlin (1976). Handbook for the IPAT Depression Scale. Institute for Personality and Ability Testing Champaign, Illinois.

Raven, J.C. (1938). Guide to Using Progressive Matrices, London: H.K., Lewis.

Raven, J.C. (1973). Advanced Progressive Matrices, P.M. 38, London H.K., Lewis & Co.

Repko, J.D., C.R. Corum, P.D Jones, L.S. Garcia (1978). The effects of inorganic lead on behavioral and neurological functions. Final Report, Washington D.C.: USDHEW (NIOSH - Publication No 78-128.

Rey, A. (1966). Les Troubles de la Memoire et leur Examen Psychometrique. Charles Dessart, Bruxelles.

Sommerhoff, G. (1974). Logic of the Living Brain. John Wiley & Sons Ltd.

Valciukas, J.A., R. Lilis (1979). Psychometric Techniques in Environmental Science Laboratory,Department of Community Medicine, Mount Sinai School of Medicine of the City University of New York 10029.

Zimmerman, I.L., J.M. Woo-Sam (1973). Clinical Interpretation of the Wechsler Adult Intelligence Scale. Grune & Stratton New York and London.

Xintaras, C., B.L. Johnson, I. de Groot (1974). Behavioral Toxicology: Early Detection of Occupational Hazards. Washington D.C.: USDHEW (NIOSH), Publication No 74-126.

Recent Neurological Developments

Chairmen: R. Boeri (*Italy*)
E. Lukas (*CSSR*)

Optacon® — A New Screening Device for Peripheral Neuropathies

M. L. Bleecker

Johns Hopkins Medical Institutes, Department of Neurology, Baltimore City Hospitals and The Johns Hopkins School of Hygiene and Public Health, Department of Environmental Health Sciences, Division of Occupational Medicine, Baltimore, Maryland, USA

ABSTRACT

Early detection of toxic neuropathies is of paramount concern to the occupational physicians. Sensory symptoms in the hands and feet are the most common presentation of peripheral nerve damage when it is still reversible if the toxic exposure is discontinued. The modified Optacon® is a simple battery powered device that measures fingertip vibration sensation. Two common peripheral neuropathies in the work place namely, alcoholic neuropathy and carpal tunnel syndrome were examined. In the alcoholic neuropathy group threshold scores were significantly elevated in all age groups. Patients with carpal tunnel syndrome tended to have markedly elevated threshold scores in the involved II digit as compared to the uninvolved V digit. The Optacon® appears to be a sensitive screening tool for these two common neuropathies.

KEYWORDS

Optacon®; sensory threshold; peripheral neuropathy

INTRODUCTION

Numerous chemicals (drugs and toxic chemicals) are known to produce sensory peripheral neuropathy. Morphological, electrophysiological and clinical studies have attempted to define the exact nature of these neurotoxic effects. Morphological studies have investigated the anatomical structure and ultrastructure of peripheral nerves. Electrophysiological studies have emphasized the importance of a functional approach to these disorders. However, electrophysiological measures do not always reflect the neurological status. Clinical studies have aimed at detecting the presence of frank disorders but do not have the refinements necessary to detect early signs of peripheral nerve damage, or to follow sensory recovery after exposure to a neurotoxic agent.

Impairment of vibration sensibility, starting distally in the limbs, may be an early sign of peripheral neuropathy (Williamson 1922; Nielson 1972), and measurement of perception thresholds can be used as an index of subclinical disease from metabolic-nutritional disorders or environmental toxic agents, where the patient themselves are unaware of the defect. Since vibration sensation is mediated by the largest afferent nerve fibres, the vibration threshold reflects the functional integrity of these sensory fibres. The clinical importance demands that vibration sensibility should be measured quantitatively. The frequency range (Rumpf 1889) or the disappearance time of vibrations from a tuning fork (Treitel 1897; Pearson 1928) used by earlier observers have been supplemented by threshold determinations by means of electromagnetic devices (Geldard and Gilmer 1924; Gregg 1951; Cosh 1953; Steiness, I. 1957; Goldberg and Lindblom 1979). Dyck et al (1978) has developed a system for sensory evaluation in which vibration is assessed with a stimulus whose waveform could be precisely defined and quantified and which would vary over a wide range and not be subject to damping, one of the major problems with the electromagnetic vibrators. Furthermore, Dyck applied the stimuli at discrete skin points rather than to bony protuberances, so that long bones would not transmit the stimulus to anatomical sites distant from the test site. Unfortunately, this elegant system is impractical for screening in the workplace due to the size of the equipment and time necessary for testing.

The Optacon® a portable electronic device that converts visual images into tactile forms, has been utilized to assess finger-tip sensation. Arezzo and Schaumberg (1980) initially used this device and were able to demonstrate a steady increase in threshold with age and no sex difference. A diabetic study group was examined and an increased threshold was found in 38% correlating with the peripheral neuropathy found on physical exam.

In a study of 150 asymptomatic government workers an attempt was made to validate the thresholds obtained with the Optacon® by comparison with a standardized neurologic exam and routine nerve conduction studies (Bleecker 1981). An age effect with increasing thresholds was again demonstrated. However, when threshold values were compared between the group having a normal physical exam and electrodiagnostic studies and the other having evidence of peripheral neuropathy, considerable overlap was observed. The routine neurologic exam and electrodiagnostic studies appeared to be more sensitive indicators of peripheral neuropathy which appeared to be idiopathic.

The Optacon® may have increased sensitivity and specificity when applied to the study of a neuropathy with a known etiology. Therefore, the alcoholic-nutritional neuropathy and the carpal tunnel syndrome, two frequently occuring neuropathies in the workplace, were examined.

METHODS

Subjects - A total of 117 subjects were tested. Fifty-six were normal with no evidence of peripheral neuropathy by exam and no history of neurological disease, diabetes mellitus, thyroid dysfunction or heavy alcohol abuse. Fifty-six patients were examined in an alcoholic outpatient clinic with only two having no evidence of peripheral neuropathy. These two individuals had been in the program for one and three years. The remainder had evidence of peripheral neuropathy on physical exam and had been involved in the clinic from two weeks to nine years. Subjects were selected solely by their availability and were tested prior to a neurologic exam by M.L.B. Patients seen in the Neurology Clinic in which the diagnosis of carpal tunnel syndrome was made and subsequently confirmed by nerve conduction study were studied with the Optacon® if there was no evidence of peripheral neuropathy or heavy alcohol use.

Apparatus - The modified Optacon® tactile stimulator is a portable, battery powered unit weighing four pounds. It was originally designed for use in concert with a camera module to enable blind individuals to read independently of Braille (Mims 1973). The stimulator consists of 144 miniature rods organized into a matrix of 6 rows with 24 pins, with a 2 mm horizontal and 1 mm vertical interod spacing. Each rod protrudes through a contoured plastic plate and contacts a discrete portion of the skin. The rods vibrate continuously at 230 H_z; the height the rods extend above the plate and the amplitude of vibration varies as a function of voltage. The second and fifth finger of the non-dominant hand was positioned within the Optacon® stimulator so that approximately 2 cm^2 of the ventral surface of the finger pad was in contact with the stimulating surface. In cases of carpal tunnel syndrome the symptomatic hand was used. The subject's forearm, wrist and hand rested on a platform which was positioned parallel to the surface of the Optacon®. Each subject was instructed to allow his finger to rest on the vibrating surface but not to exert any downward force. This was critical since the pressure of the finger on the pins not only has a possible effect on the efficacy of the touch receptors in the skin but also has a large influence on the pin movements because the piezo electric activators have no feedback control and thus the pin excursions are very much load dependent. Prior to testing, subjects were given a period of suprathreshold stimulation so that they would become familiar with the stimulus. Goldberg and Lindblom (1979) found that the variance of the vibration perception threshold was less than that of the disappearance threshold and therefore determination of perception threshold alone was felt to be sufficient. Six ascending series were obtained and the threshold was determined by averaging the six values. During testing the subjects wore earphones to mask the sounds which resulted from the higher intensity vibrations. The time of the ascending trials was varied to avoid providing a temporal clue, and the subject was prevented from viewing the intensity knob or the voltmeter to eliminate any visual artifact. On approximately one-sixth of the trials a "sham" series was run in which the stimulus was not delivered. The total testing time per subject was approximately five minutes.

RESULTS

Table 1 shows the mean voltage at threshold in the index finger and the standard deviations of these scores for the normal group and those with a peripheral neuropathy (PN). There is a systematic increase in the mean scores as a function of age in both groups. The variability of the data is homogeneous across all age groups. For each decade the PN group had a significantly elevated threshold ($p < .001$). Review of the neurologic exams in the PN group revealed 38% to have no upper extremity involvement in spite of elevated threshold scores. All these cases were essentially below the age of 40. Patients who had been enrolled in the alcoholic program for longer periods had milder peripheral neuropathies and a tendency towards lower threshold scores. For example, a 38 year old male had been enrolled for over one year and had slight vibratory loss in his feet when the threshold score was 4.27. However, a 50 year old male who had been in the program for two weeks had significant pin and vibratory loss to mid-calf and mid-forearm and absent deep tendon reflexes at the ankle. His threshold score was 6.65.

Table 1. Mean and Standard Deviation of Voltage at Threshold for Normal Subjects and Those With Neuropathy

Age	N	Normal $\bar{x} \pm SD$		N	Neuropathy $\bar{x} \pm SD$	
21 - 30	14	2.9	0.74	11	4.69	0.75 *
31 - 40	13	3.67	0.93	20	5.40	1.15 *
41 - 50	16	4.17	0.85	11	5.74	1.22 *
51 - 60	13	4.5	0.5	12	6.45	1.33 *

* $p < .001$

When the difference in threshold scores between the second and fifth digit was examined in the control group, a median of 0.52 was found with a range from .03 to 1.24. The PN group had a similar difference. This was expected since the alcoholic-nutritional polyneuropathy all the nerves in the distal extremity are affected equally in a stocking-glove distribution.

A major complaint with the Optacon® has been the large interindividual difference. It was hoped to overcome this problem by using each individual as his own control in the evaluation of carpal tunnel syndrome. In the patients with the carpal tunnel syndrome, only one out of seven had a difference in threshold scores between the II and V digit of .5. The remainder as seen in Table 2 were between 1.29 to 3.14. This demonstrates the selective involvement of the median nerve (II finger) by compression in the carpal canal as compared to the uninvolved ulnar nerve innervating the V finger.

Table 2. Threshold Scores in the Carpal Tunnel Syndrome

Age	II Digit $\bar{x} \pm SD$		V Digit $\bar{x} \pm SD$		Difference Between II & V
31	5.08	0.19	3.05	0.23	2.03
36	5.90	0.24	4.61	0.31	1.29
47	4.6	0.35	3.0	0.13	1.60
51	7.01	0.46	4.80	0.21	2.21
53	5.8	0.35	5.30	0.46	0.50
57	8.80	0.20	5.66	0.20	3.14
60	6.02	0.49	4.15	0.75	1.87

DISCUSSION

A frequent problem encountered in Occupational Neurology is toxic neuropathy. Even though neurologists have been evaluating peripheral neuropathies for a long time, this has usually occurred in symptomatic individuals. Now there exists the problem of screening asymptomatic workers for early evidence of toxic neuropathies.

The neurologic exam is a very sensitive tool, however, it requires special training and standardization, which is difficult even between trained neurologists. Since the neurologic exam requires subjective interpretation and is difficult to quantitate, routine electrodiagnostic studies have been used extensively in the evaluation of toxic neuropathies. Again there are limitations even when controlling for all of the confounding variables. Most neurotoxic substances produce axonal degeneration rather than segmental demyelmination. Consequently, there is no gross change in maximal nerve conduction velocity in the majority of toxic neuropathies. Secondly, even when symptoms and signs are predominately motor, electrophysiological abnormalities may be more marked in sensory than motor nerves. It is known that considerable loss of sensory nerve fibres can occur before there is clinically detectable sensory loss. Furthermore, when recording from sensory nervous, the amplitude of the evaluated potential reflects the number of fibres conducting in the nerve. Amplitude of the sensory response as an absolute measurement in screening for toxic neuropathies is too variable to provide the sensitivity required.

The Optacon® should be effective in detecting early changes, since it measures sensation at the location of early pathology in the "dying back" or toxic neuropathy. The Optacon® is believed to stimulate the rapidly adapting mechanoreceptors, the Pacinion corpuscle, since its maximum sensitivity occurs at 200-300 H_z. Deterioration of the Pacinion corpuscle in early acrylamide neuropathy has been observed in animal studies.(Schaumberg, 1974) This could account for the high vulnerability of vibration sense in early acrylamide neuropathy. At present, Arrezo and Schaumberg (personal communication) are using the modified Optacon® to screen workers in acrylamide plants for evidence of early neuropathy. Since they will have the ability for longitudinal measurements, the sensitivity of this tool should be greatly increased with each worker serving as his own control.

In the present study of patients seen in the outpatient alcoholic clinic, the Optacon® was able to detect sensory pathology in the upper extremities when the neurologic examination had only demonstrated evidence of peripheral neuropathy in the lower extremities. Obviously, this will not be true for all neuropathies but probably is a reflection of the underlying pathology. Behse and Buchthal (1977) examined nerve biopsies from 37 patients with alcoholic neuropathy which showed loss of small and large fibres in most nerves with rare signs of regeneration or segmental demyelimination.

A completely different pathology exists in the carpal tunnel syndrome (CTS), an entrapment neuropathy in which median nerve injury results from compression by neighboring anatomical structures. Pressure is known to affect the fastest conducting peripheral nerve fibres. Dellon (1981) believes that abnormal vibratory perception in the index finger in comparison to the ipsilateral little finger, is the earliest possible non-provocative sign in the carpal tunnel syndrome. In patients with a history compatible with the carpal tunnel syndrome, Dellon found 72% to have an abnormal perception of vibratory stimuli. Obviously, the CTS group in the present study is too small to develop meaningful statistics but if one uses the mean difference between the II and V digit in the control group ± 2.5 SD (1.5) as the upper limit, then 71% of the CTS patients had an abnormal perception of the vibratory stimulus in the index finger.

The present study describes the utility of the modified Optacon® for screening neuropathies in which the underlying pathology is known.

ACKNOWLEDGEMENT

The author wishes to thank Paula Pysik for her generous technical assistance. This work was supported by core support by NIOSH for the Center of Occupational and Environmental Health at The Johns Hopkins University.

REFERENCES

Arezzo, J., and H. Schaumberg (1980) J. Occup. Med. 22, 461-464.

Behse, F., and F. Buchthal (1977) Ann. Neurol. 2, 95-110.

Bleecker, M. (1981) Proceedings XX International Cong. on Occup. Health, Cairo, Egypt.

Cosh, J.A. (1953) Clin. Science 12, 131-151.

Dellon, A.L. (1981) Evaluation of Sensibility and Re-education of Sensation in the Hand. Williams and Wilkins, Baltimore.

Dyck, P.J., I.R. Zimmerman, P.C. O'Brien, A. Ness, P.E. Caslay, J. Karnes, and W. Bushek (1978) Ann. Neurol. 4, 502-510.

Goldberg, J.M. and U. Lindholm (1979) J. Neural Neurosurg. Psychiat. 42, 793-803.

Gelgard, F.A. and B. von Haller Gilmer (1934) J. Gen. Psychol. 11, 301-309.

Gregg, E.C. (1951) Archs. Neural and Psychiat. 66, 403-411.

Mims, F. (1973) New Outlook for the Blind 67, 407-414.

Nielsen, V.K. (1972) Acta. Med. Scand. 191, 287-296.

Pearson, G.H. (1928) Archs. Neurol. Psychiat. 20, 482-496.

Rumpf, H.T. (1889) Neurologisches Zentralblatt 8, 222-230.

Schaumberg, H.H., Wisniewski, H.M., and Spencer, P.S. (1974) J. Neuropath. and Exper. Neurol. 33, 260-285.

Trietel, L. (1897) Archiv für Psychiatrie and Nerven Krenkenheiten 29, 633-640.

Williamson, R.T. (1922) Am. J. Med. Sci. 164, 715-727.

Electro-oculography

G. Avanzini, M. Casazza and F. Girotti

Institute of Neurology "C. Besta", via Celoria 11,
20133 Milan, Italy

Electro-oculography (EOG) is a useful objective method for studying eye movements, based on the fact that the eye acts as a dipole. There is actually a potential difference between the cornea (positive pole) and the foveal sclera (negative pole), which normally ranges from 10 to 30 mV.
As the eyes rotate, the potential field moves and suitably placed peri-orbital skin electrodes can pick up the difference in voltage. These voltage changes are linearly correlated with the eye excursions, at least in the range of 30° to either side from the fixation point. The EOG recording is derived from paired skin electrodes applied to the periorbital region, either on both lateral canthi for studying horizontal conjugate movements, on the superior and inferior orbital rims for recording vertical single eye movements, or on the medial and lateral canthi of each eye for horizontal disconjugate movements. It is advisable to simultaneously record vertical and horizontal individual movements on different channels.
The EOG recording is amplified by a DC-coupled preamplifier, and the signal can be visualized on an oscilloscope and recorded on a magnetic tape and/or paper. To obtain good technical results it is important to use special electrodes that provide a stable contact with low resistance and a suitable DC amplifer with an input resistance of more than 1 megaohm and a linear dynamic range of 100:1. The maximum instrument drift should not be more than 1 μV for 60 min. EOG can be employed for recording different kinds of ocular movements, elicited by an adequate stimulation system, including moving light spots (smooth pursuit movements), randomly lit targets (saccades), black and white rotating strips covering the entire visual field (optokinetic nystagmus), and rotation of the head or heat stimulation (vestibulo-ocular movements). The data obtained from the EOG recording may be studied with automatic analysis: the EOG is converted to a digital form and the computer identifies the parameters, according to which the different kinds of ocular movements can be characterized (Fig. 1). The aim of the systems that control ocular motility is to obtain a perfect matching between fovea and retinal image of meaningful objects.
This is achieved through the combination of different systems that control ocular movements: the main ones, which control conjugate ocular motility, are saccadic, pursuit, optokinetic and vestibulo-ocular systems. Many studies have been devoted to define their anatomo-physiological basis and their functions, and a significant contribution to this complex problem

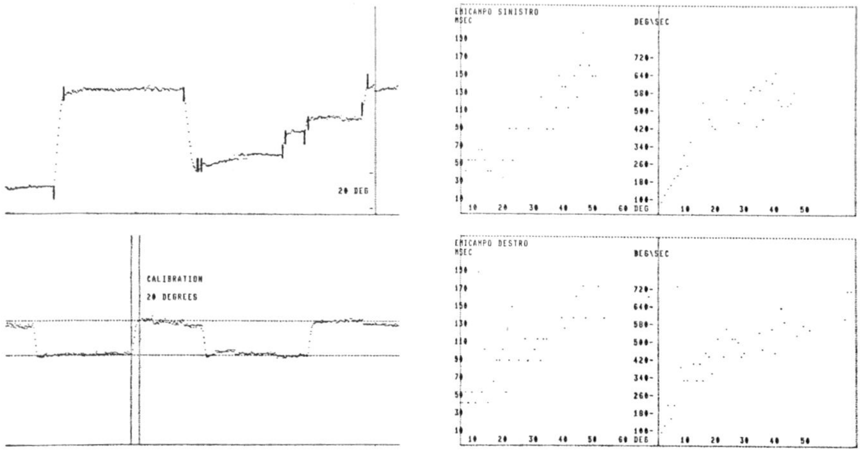

Fig. 1. Example of automatic analysis of saccades according to the SACCAD program (Cefalà et al., 1979). Left: top, the start and end of each saccade is automatically identified by computer on digital EOG; amplitudes and durations are calculated with reference to the calibration movement, as shown at the bottom. Right: durations (msec) and velocities (deg/sec) of individual saccades are plotted against amplitude (deg) for both hemifields (emicampo sinistro = = left hemifield; emicampo sinistro = right hemifield).

has been recently provided by mathematical models (Young and Stark, 1963; Robinson, 1977). These results emphasize the crucial role of some brainstem structures, in particular the vestibular nuclei and their projections to the oculomotor nuclei through the medial longitudinal fasciculus, and the paramedian pontine reticular formation (PPRF), which also acts on oculomotor neurons according to an appropriate motor code. Many other influences seem to converge on these structures, particularly from the cerebral cortex, basal ganglia, thalamus, and cerebellum, suggesting the complexity of the oculomotor system and the involvement of many central structures in its correct functioning. It appears therefore interesting to examine each system in more detail, with particular regard to the respective pathological alterations. In the present context we will consider only the most important kinds of conjugate movements, leaving out disconjugate movements physiologically elicited by retinal image disparity (vergence movements).

The saccadic system generates rapid eye movements, used to catch an object that suddenly appears in the peripheral visual field and bring its image onto the fovea. The saccadic movement starts after an initial latency of about 200 msec, velocity can reach 700°/sec, and is proportional to the movement amplitude (Fig. 2). Both these parameters may be prolonged in hemispheric lesions (Sharpe et al., 1979) and in basal ganglia diseases (Shibasaki et al., 1979). The slowing-down of the saccades is a typical feature in degenerative diseases involving brainstem structures; it is particularly evident in olivo-ponto-cerebellar atrophy and in progressive supranuclear palsy (Fig. 3). In these cases, the PPRF is altered, which is consistent with its well-known role in programming and executing the saccade (Cohen and Henn, 1972; Luschei and Fuchs, 1972).

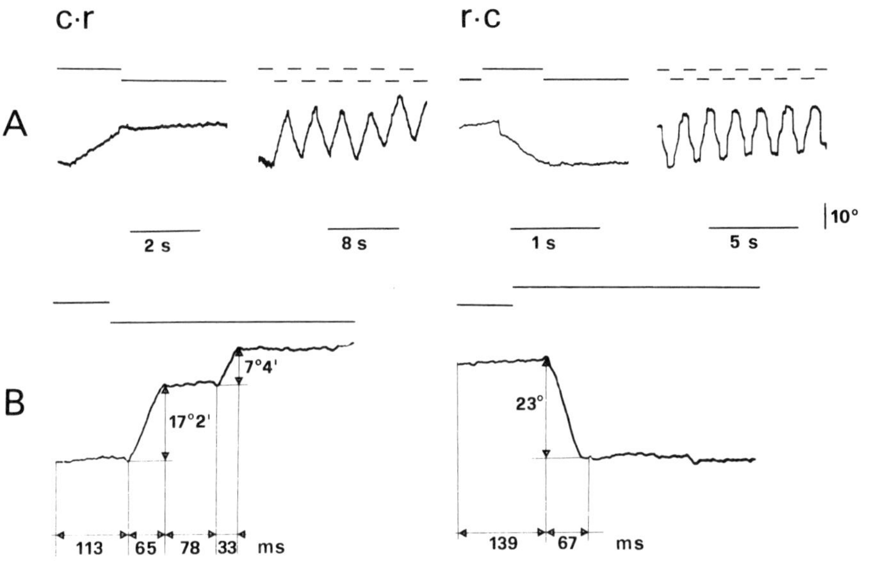

Fig. 2. Normal subject. (A) Single and repeated smooth pursuit movements at two different tracking velocities. C-r = center to right; r-c = = right to center. (B) Refixation saccadic movements. In each recording, the upper trace indicates the beginning and end of the spot track. Internal latency of the stimulation system, 60 ms. s = seconds; ms = milliseconds. The amplitude of the eye movements is expressed in sexagesimal degrees (°) and minutes ('). (Reprinted from Avanzini et al., 1979a.).

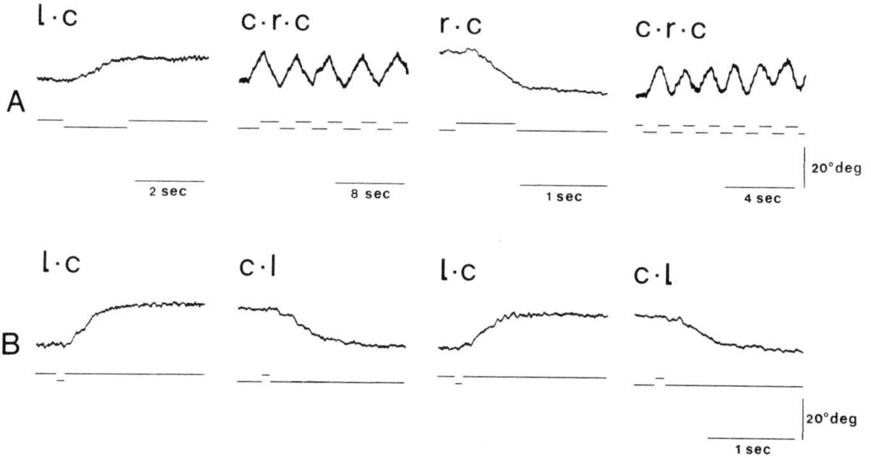

Fig. 3. Smooth pursuit movements (A) and saccades (B) performed by a patient with progressive supranuclear palsy. Note the extreme slowing-down of saccades. (Reprinted from Avanzini et al., 1978).

Saccade accuracy may also be impaired, leading to metric alterations: dysmetric saccades, either hypometric or hypermetric, occur in several pathological conditions and even in normal subjects, particularly in large movements (more than 20°). Oculomotor disturbances of this type are the expression of a mismatch of the control signals encoded in the PPRF, as Bahill and Troost (1979) have exhaustively discussed. Dysmetric (usually hypermetric) saccades are often found in patients with cerebellar diseases (Zee *et al.*, 1976; Avanzini *et al.*, 1979b); they usually consist of a main hypermetric saccade, followed by a single corrective saccade (dynamic overshoot) or by a slow drift movement (glissadic overshoot) (Fig. 4). In other cases, the target is reached by a sequence of small saccades giving rise to a "fragmented" or, more appropriately, "multiple-step saccadic refixation", which may be observed in mesencephalic tectal lesions (Girotti *et al.*, 1980), basal ganglia diseases (DeJong and Jones, 1971), and cortical lesions involving parietooccipital cortex (Meienberg *et al.*, 1981; Girotti *et al.*, 1982).

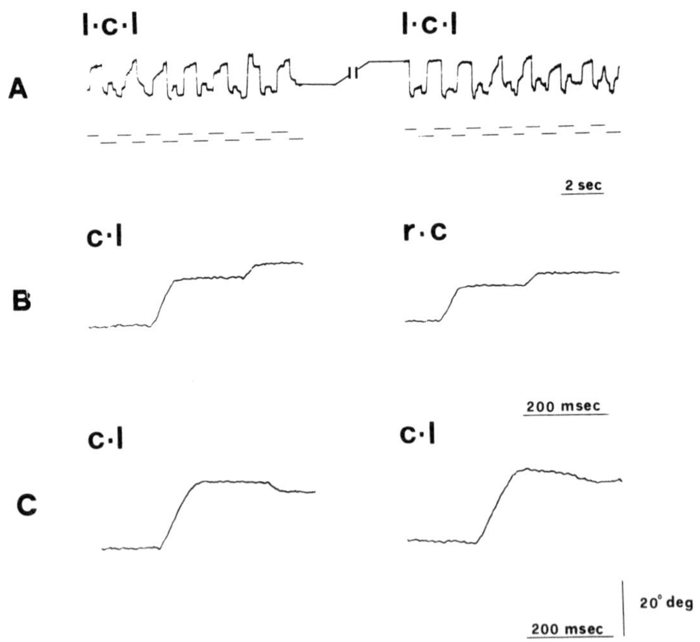

Fig. 4 (A) Irregular smooth pursuit movements recorded from a patient with Holmes' cerebellar atrophy. (B) Hypometric saccades recorded from the same patient. (C) Same patient, saccadic and glissadic overshoot. (Reprinted from Avanzini *et al.*, 1979b).

Smooth pursuit movements are used to follow an object moving slowly in the visual field, while keeping its image on the fovea. If the target's velocity is between 40-45°/sec, the matching between eye and target is almost perfect: in these cases, the gain of the pursuit system, i.e., the ratio between eye and target velocities, is close to one. When the target's velocity exceeds these limits the oculomotor strategy appears completely different: the eyes cannot "follow" the object, they "lose" it and "catch" it again with one or more saccadic movements. The same strategy is also observed in pathologic situations, resulting in a loss of pursuit gain, as in diseases involving the cerebellum (Zee et al., 1976; Kirkham et al., 1979), cerebral cortex (Sharpe et al., 1979), basal ganglia (Avanzini et al., 1979 a; Shibasaki et al., 1979) (Fig. 5), and in drug intoxication (Bittencourt et al., 1980).

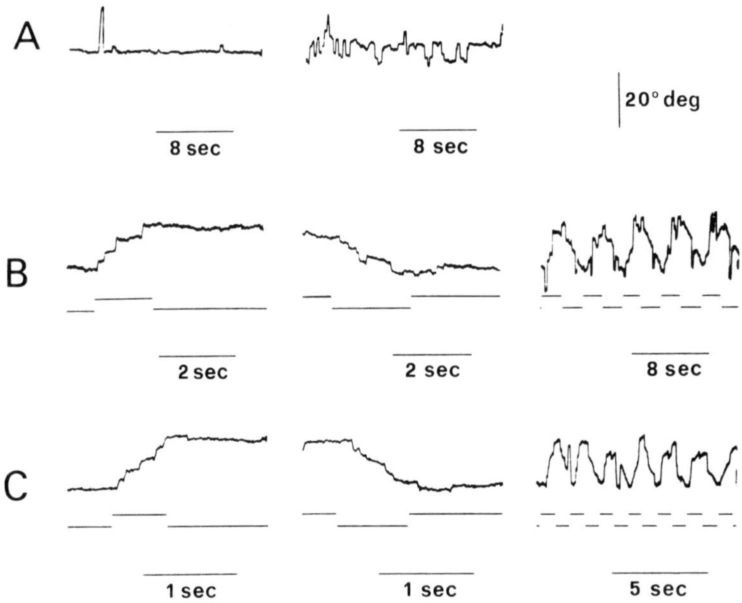

Fig. 5. Fixations (A) and pursuit movements (B, C) of a patient with Huntington's chorea. Note the instability of fixation and the "staircase" pattern of pursuit movement due to the interposed small "catch-up" saccades. (Reprinted from Avanzini et al., 1978).

The movement of the entire visual field evokes an optokinetic nystagmus (OKN), consisting of an ipsiversive slow phase, which represents the specific response of the optokinetic system, followed by a quick contraversive phase, which is a regular saccade. The OKN can be elicited by the rotation of a visual pattern (black and white strips or even irregular patterns) projected on a screen, which occupies the entire visual field of the subject. It should be pointed out that the pursuit system is usually coactivated by these kinds of stimulation; however, the specific contribution of the optokinetic system can be evaluated when the stimulus rotation velocity exceeds the maximum critical value for the pursuit system. In fact, a gain close to one is maintained for optokinetic response up to 60-65°/sec in man. The OKN may be altered and irregular, or totally absent in several pathological conditions: cerebellar and brainstem diseases, and occipital cortex lesions (Dichgans and Jung, 1974; Dichgans, 1979).

A fourth kind of ocular movement is vestibulo-oculomotor. It can be elicited by rotation of the head, which stimulates vestibular receptors evoking a reflex contraversive ocular movement, followed by a compensatory saccade, resetting the eyes in their primary position. The sequence of these two phases represents the vestibular nystagmus. Changes in excitability of vestibular receptors, lesions of vestibular nuclei and vestibulo-oculomotor pathways may all induce alterations in rhythm and/or morphology of vestibular nystagmus. The analysis of vestibular nystagmus has therefore been recently proposed as a quick, sensitive method to evaluate the effects of alcohol, diazepam, opiates, barbiturate, cocain, marijuana and hallucinogenic drugs on the central nervous system (Westerman and Gilbert, 1981). It should be pointed out that the vestibulo-oculomotor projections represent the main system that secures the proper pattern of reciprocal innervation to the oculomotor muscles, resulting in conjugate ocular motility. Lesions involving the medial longitudinal fasciculus usually manifest themselves by a disconjugate eye movement syndrome, termed internuclear ophthalmoplegia.

EOG provides a precise qualitative and quantitative evaluation of the performances of different systems that control ocular motility. As previously considered, each system can be altered by various neurological diseases and different neurotoxic substances. Alcohol reduces saccade velocity, probably acting on the synaptic transmission in the PPRF (Lehtinen et al., 1979). Similar effects have been reported for drugs such as barbiturate and diazepam (Gentles and Thomas, 1971). In subjects taking phenobarbital and phenytoin, a loss of gain of the smooth pursuit system, correlated with blood levels, has recently been reported (Bittencourt et al., 1980). Methadone hydrochloride is reported to depress vestibulo-ocular reflex (Westerman and Gilbert, 1981). All these data stress the existence of oculomotor abnormalities determined by several neurotoxic agents that act on brainstem structures, whose effects deserve further investigation.
EOG could thus be an interesting tool in occupational medicine to assess or exclude brainstem involvement after prolonged contact with toxic agents.

Another mechanism which should be taken into account in the evaluation of ocular movements is attention, since changes in attention may affect psychomotor reactions and performances. It has been demonstrated that repeated trials lead to a marked slowing-down of saccades with a variety of anomalous-looking saccadic eye movements. According to Bahill and Stark (1975), these aspects can be related to a "fatigue" phenomenon, considered as a time-correlated decrement in efficiency and skill, which may be observed while performing a monotonous task that requires a high level of arousal. EOG may therefore be viewed as an ideal means to quantitatively assess the general psychophysiological state of individuals who must perform critical tasks under such particular conditions.

ACKNOWLEDGMENT

We wish to thank Ms. B. Johnston for editing and preparing the manuscript. Dr. Marina Casazza was supported by a fellowship (*assegno di formazione professionale*) from the Consiglio Nazionale delle Ricerche, Rome.

REFERENCES

Avanzini, G., F. Girotti, T. Caraceni and R. Spreafico (1979a). J. Neurol. Neurosurg. Psychiatry, 42, 581-589.

Avanzini,G., F. Girotti, P. Crenna and S. Negri (1979b). Arch. Neurol., 36, 274-280.
Avanzini,G., F. Girotti, R. Spreafico and T. Caraceni (1978). Rev. Otoneuroophtalmol., 50, 211-217.
Bahill, A.T. and L. Stark (1975). Exp. Neurol., 48, 95-106.
Bahill, A.T. and T.B. Troost (1979). Neurology, 29, 1150-1152.
Bittencourt, P.R.M., M.A. Gresty and A. Richens (1980). J. Neurol. Neurosurg. Psychiatry, 43, 1119-1124.
Cefalà, A., F. Girotti, M. Pederzoli and G. Avanzini (1979). Riv. It. EEG Neurofisiol. Clin., 2, 7-10.
Cohen, B., and V. Henn (1972). Brain Res., 46, 403-410.
DeJong, J.D. and G.M. Jones (1971). Exp. Neurol., 32, 58-68.
Dichgans, J. (1979). Visual-vestibular interaction in the control of eye movements. Eye Movement Analysis in Neurological Diagnosis, Schmid, R., and Zambarbieri, D., eds., Center of Bioengineering, University of Pavia, Pavia.
Dichgans, J. and R. Jung (1974). Oculomotor abnormalities due to cerebellar lesions. Basic Mechanisms of Ocular Motility and Their Clinical Implications. Lennerstrand, G., and Bach-y-Rita, P., eds., Pergamon Press, New York, pp. 281-298.
Gentles, W. and E.L. Thomas (1971). Clin. Pharmacol. Ther., 12, 563-574.
Girotti, F., M. Casazza, C. Milanese, F. Corridori, G. Avanzini and A. Allegranza (1982). Cortex, in press.
Girotti, F., A. Cefalà, A. Sghirlanzoni, M. Savoiardo and G. Avanzini (1980). It. J. Neurol. Sci., 1, 139-148.
Kirkham, T.H., D. Guitton, A. Katsarkas, L.B. Kline and E. Andermann (1979). Can. J. Neurol. Sci., 6, 167-172.
Lehtinen, I., A.H. Lang, V. Jantti and E. Keskinen (1979) Psychopharmacology, 63, 17-23.
Luschei, E.S. and A.F. Fuchs (1972). J. Neurophysiol., 35, 445-461.
Meienberg, O., W.H. Zangemeister, M. Rosenberg, W.F. Hoyt and L. Stark (1981). Ann. Neurol., 9, 537-544.
Robinson, D.A. (1977). Vestibular and optokinetic symbiosis: An example of explaining by modelling. Control of Gaze by Brainstem Neurons, Developments in Neuroscience, vol. 1, Baker, R. and Berthoz, A., eds. Elsevier- North Holland Biomedical Press, Amsterdam, New York.
Sharpe, J.A., A.W. Lo and H.E. Rabinovitch (1979). Brain, 102, 387-403.
Shibasaki, H., T. Sadatoshi and K. Yoshigoro (1979). Arch. Neurol., 36, 360-364.
Westerman, S.T. and L.M. Gilbert (1981). Laryngoscope, 91, 1536-1547.
Young, L.R. and L. Stark (1973). IEEE Trans. HFE, 4, 28-51.
Zee, D.S., R.D. Yee, D.G. Cogan, D.A. Robinson and W.K. Engel (1976). Brain, 99, 207-234.

Electroencephalographic Spectral Analysis

P. Genta* and I. Bruni**

*Institute of Occupational Health, Department of Preventive Medicine, University of Milan, Italy
*OTE Biomedica, Florence, Italy

ABSTRACT

While the vast amount of information supplied by conventional EEG is of paramount importance in clinical neurology, some problems may arise with EEG studies of large working populations: 1 - the content of the raw EEG trace is qualitatively adequate for specific demands, such as the identification of epileptogenic activity, but supplies limited quantitative information; 2 - the interpretation of the EEG tracings is to a great extent subject to individual evaluation.
The introduction into EEG of the Fast Fourier Transform (FFT) and the availability of special purpose computers have made an entirely new approach to the quantification of the EEG curve possible. After a review of the classical methods of EEG analysis (amplitude analysis, interval analysis) the importance of frequency analysis is emphasized and its main features discussed.
The Fast Fourier Transform is evaluated in relation to the most recent Berg Transform. The parameters apt to characterize the EEG curve as well as their statistical evaluation are then discussed. Since it is now possible to work on continuous variables also in the field of EEG interpretation, it is desirable that this new method be utilised also in neuroepidemiological surveys of populations at high risk of exposure to neurotoxic workplace chemicals. Preliminary validation testing is being carried out and should be concluded soon.

KEYWORDS

Neurophysiology; neurotoxicology; conventional EEG; Fast Fourier Transform; Berg Transform.

INTRODUCTION

Electroencephalography (EEG) is a technique of recording electrical activity of the brain registered by means of electrodes applied to the scalp and provides information on the cerebral electrophysiological conditions. Independently of where and how the brain electrical activity originates, it is transformed into continuous signals that are conventionally reported on paper. The semantic content of the recordings is not fully understood; the electroencephalographer draws from the EEG reading only what he was taught to read and look for. Once the electroencephalographer is provided with the information concerning age, general conditions and the level of vigilance of the subject, he takes into account the following seven parameters: frequency or wave length, amplitude, distribution, wave shape, interhemisphere coherence, pattern of recurrence of the wave shape, reactivity of the tracing. However, the interpretation of each of these parameters is loaded with the so-called "human factor". In other words, the evaluation of the tracings is influenced by education and professional background, experience, emotional conditions, etc., of the electroencephalographer.

Even when the influence of these confounding factors can be reduced, a drawback of direct interpretation of the EEG trace is represented by the limited resolving power of visual analysis. In fact a number of information contents of the EEG tracing are not perceived even by the most experienced examiner.

It has been shown, in fact, that there is a good correlation between the evaluations of different electroencephalographers only when the tracings are normal; however such correlation falls rapidly for borderline tracings and rises again, but to an insufficient level, for pathologic traces. It is known from the literature that the EEG parameters that are more likely to lose power of information are frequency, amplitude and interhemisphere coherence. A great amount of research, over the last twenty years, has concentrated on the analysis of the EEG signal along the following lines: spectral analysis by filtering; spectral analysis by the Fourier Transform; correlation analysis. In general, spectral analysis has given results that can be transferred from research into clinical practice and is considered more applicable. Spectral analysis is a methodology that permits not only analysis and evaluation of some contents of the EEG trace, but also, and this is of crucial importance, quantitation of these contents.

Kelvin's famous remark "when you can measure what you are speaking about and express it in number, you know something about it", is incredibly pertinent to neurophysiology and especially in the case of neuroepidemiological surveys of occupationally exposed populations. EEG spectral analysis, in our opinion, can meet the demands of providing quantitative data for further statistical processing. In addition to this advantage, EEG spectral analysis is an important clinical adjunct for the interpretation of vague and borderline EEG abnormalities so frequently observed in practice. Moreover, the distribution of the abnormalities in the population can be studied using a rigorous method and their pattern of evolution can be followed.

METHODOLOGIES

The term "spectrum" indicates a mathematical function or the diagram that represents how the relative quantities are distributed within a group of homogeneous entities with respect to a certain magnitude that differentiates them. In the domain of physics the term spectrum is used to indicate a diagram giving the distribution of a magnitude, usually intensity as a function of another magnitude, usually energy. It is possible, with the spectroscopic method, to separate a light radiation produced by a flame into the constituent frequencies so as to determine the chemical nature of the burned components. In electroencephalography, it is possible to create a similar situation: the separation of the EEG signal into quantities so as to permit the construction of a histogram signal representing qualitatively as well as quantitatively the given. As is known, the EEG trace is obtained by recording the trend of the brain potentials against time. The parameter frequency has acquired such importance for EEG evaluation that it is considered to be one of the primary elements of judgement. In fact, it is common to speak of more or less "slow" tracings and the classical EEG rhythms are exactly identified and classified on a frequency basis.

The first attempts to supply information on the existence, organization and distribution of the dominant frequencies in the EEG, was based on the measurement of the distance between zero points or on the equivalent method of the null derivative.

In the first case (fig. 1), the problem lies in the identification of the instants when the EEG signal crosses the zero line; in the second case, since definition of zero is vague and questionable, the problem lies in the identification of the instants when the slope takes a different direction (fig. 2).

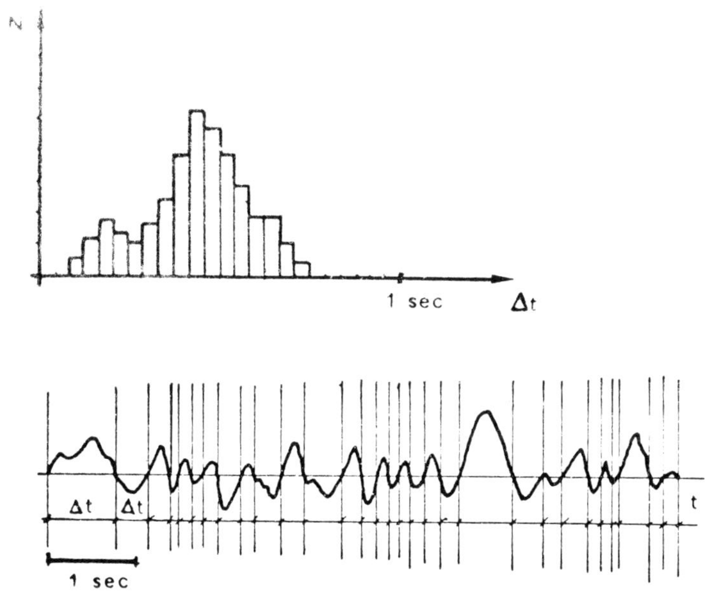

Fig. 1. Method of zeros: histogram constructed by plotting against each Δt the number of intervals between two consecutive zero points of Δt duration.

When these points are obtained and the intervals between consecutive points are measured, the histogram is constructed by plotting against each duration the number of intervals having that duration. These two methods are subject to great variability due essentially to the signal amplification factor or, better still, to the resolution by which that signal is observed (fig. 3).

The electric signals can be classified in two broad categories according to their nature: deterministic and random signals.

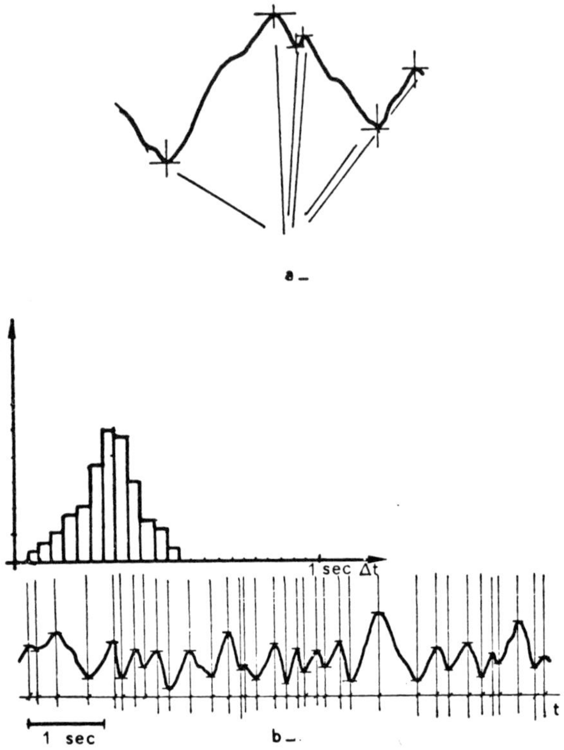

Fig. 2. Method of the null derivative: a) identification of null derivative points by means of the tangent; b) histogram constructed by plotting for each Δt the number of intervals of Δt duration between two null derivative points.

Deterministic signals can be expressed analytically by a function of time; for these signals it is possible not only to know their trend instant by instant, but also to predict their behavior for time intervals not yet elapsed. The pure sinusoidal signal is a classical example of a deterministic signal.

The random signal is more complex in that the prerogative of a definite relation with time does not exist any longer: a factor of casualness is present and no predictive criteria are offered even if this signal can still be defined with representative parameters. The height of the sea waves gives an idea of the random signal.

Let us now suppose we have two sinusoids with different frequency and phase; by associating them we will obtain a signal where the two original components keep their previous characteristics, but the generated signal will have different shape and frequency. This can be repeated by an n number of sinusoids with different frequency to produce a signal that will result from the summation of n sinusoids.

When analyzing the EEG signal the problem lies in having a method available that permits this signal to be observed by the reverse procedure: resolve a

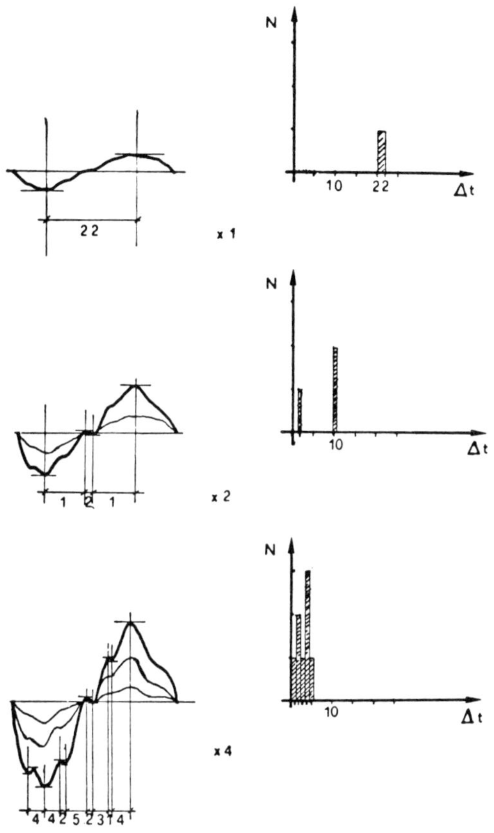

Fig. 3. Example of dependence on the amplitude of the histograms calculated with null derivative method: signal amplification x1, x2, x4.

signal into its components with different frequencies.
One could imagine a series of sieves where each sieve prevents the signal with a particular grain from passing through. If we speak of filters and frequencies instead of sieves and grains we obtain the system of analysis based on the differences of frequencies (fig. 4).
If the mean value for each frequency in a given time interval called epoch is diagramatically represented, the spectrum of the signal during the sampled epoch is obtained (fig. 5). The method of the battery of filters, however, has inherent limitations in terms of practical feasibility and reliability; a solution to this problem was offered by the Fourier analysis. J.B.J. Fourier has supplied the mathematical instrument to separate a signal into its constituents at different frequencies so that the filtering method can be replaced by this mathematical analysis (fig. 6).
With this analysis, the EEG curve is converted from analogue to digital and a computer handles the signal with the Fast Fourier Transform (FFT). The result obtained is a diagram termed spectrum with the frequencies on the x-axis and the relative energy contents on the y-axis (fig. 7).

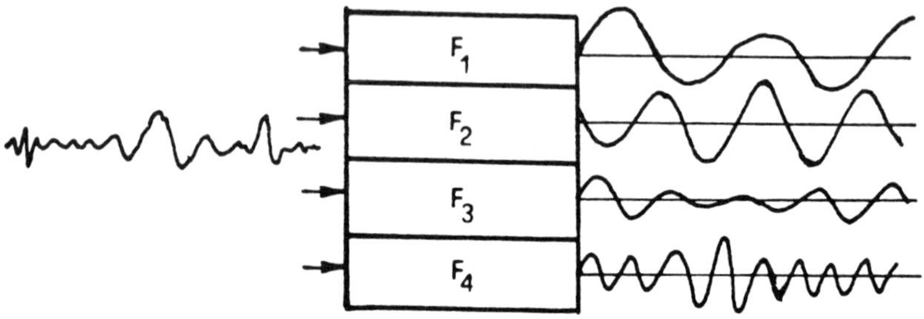

Fig. 4. A battery of filters separates a signal into the harmonic constituents.

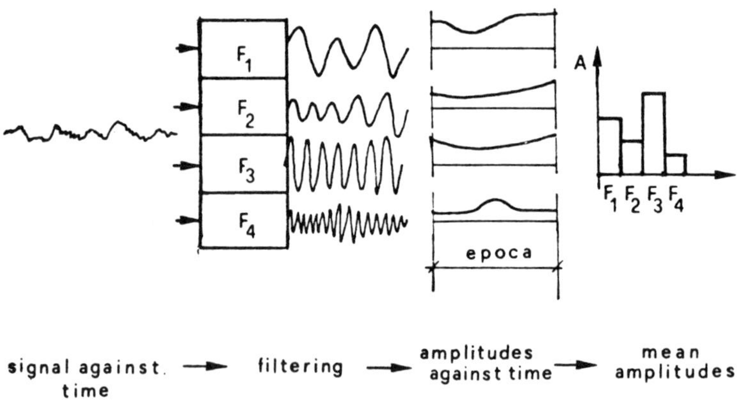

Fig. 5. Spectral analysis obtained with the filtering method.

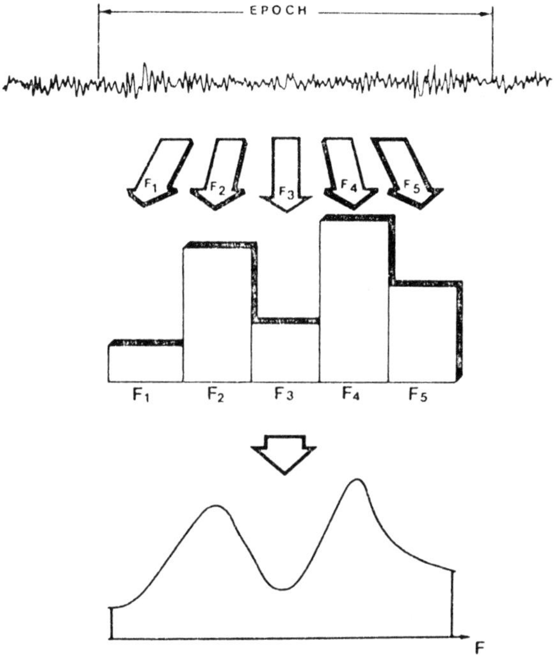

Fig. 6. Spectral analysis by the Fourier Transform. The spectrum of a length (epoch) of EEG recording is obtained by selecting a certain number of frequencies and plotting against each one the relevant energy content. Then the spectrum is made continuous by smoothing processes.

The sampling interval is a few seconds while in conventional recording the EEG trace is evaluated for long periods, from 20-30 minutes to several hours. This limitation in the application of the spectral analysis was overcome bearing in mind that, by observing the trend of the EEG curve, it is possible to study how the spectrum evolves over time, that is, in a continuous succession of sampled and processed epochs.

An elegant presentation of this principle was proposed by R.G. Bickford under the name of compressed spectral array (CSA) in which the spectra of adjacent epochs are displayed prospectively aligned and arranged, thus achieving a dynamic spectral analysis (fig. 8) of the EEG.

This type of presentation offers numerous advantages, ranging from the possibility of monitoring to the construction of spectral maps. Another advantage is the high information content of the spectra due to the possibility of catching even small variations in frequency or the harmonic components usually undetected by conventional EEG.

FFT analysis has two inherent limitations: the energy disproportion and the loss of positional information.

The energy disproportion is due to the fact that in physics, and in EEG in particular, an inverse proportion exists between the parameters "amplitude" and "duration" which are involved in defining the amplitude of the spectral peak and its natural frequency. This means that slow phenomena are generally

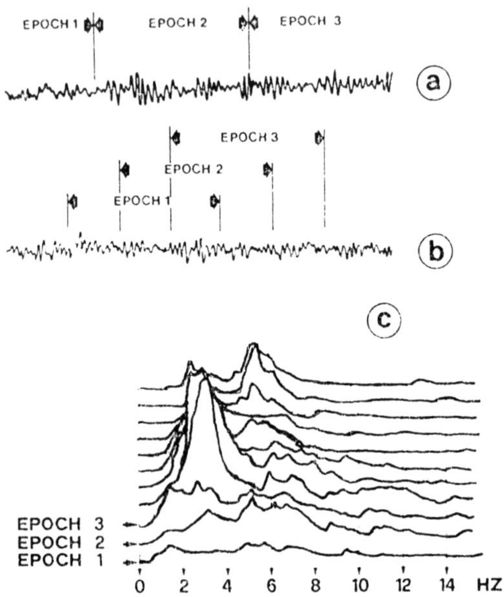

Fig. 7. Dynamic spectral analysis. By recording in sequence the spectra related to (a) contiguous or (b) partially overlapping epochs, a representative (c) is obtained of the evolution in time of the spectral content.

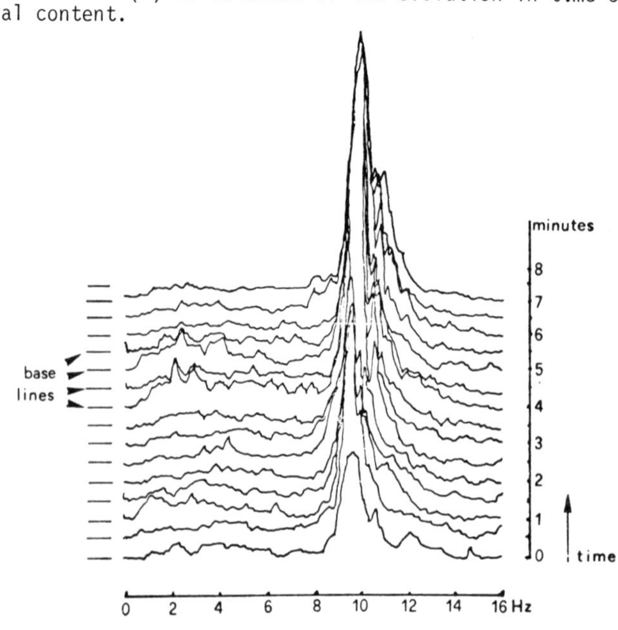

Fig. 8. Dynamic spectral analysis of the EEG

of large amplitude and long duration while high frequency phenomena are more rapidly damped and generally present a smaller amplitude (fig. 9).

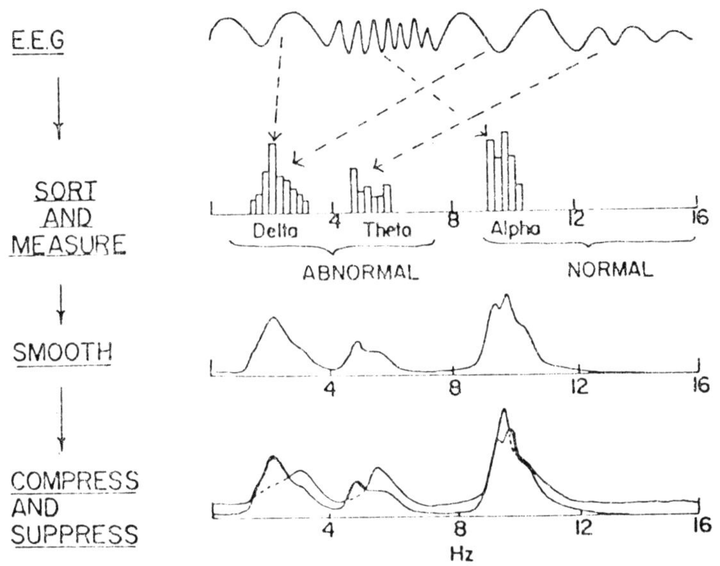

Fig. 9.

Therefore, the electroencephalographer instinctively ascribes the same importance to sequences which, broadly speaking, have the same number of cycles.
Fourier analysis, on the contrary, being a rigorous mathematical method, impartially calculates the energy content at the various frequencies. When applied to EEG, Fourier analysis heavily favors the slow frequencies so as to limit the possibility of detecting and evaluating quickly-varying phenomena.
The second limitation of conventional spectral analysis is the impossibility of collecting information on the phenomenon position within the analyzed epoch, in spite of the phenomenon presenting typically differentiated spectral contents. On the other hand, since Fourier Transform is structural analysis, it is rightly independent of the position of the episode within the analyzed epoch (fig. 10).
To compensate for these limitations and to preserve the possibilities offered by dynamic spectral analysis, a different transformation algorithm has now been proposed by Sciarretta and Erculiani (1975) called Berg Transform which is based on the following premises.
The presentation at successive spectra provides information on the spectral content and on its evolution in time; the resolving power of the latter increases as the epoch duration decreases, so much so that, assuming as a limit that the spectra could be calculated over short intervals with respect to the variation time of the spectral characteristics, even the function of time evolution would be fully preserved.
Unfortunately, this is not the case with the EEG signal which rapidly and continuously varies in time and frequency. We then consider transferring the human criterion to the Fourier Transform to adjust to the extension of the

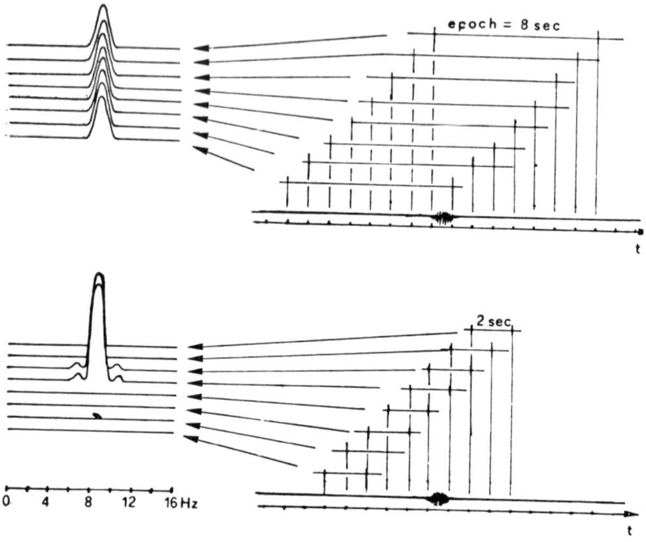

Fig. 10. CSA of a 9-Hz bouffée of 1 sec. duration.
a - Fourier Transform with 8-sec., 7/8 overlapped epochs;
b - Berg Transform with K_2= 16 K_1 = 2.

analyzed interval to the value of the frequency being calculated.
This can be achieved, for instance, by varying the width of the temporal window ininverse proportion to the analyzed frequency so as to construct a constant percent resolution spectrum (fig. 11).
The application of this transform on commercial apparatus (BFA: Berg-Fourier analyser (fig. 12) has made it possible to study the EEG by means of an extremely reliable and sophisticated method (see also Gilioli et al, in this volume).
The possibility of applying a post-elaboration system greatly enhances the capacity for qualitative analysis and application of this method. With the post-elaboration unit the following advantages can be obtained: 1 - the possibility of constructing mean spectra on selected epochs (figs.13 and 14); 2 - the analysis of single trends of selected frequency ranges for predetermined epochs (fig. 15); 3 - the possibility of percent quantitation of the power within single frequency ranges (fig. 16); 4 - spectral maps can be constructed (fig. 17).

CONCLUSIONS

The methodology of spectral analysis is a recent advancement in neurophysiology and has therefore been described in detail.
Spectral analysis has been extensively validated in studies on disorders of vigilance and level of consciousness, in anesthesiology, neuropharmacology and in intrasurgical monitoring procedures of cerebral functions.
Validation studies of this method for the purposes of occupational health and, in particular, neuroepidemiology surveys of subjects at high risk of exposure to neurotoxic workplace chemicals are being carried out and the preliminary results are encouraging. Other possible applications in occupational health and ergonomics are monitoring of the conditions of vigilance of working populations so as to prevent accidents and better understand work rhythms.

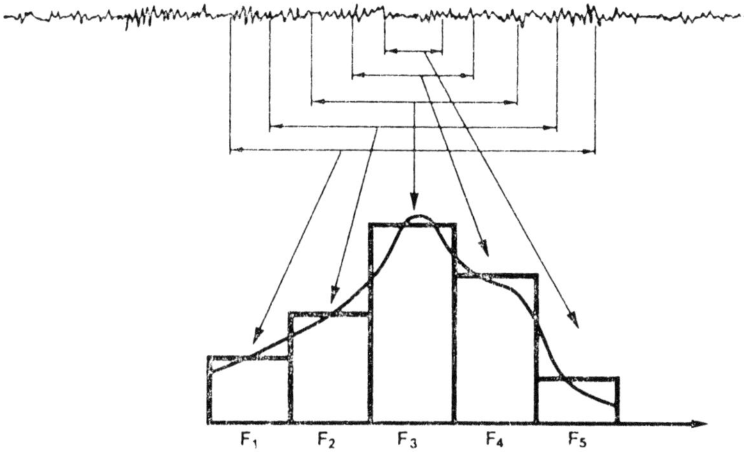

Fig. 11. Spectral analysis by the Berg Transform. With the Berg Transform the contributions to the various frequencies are calculated on epochs decreasing in duration as the value of each analyzed frequency rises. Thus, the "observation range" of the analysis fits the normal extent of the phenomenon, so that isolated episodes of peculiar shape can be emphasized.

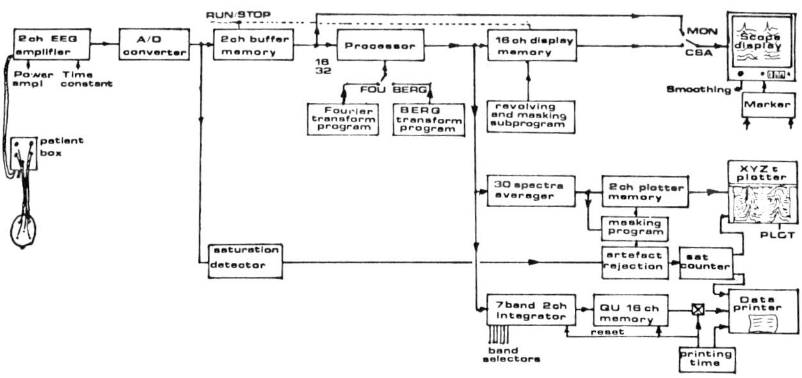

Fig. 12. BFA: function diagram

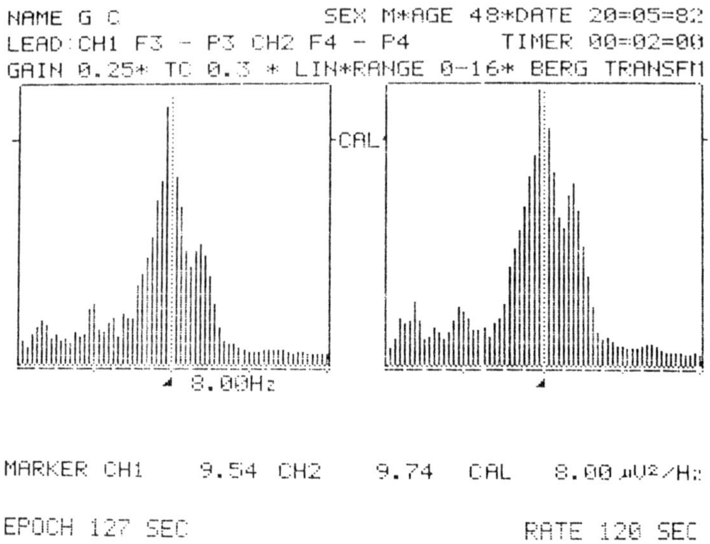

Fig. 13. Mean spectrum on a EEG epoch of 127 seconds.

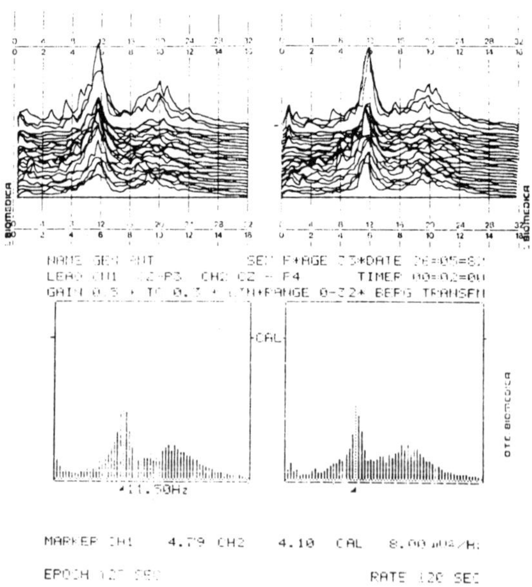

Fig. 14. Mean spectrum on 127 seconds with relative CSA.

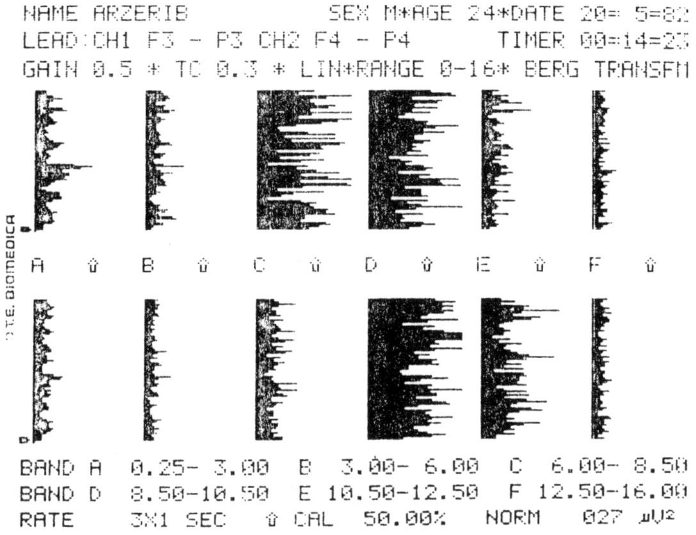

Fig. 15. Analysis in trends.

NAME GEN ANT SEX F*AGE 33*DATE 26=05=82
LEAD:CH1 CZ-P3 CH2 CZ - P4 TIMER 00=02=08
GAIN 0.5 * TC 0.3 * LIN*RANGE 0-32* BERG TRANSFM

O.T.E. BIOMEDICA

BAND A <%>	013	018
BAND B <%>	060	048
BAND C <%>	051	037
BAND D <%>	021	019
BAND E <%>	012	015
BAND F <%>	006	005
ARTIFACTS		000

BAND A 0.50- 6.00 B 6.00-12.00 C 12.00-17.00
BAND D 17.00-21.00 E 21.00-25.00 F 25.00-32.00
RATE 2X1 SEC CAL 50.00% NORM 036 μV²

Fig. 16. Percent trend quantitation.

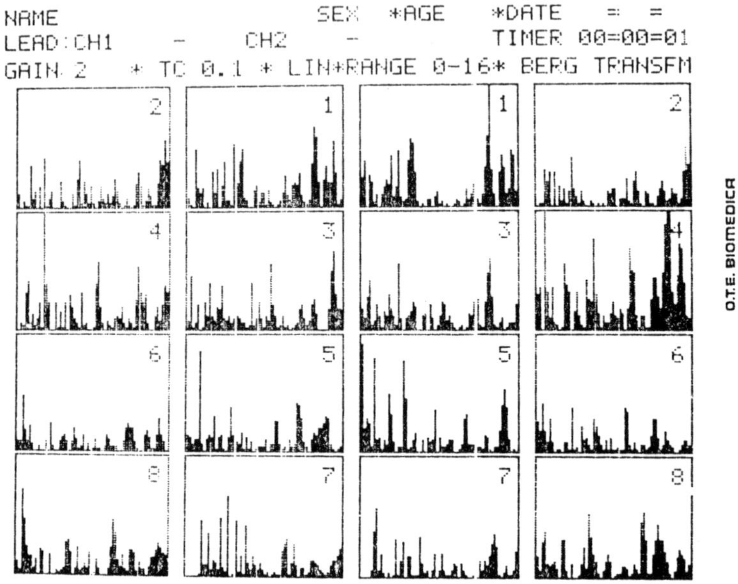

Fig. 17. Spectral map.

ACKNOWLEDGEMENT

The Authors gratefully acknowledge the permission granted by G. Sciarretta to use graphic material from the book "An introduction to EEG spectral analysis", Florence, 1977.

REFERENCES

Aminoff, M.J. (1980). Electrodiagnosis in clinical neurology. Churchill Livingstone.
Binnie, C.D., A.J. Rowan and Th. Gutter (1982). A manual of electroencephalographic technology. Cambridge University Press.
Dolce, G. and H. Kunkel (1975). CEAN computerized EEG analysis. Fisher.
Genta, P., S. Bonini and I. Bruni (1982). EEG - Acquisizione del segnale elettroencefalografico. Libreria Scientifica già Ghedini, Milano.
Kellaway, P. and I. Petersén (1973). Automation of Clinical EEG, Raven Press
Matejcek, M. and G.K. Schenk (1975). Quantitative analysis of the EEG. From Symposium of the Study group for EEG methodology. Vol. 1, 1973, vol. 2, 1975.
Matousek, M. (1973). Frequency and correlation analysis. Handbook of EEG and clinical neurophysiology. Vol. 5, part. A, Elsevier.
Niedermeyer, E. and F. Lopes da Silva (1982). Electroencephalography. Urban & Schwarzenberg.
Nunez, P.L. (1981). Electric field of the brain. The neurophysics of EEG. Oxford University Press.
Prior, P.F. (1979). Monitoring cerebral function. Elsevier.
Randall, R.B. (1977). Frequency Analysis. Bruel & Kjaer.

Rémond, A. (1977). EEG informatics. A didactic review of methods and application of EEG data processing. Elsevier.
Sciarretta, G. and P. Erculiani (1977). Introduction to EEG spectral analysis. OTE-Biomedica, Firenze.

The Application of Event-related Slow Brain Potentials in Occupational Medicine

D. A. Otto

Neurotoxicology Division, U.S. Environmental Protection Agency, Research Triangle Park, NC 27711, USA

ABSTRACT

Workers exposed to solvents, pesticides, heavy metals and other neurotoxic chemicals frequently exhibit symptoms of mental confusion, impaired attention, and memory loss. These symptoms suggest the impairment of several aspects of cognitive function that can be assessed readily by noninvasive electrophysiological tests. Event-related slow potentials (ERSPs) of the brain constitute a family of scalp-recorded electrophysiological measures of cognitive and/or motor function. Applications of CNV, P300 and other ERSPs, as well as criteria for the selection and evaluation of neurotoxicity tests, in occupational health assessment are discussed.

KEYWORDS

Occupational health, neurotoxicology, event-related slow potentials, CNV, P300.

INTRODUCTION

Workers today are exposed to a growing array of chemicals, some of which are known to produce neurotoxic effects if ingested, inhaled, or absorbed through the skin. In the case of lead poisoning the symptoms of overt neurotoxicity are clearly defined and body burden can be monitored to prevent excessive accumulation of the neurotoxic substance. For other chemicals developed during the past few decades, far less is known about the consequences of worker exposure and biological markers of body burden are seldom available. In the face of these uncertainties, the search continues for more sensitive and reliable methods to detect and characterize the neurobehavioral effects of chemical exposure in the workplace.

The electroencephalogram (EEG) recorded non-invasively from the scalp is a direct measure of CNS function that is used in neurotoxicological assessment. Clinical analysis of spontaneous EEG, however, is a relatively insensitive index of neurotoxicity, particularly when exposure levels are moderate to light. Several computer-based methods of EEG analysis are available that offer more sophisticated and, presumably, more sensitive measures. Gilioli et al (this volume) describes one of these methods--EEG spectral analysis, and Cosi (this volume) describes another--sensory evoked potential analysis. This paper will review event-related slow potentials (ERSPs) of the brain, a third type of EEG measure that has proven to be useful in assessing the effects of psychoactive drugs (Tecce et al, 1978). The purpose of this paper is to explore the potential utility of ERSPs in occupational medicine.

DEFINITION OF EVENT-RELATED POTENTIAL (ERP)

ERPs are electrical responses of the brain to *external* stimulus events (i.e., auditory, visual, and somatosensory evoked potentials) or electrical signs of *internal* events (or processes) initiated in the brain. Internal events include efferent motor commands (*ergo* motor-related potentials) and a broad spectrum of information processing, associative, or cognitive processes that originate in the brain (e.g., assocciation potentials and the expectancy wave). Two general classes of ERPs may be distinguished: *exogenous* and *endogenous*. Exogenous potentials are evoked by external stimulus events, while endogenous potentials reflect events or processes that originate in the brain.

If subjects are required to attend, process, or respond to external stimuli in some manner such as selectively counting or button pressing when a particular signal occurs, the resultant waveform will include early (sensory-evoked) exogenous components as well as later (cognitive) endogenous components. The term event-related "*slow*" potentials refers to

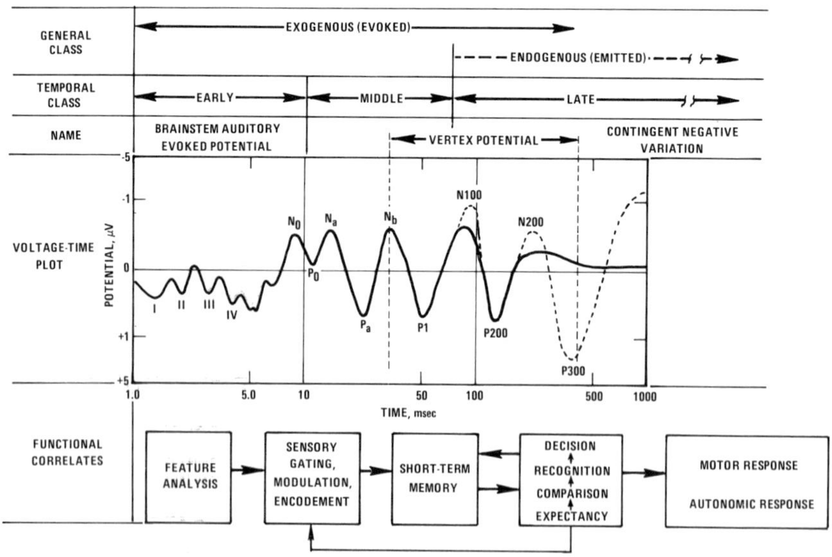

Fig. 1. Temporal and functional characteristics of event-related brain potentials evoked by or emitted in response to a brief auditory click. Functional correlates are hypothetical. (Adapted from Hillyard and Woods, 1979).

later endogenous components associated with information processing, cognition, and memory. The latency of endogenous potentials is usually 100 msec or more after stimulus onset. Endogenous potentials may persist for several hundred msec, while exogenous components are usually much briefer. Fig. 1 illustrates the temporal and functional characteristics of these two classes of potentials in the auditory modality (cf. Hillyard and Woods, 1979 for an excellent review).

The three most prominent ERSPs are: (1) Contingent Negative Variation (CNV)--a slow vertex-negative shift in scalp-recorded EEG that arises during a signalled reaction-time foreperiod (Walter et al, 1964); (2) Readiness Potential (RP)--a slow vertex-negative shift in EEG baseline that arises about one second prior to a stereotyped movement (Kornhuber and Deecke, 1965); and (3) P300--a large positive wave that peaks 300-600 msec following a stimulus that delivers relevant information to the perceiver (Sutton et al, 1965). Figure 2 demonstrates the remarkable sensitivity of CNV and P300 components to the behavioral contingencies of a continuous performance task. CNV and P300 are clearly defined in hit trials, but are absent in miss trials.

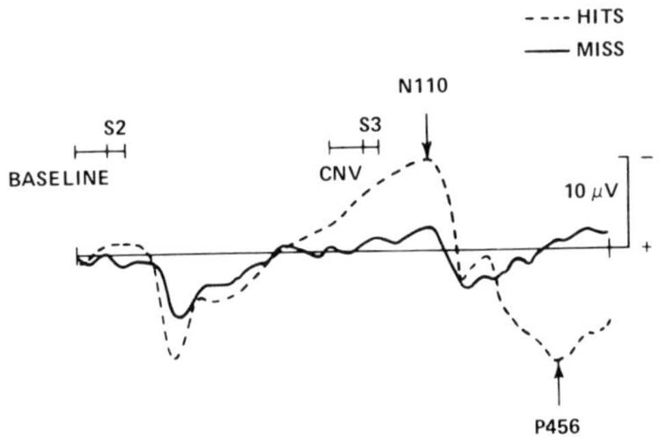

Fig. 2. Summary waveforms of hit and miss trails from seven subjects. Note the absence of CNV and P300 (labelled P456 in this example) components during miss trails. (From Otto, 1977 p 10-23.)

SENSITIVITY OF ERSPs TO DRUGS AND TOXICANTS

Considerable evidence is available concerning the effects of psychotropic drugs on the CNV. The prospective sensitivity of ERSPs to toxicant substances may be inferred from the effects of commonly used psychotropic drugs. In general, CNV amplitude is enhanced by stimulants such as amphetamine (Kopell et al., 1974) and caffeine (Ashton et al., 1974) and decreased by depressants and tranquilizers and nitrazepam (Ashton et al., 1974). The divergent effects of stimulants and depressants on CNV amplitude are illustrated in fig. 3. In cases where the behavioral effects of drugs has been paradoxical, CNV changes have consistently mirrored behavioral observations. That is, when a given drug enhanced performance in some subjects, but depressed performance in others, CNV amplitude varied in the same direction as performance. Paradoxical effects have been reported with d-amphetamine (Tecce et al., 1974) and nicotine (Ashton et al., 1973). This evidence suggests that the CNV should provide a useful index of central-acting neurotoxicant effects.

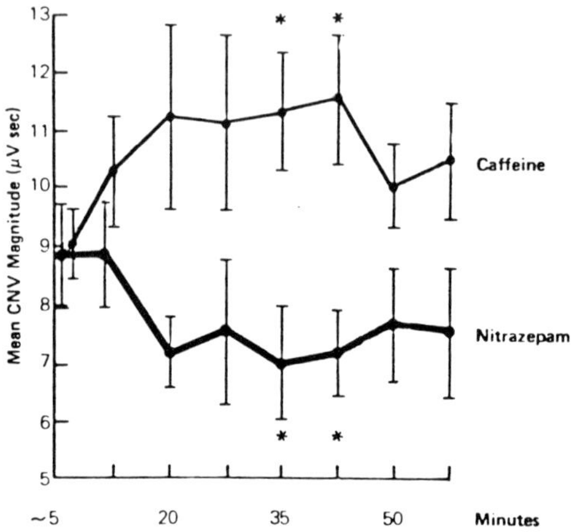

Fig. 3. Effect of stimulant (300 mg caffeine citrate, 13 subjects) and tranquilizer (2.5 mg nitrazepam, 12 subjects) drugs on CNV amplitude. Stimulants generally increase CNV amplitude, while depressants or tranquilizers usually decrease CNV amplitude. (Redrawn from Ashton et al, 1974).

Few efforts have been made to apply ERSP methods in neurotoxicological research. Groll-Knapp et al (1972) reported a dose-related decrease in CNV amplitude in subjects exposed to low concentrations of carbon monoxide, although the same laboratory (Groll-Knapp et al, 1978) was unable to replicate these results. Otto et al (1978) reported a paradoxical increase in CNV amplitude in subjects exposed to 100 ppm CO, but a decrease in subjects exposed to 200 ppm CO. P1 amplitude (following the CNV) increased proportional to CO exposure. These findings suggest a disinhibitory or excitatory effect on CNS function at low CO concentrations followed by the expected depressant effect at higher doses. Further study is required to clarify the extensive contradictions in reported neurobehavioral effects of low-level CO exposure.

A more convincing demonstration of the sensitivity of ERSPs to neurotoxicant exposure was recently reported by our laboratory (Otto et al., 1981). A linear relationship between slow wave voltage and blood lead level (figure 4) was observed in young children during sensory conditioning. Slow wave voltage varied systematically across the entire range of blood lead levels observed in these children (6-52 μg/dl) -- evidence that Pb exposure alters CNS function at levels considerably below the 30 μg/dl limit that is generally considered to be "safe" (CDC 1978). Although the functional significance of this finding is not clearly understood -- i.e., we cannot state with certainty that altered slow wave voltage reflects neural damage or even impaired sensory conditioning, the data indicate that this ERP parameter is extremely sensitive to body lead burden in children. Results of this work are reviewed in more detail elsewhere in the volume, (Otto et al.).

While the latter report provides encouraging evidence of the utility of ERP measures in neurotoxicology, ERSPs must be considered as experimental measures for several reasons. Although the relationships among ERSPs, behavior, and cognitive processes have been exhaustively explored during the past decade (cf. Otto 1978), very little is known

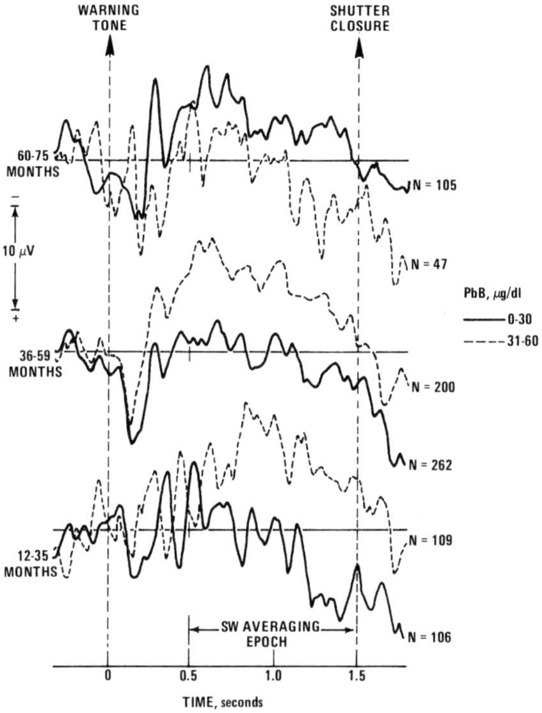

Fig. 4 Interactive effects of age and PbB levels on slow wave voltage during tone-shutter conditioning. Summary averages of all artifact-free trials were constructed for children in three age groups at two different PbB levels. N=number of trials included in average. (Redrawn from Otto et al., 1981).

about the anatomical, physiological, or chemical substrate of these intriguing waveforms. Even the psychological and behavioral significance of ERSPs remains a subject of lively debate. Little serious effort has been made to standardize testing paradigms or to construct population norms for any ERSP paradigm.

Other methodological problems limit the application of ERSP methods in occupational medicine. Eye and body movements produce considerable artifact during recording. Trials in which eye or body movements occur must be rejected from signal averages. Eye fixation on the stimulus generator is usually required to minimize eye movement artifact. The complexity and expense of computerized data acquisition and analysis systems also limit the application of ERSP methods.

EVALUATION OF NEUROTOXICITY MEASURES

Salient features and limitations of several event-related slow potentials that are promising experimental measures have been described. In order to compare the relative merits of different electrophysiological measures for neurotoxicological assessment of workers, a set of evaluation criteria must be specified. Twelve such criteria are listed and defined in Table I.

Neurotoxicity tests should meet the criteria required of psychometric tests -- i.e., each test should be valid, reliable, sensitive, and objective. Although not an essential requirement, tests should also be non-invasive and non-threatening, culture-free and as age-independent as possible. If the objective is population screening (as opposed to exhaustive clinical evaluation), three additional criteria should be considered in test evaluation. Tests should be rapid (as a rule of thumb, the battery should be limited to about 60 minutes in length -- that is only 8 subjects/day/ testing station), simple, and inexpensive.

TABLE I. EVALUATION CRITERIA FOR HUMAN NEUROTOXICITY TESTS

1.	STANDARDIZED	standard procedures for test administration and measurement have been developed, published, and accepted by user community; normative data available for evaluation of clinical significance of data
2.	VALID	test measures the dimension or function that it was selected to assess
3.	RELIABLE	test provides comparable results when administered on repeated occasions
4.	SENSITIVE	test discriminates between exposed and non-exposed populations; detects change at minimal effect threshold
5.	OBJECTIVE	test provides a quantitative measure that is not susceptible to subjective bias
6.	NON-INVASIVE	test does not require skin penetration
7.	NON-THREATENING	test does not frighten or upset subject
8.	AGE-INDEPENDENT	test scores do not vary with age
9.	CULTURE-FREE	test scores do not vary with ethnic background or social class
10.	RAPID	for population testing, individual measures should require no more than 10 minutes to administer; rapid scoring of response is also important
11.	SIMPLE	test should be simple to administer and simple for the subject to perform
12.	INEXPENSIVE	cost of test materials and scoring less than $100

In Table II a simple binary rating scheme has been used to compare the relative merits of several ERSP measures and three sensory evoked potentials used widely in clinical neurology. The sensory measures are brainstem auditory (BAEP), pattern-reversal visual (PREP), and somatosensory (SEP) evoked potentials. Plus (+) indicates that the measure meets the criterion; minus (-) indicates that the measure does not; a question mark (?) indicates that insufficient evidence is available to assign a rating. While this rating scheme is crude and little evidence exists to evaluate the sensitivity or utility of these tests in populations exposed to industrial toxicants, Table II does provide a systematic basis to select appropriate electrophysiological measures for neurotoxicity assessment. Clinically validated tests of sensory dysfunction and non-specific demyelinating diseases (BAEP, PREP, and SEP) must be considered as prime candidates. Event-related slow potentials including the CNV and P300 have considerable promise, but require modification, standardization, and validation in order to meet many of the selection criteria for neurotoxicity testing in occupational medicine.

TABLE II. COMPARISON OF TESTS USED FOR NEUROTOXICITY ASSESSMENT

		BAEP	PREP	SEP	SW*	CNV	P300
(1)	STANDARDIZED	+	+	+	-	-	-
(2)	VALID	+	+	+	?	?	?
(3)	RELIABLE	+	+	+	?	?	?
(4)	SENSITIVE	+	+	+	+	?	?
(5)	OBJECTIVE	+	+	+	+	+	+
(6)	NON-INVASIVE	+	+	+	+	+	+
(7)	NON-THREATENING	+	+	-	+	+	+
(8)	AGE-INDEPENDENT	-	-	-	-	-	-
(9)	CULTURE-FREE	+	+	+	-	?	?
(10)	RAPID	+	+	+	-	+	+
(11)	SIMPLE	+	+	-	-	-	-
(12)	INEXPENSIVE	-	-	-	-	-	-

(*Slow wave voltage during sensory conditioning)

ACKNOWLEDGEMENT

The able assistance of C. Keadle in the preparation of figures and J. Nichols and staff in manuscript preparation is gratefully appreciated.

REFERENCES

Ashton, H., J. Millman, R. Telford, and J. Thompson, (1973. *Brit. J. Pharmacol.*, *48*, 715-717.

Ashton, H., J. Millman, R. Telford and J. Thompson (1974). *Electroenceph. Clin. Neurophysiol.*, *37*, 59-71.

Center for Disease Control (1978). *Preventing Lead Poisoning in Young Children*. USDHEW.

Groll-Knapp, E., H. Wagner, H. Hauck and M. Haider (1972). *Staub-Rheinhalt. Luft*. *32*, 64-68.

Groll-Knapp, E., M. Haider, H. Hoeller, H. Jenkner and H. Stidl (1978). In D. Otto (ed.) *Multidisciplinary Perspectives in Event-Related Brain Potential Research*. EPA-600/9-77-043, 424-430.

Hillyard, S. and D. Woods (1979). In: M. Gazzaniga (ed.) *Handbook of Behavioral Neurobiology*, *vol. 2 Neuropsychology*, Plenum Press, New York, 345-378.

Kopell, B., J. Tinklenberg, and L. Hollister (1972). *Arch. Gen. Psychiat.*, *27*, 809-811.

Kopell, B., W. Wittner, D. Lunde, L. Wolcott and J. Tinklenburg (1974). *Psychopharmacologia* (Berl.), *34*, 55-62.

Otto, D. (1977). In: H. Zenick and L. Reiter (eds.) *Behavioral Toxicology: An Emerging Discipline*. EPA-600/9-77-042, 10.1-10.31.

Otto, D., ed. (1978). *Multidisciplinary Perspectives in Event-Related Brain Potential Research*. EPA-600/9-77-043.

Otto, D., V. Benignus, J. Prah and B. Converse (1978). *ibid*, 440-443.

Otto, D., V. Benignus, K. Muller and C. Barton (1981). *Electroenceph. Clin. Neurophysiol.*, *52*, 229-239.

Sutton, S., M. Braren and J. Zubin (1965). *Science*, *150*, 1187-88.

Tecce, J., and J. Cole (1974). *Science*, *185*, 451-453.

Tecce, J., J. Savigno-Bowman and J. Cole (1975). *Psychoparmacologia* (*Berl.*), *43*, 293-295.

Tecce, J., J. Savigno-Bowman and J. Cole (1978). In: M. Lipton, A. DiMaschio and K. Killam (eds.), *Psychopharmacology: A Generation of Progress*, Raven Press, New York, 745-758.

Stimulus-related Cerebral Evoked Potentials in Occupational Medicine

V. Cosi

Institute of Neurology, University of Pavia, Pavia, Italy

ABSTRACT

Description of the most frequently employed Evoked Potentials (EPs) and discussion of their utilisation in occupational medicine.

KEYWORDS

Evoked Potentials (EPs); AEP; BAEP; Middle latency AEP; SEP; VEP-flash; VEP-pattern; occupational medicine.

INTRODUCTION

Stimulus-related cerebral evoked potential, or evoked potential (EP), is any so called "exogenous" sensory evoked potential or the obligatory potential evoked by a specific stimulus than can be auditory (AEP), somatosensory (SEP), visual (VEP). By any stimulus modality, the early components are mainly stimulus-related, i.e., sensitive to the stimulation parameters, while the late components are affected by many variables.

The study of EPs permits an objective evaluation of the functional state of the afferent sensory pathway from the receptor through the nervous system up to the specific sensory cortex and non-specific cortex and the localisation of the anatomical site of the disorder. Moreover, with this method it is possible to obtain further information on the presence of a general neurophysiological brain dysfunction. If compared with EEG, the stimulus-related EPs give a more selective and dynamic set of information which is of a neurophysiological nature, that is to say, little or not at all influenced by attentional and cognitive factors. On the contrary, slow potentials (CNV, P300) as well as late components of EPs, which are largely event-related, provide a widely psychophysiological set of information.

Only a few experimental and human data are at present available on EPs in the field of occupational medicine. They include EP alterations by *noise* (Bergamasco et al, 1975: AEP, VEP), *vibration* (Ohta et al, 1980: SEP), *microwaves* (Edelwejn, 1976: SEP, rabbits), *radiation* (Minamisawa et al, 1980: VEP, rabbits), *solvents* (Seppäläinen et al, 1979: n-hexane, VEP and ERG;

Serra et al, 1979: VEP and SEP; Johnson et al, 1979: MnBK, VEP, monkeys; Seppäläinen et al, 1981: VEP and SEP), *metals* (Fox et al, 1977: Pb, VEP, rats; Mattsson et al, 1981: methylmercury, VEP, dogs; Signorino et al, 1981: Pb, VEP and ERG; Otto et al, 1981: Pb, slow potentials).

However, the main field of application of EPs studies is probably that of neurotoxic agents.

We can in some way compensate for the scanty experience in this field by employing the setup of these methods for the study of the effects of alcohol, drugs or anesthetics. Some data are available on acute intoxications and late brain disorders (e.g., chronic alcoholics, uremics). Reports that can be quoted here concern *alcohol*: Squires et al, 1978: BAEP; Lewis et al, 1977: VEP and SEP; *drugs*: Lader, 1977: psychotropic drugs, AEP; Saletu, 1977: psychotropic drugs, VEP, SEP and AEP; Yiannikas et al, 1981: ethambutol, VEP; Green et al, 1982: phentoin, SEP and BAEP; *anesthetics*: Uhl et al, 1980: halothane, VEP; *chronic alcoholics*: Dustman et al, 1979:VEP; Begleiter et al, 1981: BAEP; Psthuma et al; 1982: VEP; Cosi et al, 1982a: VEP; Cosi et al, 1982b: VEP and SEP; *uremics*: Hamel et al, 1978: VEP; Rossini et al, 1981: VEP; Knoll et al, 1982: BAEP and middle AEP.

At the present time EPs studies are rather common and consolidated in many fields but not yet in occupational medicine. We therefore consider it important to explain some fundamental characteristics of EPs in order to point out the possibilities this method has to offer.

Evoked potentials are obtained by techniques averaging the responses to a sensory stimulation, which are familiar to neurophysiologists; their illustration will of course be omitted here.

VISUAL EVOKED POTENTIALS (VEPs)

VEPs can be obtained by a flash stimulation (VEP-flash) and by patterned stimuli (VEP-pattern).

VEP-FLASH

VEP-flash (Fig. 1) is recorded by electrodes placed in occipital regions in bipolar or referential derivations (01, 02 to A1 or A2) to stimulation by diffuse light flashes presented at a frequency below 1 Hz, bandpass 1.6-70 Hz, time base 500 msec, 128 responses. Eight peaks can be identified within the first 300 msec following the stimulus; they may be labelled from 0 to VII according to Cigânek (1961). Only peak III can be detected in all cases, as the other first three peaks ("early components") have small amplitudes and are often lost. The subsequent peaks IV to VII ("late components") are the most constant constituent of VEPs.

The visual evoked potential is a specific response that reflects the visual signal and probably also its processing within the visual system and related brain structures. Two main sources contribute to cortical VEP: a) activity originating from the retina that travels to the primary visual cortex through the lateral geniculate body; b) retinal activity passing through the reticular formation and diffuse thalamic projections. Early (0-III) and late (IV-VII) complexes are relatively independent units originating perhaps in different cerebral structures, the early ones showing less variability than the late ones. This variability is chiefly interindividual, but can also be intraindividual. Early components are usually more steady, but peaks 0 and I often cannot be identified. Amplitudes are affected by variability more than latencies.

Many factors must be taken into account to avoid this variability: they are methodological and technical, such as intensity, the monocular versus binocular stimulation, etc., but can be psychophysiological as well, such as alertness, attention, distraction. Beside these factors, there are others which are, however, difficult to control and about which knowledge is still insufficient.
The commonly considered parameters are the latencies of peaks III and IV and the amplitude peak III to peak IV.

VEP-PATTERN

VEP-pattern can be detected in different ways (reversal pattern, pattern onset, checkerboard, bars, etc.) but the most frequently employed method is the VEP by reversal pattern of a checkerboard of different size shown by the stimulator on a TV monitor. VEP-pattern (Fig. 2) is usually recorded in bipolar derivation (Oz-Cz or Fz) by monocular stimulation (reversal rate 2 Hz, band-pass 1.6-300 Hz, time base 200 msec, 128 responses). Many stimulation parameters should be taken into account: spatial frequency (check size), temporal frequency (reversal at 2 Hz or more), luminance, contrast (more than 50%), field size, pattern orientation. VEP by checkerboard reversal consists of three major peaks (N1, P1 or P100, N2) sometimes preceded by a smaller positive (Po). The parameters usually considered are the absolute latencies of N1 and P1 and their respective inter-ocular difference. The amplitude mainly considered is between peak N1 and peak P1; both peak latencies and amplitude are closely related to the check size: the latencies decrease exponentially and amplitudes vary nonmonotonically with check size (Ristanovic et al, 1981). The variations in amplitude and, to a lower degree latency, depend on the fact that, when larger checks are presented, parafoveal responses increase and foveal responses decrease.

The patterned stimulus has to be sharply focused and any refractive error corrected, as blurring of the image consequently modifies the amplitudes of response. A monocular stimulation is needed, because a binocular stimulation causes different interactions on the amplitude such as inhibition, zero summation, partial summation, summation, facilitation, which can be partially explained by the high degree of VEP specificity to varying stimulus determinants, in particular to spatial and temporal frequency, contrast and orientation (Apkarian et al, 1981). Interactions due also to latency are possible with binocular stimulation (Hoeppner, 1980). Some differences between dominant eye (shorter latency and higher amplitude) and non-dominant eye (Seyal et al, 1981) appear less important in practice.

The most probable origin of each component of VEP reversal pattern from various occipital areas is still under investigation: N1 seems likely to reflect the activity in the part of area 17 emerging on the convexity whereas P1 is supposed to originate in area 19, and the midline component N2 in area 18 (Lesèvre et al, 1979).

Intraindividual variability of VEP reversal pattern is negligible and interindividual variability is moderate. In similar experimental conditions, the data obtained prove highly reliable and can be easily reproduced in different laboratories. On this basis, it can be recommended as a reliable method of functional investigation of visual pathways, even apart from neurological clinical investigation, although the information that can be obtained thereby is probably less than what we can get with the VEP-flash and also somehow different.

The use of VEP-flash and pattern in occupational medicine can be supported by the following facts: these methods can be usefully applied in different fields of investigation other than neurology and ophtalmology, that is, for subjects with no alterations of visual pathways. Studies on this subject, which are too numerous to be mentioned here, cover psychiatry (see Shagass, 1972), encephalopathies, degenerative diseases of CNS, psychopharmacology, etc. Experience is particularly vast for VEP-flash, but still limited for VEP pattern (Rossini, 1981; Yiannikas, 1981; Cosi, 1982a; Posthuma, 1982).

Moreover, VEP permits investigation of the functional state of several pathways and brain structures with a very large representation of the responses on the cerebral cortex, as demonstrated by spatio-temporal mapping on the recorded scalp responses.

In conclusion, this investigation is easily performed and can be repeated at any time; it is also well accepted as it causes no discomfort. Limitations and cautions arise from the possible existence of ocular alterations such as opacity, retinopathy, refractive errors, etc. Unfortunately no researches with reversal pattern stimulation are available at present in occupational medicine and those with flash are very few (Serra, 1979; Seppäläinen, 1979 and 1981).

<u>Electroretinogram</u> (ERG) is recommended as a recording device jointly with VEP-flash and pattern in order to exclude previous pathological conditions at the receptor level. But a VEP alteration by neurotoxic agents could also be due, at least partially, to an effect at retinal level caused by an interference with neurotransmitters. In particular, the dopaminergic activity which occurs in the retina has been studied by ERG with administration either of L-Dopa or neuroleptic drugs (Fornaro <u>et al</u>, 1980).

fig.1

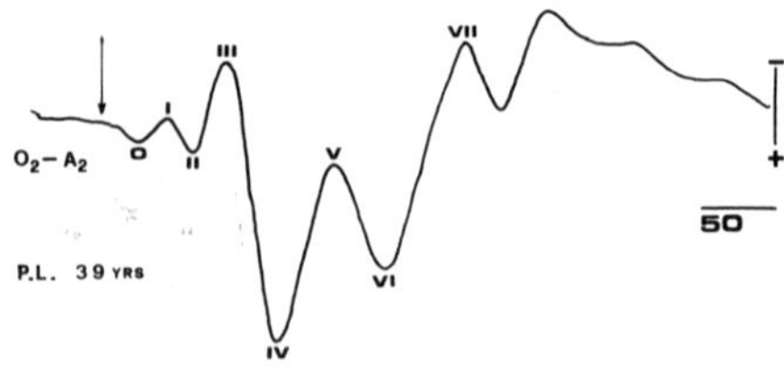

fig. 2

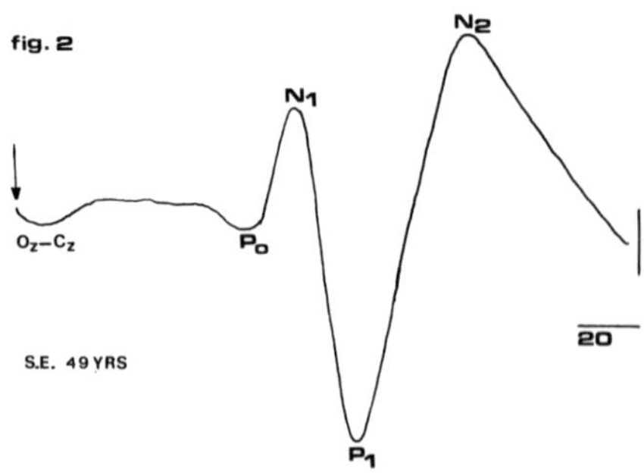

SOMATOSENSORY EVOKED POTENTIALS (SEPs)

SEP is usually obtained by electrical stimulation of a sensory or mixed nerve at a frequency rate of 2-4 Hz. Cortical SEP is recorded by electrodes placed at C3, C4 to Fz or A1+A2 for upper limb nerves and at Cz for the stimulation of lower limb nerves. In the SEPs evoked by stimulation of the median nerve at the wrist (Fig. 3), at least 6 peaks can be identified (bandpass 1.6-1000 Hz, time base 100 msec, 256 responses). The first peak P1 or P15 is subcortical in origin, while N1 or N19 is the earliest cortical component.
This complete response can be obtained by supramaximal electrical shocks: the stimulus travels through the nerves, roots, dorsal columns and medial lemniscal pathway to the sensory cortex. In addition to cortical responses, the study of spinal responses and subcortical components detected by farfield recording techniques provides objective measures of somatosensory function from peripheral nerves to cortex. More than 50% of the latency of cortical SEP as measured at the peak of N1 is due to the peripheral conduction time in nerves and roots and it therefore provides information obtainable by other methods of investigation like electroneurography. The latency of cortical SEP by stimulation of the peroneal nerve at the ankle includes not only a large peripheral conduction time but also the conduction time through the whole spinal cord.

The origin of early SEP components has been investigated more extensively than for other SEP components, except for Goff _et al_ (1977b), who demonstrated the neurogenic or myogenic origin of the various components up to 500 msec. In neurological investigations the latencies and amplitudes of both N1 and P2 are usually considered.

The somatosensory potential evoked on the median nerve by supramaximal stimuli is the cortical evoked potential that shows the lowest degree of both intraindividual and interindividual variability, at least in the stimulus-related components recorded during the first 60-80 msec, while later components show a wide variability. Cortical SEP (stimulus-related) is recommended for its stability and for the fact that it has no limitations as regards receptor alterations, as do VEP and AEP. The fact that it provides data on a largely extracerebral sensory pathway may be a limit, but also an advantage in some cases. Actually, electrical stimulation is not generally well accepted by subjects under examination and this is undoubtedly a weak point in the method.

In the field of occupational medicine SEP has been used by some of the quoted authors (Otha, 1980; Serra, 1979; Seppäläinen, 1981).

fig.3

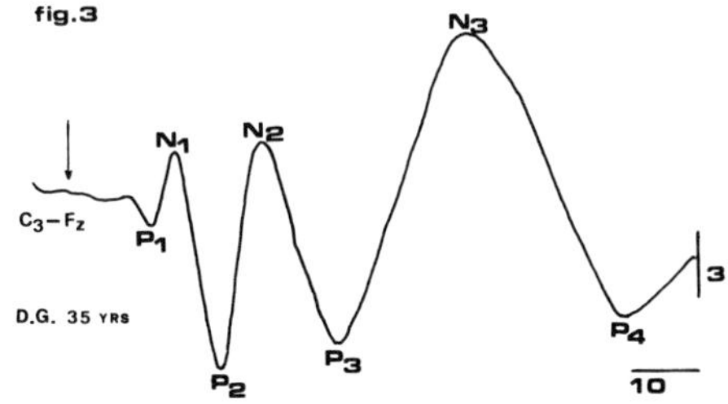

AUDITORY EVOKED POTENTIALS (AEPs)

Early, middle and late components of AEPs are classified by their latency: they must be examined separately and under different technical conditions. Only the recording electrodes can be placed in the same position at Cz with reference to the ipsilateral mastoid.

"Early components" (Brainstem Auditory Evoked Potential or BAEP): are recorded in the first 10 msec by click stimulation (intensity 60-70 dB S.L.), at high frequency repetition rate (10 Hz or more), binaural and/or monoaural stimulation bandpass 160-3000 Hz, 2048-4096 responses. Seven vertex positive peaks can be identified (Fig. 4A) and they are often indicated by Roman numerals I to VII; they are regarded as the far-field reflection of activity arising in the auditory pathway, according to the following locations: I - eight nerve; II - cochlear nucleus; III - superior olivary complex; IV - lateral lemniscus; V - region of inferior colliculus; VI - medial geniculate (?), thalamus; VII - thalamo-cortical acoustic radiation (?)

The most prominent and constant components are waves I, III and V. Absolute latencies of the first five peaks and inter-peak times I-III, III-V and I-V are usually considered. Latencies have a very high intra- and interinvidi vidual stability. Amplitudes, on the contrary, are more variable but the IV-V/I ratio is useful for clinical purposes (it is greater than 0.5 for signal intensity 65-75 dB S.L.).

In other than brainstem lesions (demyelination, vascular lesions, tumors) BAEP seems poorly affected, even in comatose patients when coma is due to metabolic causes, such as drug overdose, uremia, hepatic encephalopathy (Starr, 1977). This fact yet does not support the use of BAEP in the study of early toxic effects, but this statement may need reconsidering in the light of the recently reported effect of phentoin (Green, 1982) on BAEP. It must be stressed that experience in other than brainstem lesions and hearing assessment is still scant.

In conclusion, though the possibilities of this kind of investigation are still far from being wholly explored and definitely established, yet this simple method holds interesting potentialities and the results it supplies are easy to compare. Caution must be used and limitations arise only in cases of severe hearing loss.

Middle Latency Auditory Evoked Potentials (Fig. 4B) are those appearing between 10-80 msec after the stimulus. They are recorded by click stimulation (0.2 msec) at a rate of 5 Hz, an intensity of 65-70 dB S.L., bandpass 16-1000 Hz, 1024 responses. Several waves can be identified: the early ones up to 20 msec, are myogenic in origin (Post-Auricular Response), but the subsequent Pa-Nb-Pb are neurogenic. Primary acoustic response is not identifiable, as it is masked by myogenic potential; Pa is not the primary positivity, since it is suppressed by barbiturate anesthesia; Pb is too late to be primary and is suppressed, too (Goff *et al*, 1977a).

So far, middle latency AEPs have seldom been employed in every field of investigation; this is probably due to the variability of components 10÷20 msec, myogenic in origin, which are modified by relaxation of the neck muscles. In our opinion, however, the study of middle latency AEP can be a useful tool of investigation that, together with BAEP, allows the evaluation of auditory pathways through the brainstem to the cortex. In fact, the complex Pa-Nb-Pb has good intra- and interindividual stability (see also Robinson *et al*, 1977 and 1978). Moreover, the examination is easy to perform and causes no discomfort. The use of middle latency AEP has proved useful in the evaluation of uremic brain dysfunction (Knoll et al, 1982).

fig. 4

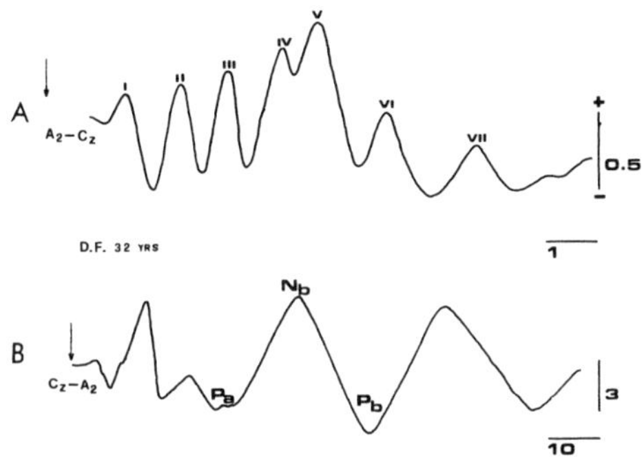

Late AEP - main components N1-P1-N2 with latencies respectively about 83-147-276 msec (Robinson, 1977) - is not considered here: in fact it is a non-specific response. It is considered as a V (vertex) potential, i.e., a slow diffuse response to any sudden sensory stimulus. Late AEP's exhibit a very wide variability according to psychophysiological conditions, inter-stimulus interval, etc.

OTHER INVESTIGATIONS

Other techniques have been developed or are under investigation: they may overcome some of the averaging limitations and provide new additional parameters to evaluate the sensory functions. Among them are the "*Critical Frequency of Photic Driving*" method or CFPD by Celesia et al (1977) and the "*Visual Steady State Evoked Potentials*" method by Regan (1972) for the evaluation of visual function. The "*Frequency Following Potentials*" or FFP by Moushegian et al (1973), evoked by a low frequency tone burst, depend on the integrity of pathways originating in the apical turn of the cochlea (Sohmer et al, 1976), thus completing the data from BAEP's evoked by clicks, which principally excite the basal turn of the cochlea. At present, however, experience with these methods is still too fragmentary and limited to permit a sound judgement on possible applications in the field of occupational medicine.

DISCUSSION

We have so far illustrated the most commonly used potentials and relative parameters, the acknowledged data about the origin of responses, the main variability factors, the methodological and non-methodological features which recommend or limit their use. EP recording and stimulation techniques differ considerably and often make it difficult to compare data.
We have presented those we use in our laboratory: they are at present the most frequently used techniques in routine investigations and can be recommended for application also in occupational medicine.

We would conclude with some general remarks. The rather limited experience with neurotoxics and the more extensive experience with drugs and chronic intoxications show that EP studies can be considered as a useful tool of investigation, at least in the field of neurotoxic agents. Since the information provided by EP studies are quantifiable, these researches can be useful to evaluate not only the effects of chronic exposures, but also early neurobehavioral effects.

It should, however, be borne in mind that, since it is a question of evaluating the presence of a brain dysfunction rather than the functional state of a sensory pathway, a multimodality stimulation seems more useful. All modalities (AEPs, SEPs, VEPs) can be employed, but their effectiveness is quite different. Consequently, the research is inevitably extensive, time-consuming and expensive: therefore, at least for the present time, it can be recommended only for preliminary or "pilot" studies.

The choice of the suitable stimulation modalities must be accurately made on the basis of general principles and after carefully considering the kind of information obtainable from each EP, as well as the limits, precautions and practical applicability of each method. It is also important, as far as the choice of the proper test set is concerned, to be acquainted with the mechanism of action of each neurotoxic agent and the relative sites affected in the nervous system (e.g., simultaneous effect on CNS and PNS), also considering the electrophysiological modifications observed with other methods.

The problems EP research raise in occupational medicine are not unlike those inherent in other neurophysiological studies concerning experimental design. However, we would emphasize that a longitudinal investigation,which is always recommended, is absolutely necessary with VEP flash, while with other stimulation modalities presented above, a comparison with a control group is definitely possible.

In the study of EPs for clinical purposes, only a few parameters are taken into account, as specified above, for each potential. Generally the latency values of only a few peaks are considered, whereas less attention is devoted to amplitudes. The peaks taken into consideration are most constant, regardless of their significance, in the case of VEPs (flash peak IV, pattern P1) while N1 and P1 latencies of SEP reflect the arrival of the afferent volley at the cortex. For clinical purposes, EPs are classified as abnormal when absent, when a marked waveform alteration makes the potential unrecognizable, or when the considered latencies are delayed more than 2-3 S.D. The effort of interpretation of the results is made chiefly from a clinical standpoint and is often reduced to an attempt to locate the lesion sites.

Such measurements are unprofitable and self-limiting in the field of occupational medicine. Since we are seeking the electrophysiological correlates of a hypothetical pathological process or dysfunction, the use of EPs as a neurobehavioral method requires a more extensive and thorough investigation of the signal. Many problems arise in this sort of evaluation by EPs which requires specific statistical competence. However, an approach of this kind seems really valuable: it can also avoid some disappointing results. We can find ample information on the signal classification procedures in the report by Roy John et al (1978). Besides a sophisticated evaluation of the results, we would propose a rigorous standardization of methods as far as recording, stimulation and averaging parameters are concerned, in order to reduce any variability in the data collection stage and make the results easy to compare.

On the basis of the above specified factors which recommend or limit the use of each EP, we would emphasize the utility of a particular model of investigation which includes mainly the study of VEPs (pattern) and AEPs (BAEPs and middle latency AEPs). SEPs, on the contrary, seem less effective, since they investigate largely extracerebral pathways: on the other hand, any interference at receptor level is avoided.

Finally, at least one aspect is still open to discussion. Possible alterations revealed by EPs studies in humans exposed to neurotoxics are naturally non-specific. First of all, they must be evaluated to take account of possible artifacts and factors of error originating from both technical procedures and experimental design. The interpretation of such alterations is usually rather complex. EPs data must of course be evaluated jointly with the results of all other kinds of investigation.

REFERENCES

Apkarian, P.A., K. Nakayama and C.W. Tyler (1981). Electroenceph. clin. Neurophysiol., 51, 32-48.

Begleiter, H., H. Porjesz and C.L. Chou (1981). Science, 211, 1064-1066.

Bergamasco, B., M. Gilli and G. Rossi (1975). Acta Neurol., 30, 24-29.

Celesia, G.G., and R.F. Daly (1977). Arch. Neurol., 34, 403-407.

Cigànek, L. (1961). Electroenceph. clin. Neurophysiol., 13, 165-172.

Cosi, V., G. Mazzella, R. Callieco and E. Sinforiani (1982a). Riv. It. EEG Neurofisiol. Clin., in press.

Cosi, V., R. Callieco and L. Mazzini (1982b). Progetto HUSPI, Università di Pavia, vol. 7, in press.

Dustman, R.E., E.W. Snyder, D.A. Callner and E.C. Beck (1979). In: Evoked Brain Potentials and Behaviour - Begleiter H. (Ed.), Plenum Press, 321-363.

Edelwejn, Z. (1976). Proceedings 2nd Industrial and Environmental Neurology Congress, Univerzita Karlova, Praha, 306.

Fornaro, P., G.F. Placidi, P. Castrogiovanni, M. Perossini and G. Cavallacci (1980). Acta Neurol., 2 (XXXV), 293-299.

Fox, D.A., J.P. Lewkowski and G.P. Cooper (1977). Toxicol. Appl. Pharmacol., 40, 449-461.

Goff, W.R., T. Allison, W. Lyons, T.C. Fischer and R. Conte (1977a). In: Auditory Evoked Potentials in Man - Psychopharmacology Correlates of Evoked Potentials,- Desmedt J.E. (Ed.), Karger, 30-44.

Goff, G.D., Y. Matsumiya, T. Allison and W.R. Goff (1977b). Electroenceph. clin. Neurophysiol., 42, 57-66.

Green, J.B., M. Walcoff and J.P. Lucke (1982). Neurology, 32, 85-88.

Hamel, B., J.R. Bourne, J.W. Ward and P.E. Teschan (1978). Electroenceph. clin. Neurophysiol., 44, 606-616.

Hoeppner, T.J. (1980). Electroenceph. clin. Neurophysiol., 47, 49-58.

Johnson, B.L., J.V. Setzer, T.R. Lewis and W.K. Anger (1977). Am. Industr. Hyg. Ass. J., 38, 567.

Knoll, O., U. Harbort, K. Schulte and F. Zimpel (1982). In: Clinical Application of Evoked Potentials in Neurology, - Courjon J.,et al., (Eds.), Raven Press, 227-232.

Lader, M. (1977). In: Auditory Evoked Potentials in Man - Psychopharmacology Correlates of Evoked Potentials - Desmedt J.E. (Ed.), Karger, 142-159.

Lesèvre, N., and J.P. Joseph (1979). Electroenceph. clin. Neurophysiol., 47, 183-203.

Lewis, E.G., R.E. Dustman and E.C. Beck (1977). In: Auditory Evoked Potentials in Man - Psychopharmacology Correlates of Evoked Potentials - Desmedt J.E. (Ed.), Karger, 160-174.

Mattsson, J.L., E. Miller, J.P. Alligood, J.E. Koering and S.G. Levin (1981). Neurotoxicology, 2, 499-514.
Minamisawa, T., K. Yamamoto and T. Tsuchiya (1980). Electroenceph. clin. Neurophysiol., 50, 59-70.
Moushegian, G., A.L. Rupert and R.D. Stillman (1973). Electroenceph. clin. Neurophysiol., 35, 665-667.
Ohta, Y., H. Nanno, O. Fujimoto, et al. (1979). Japan J. Industr. Health, 21, 433-441.
Otto, D.A., V.A. Benignus, K.E. Muller and C.N. Barton (1981). Electroenceph. clin. Neurophysiol.,52, 229-239.
Posthuma, J. and S.L. Visser (1982). In: Clinical Applications of Evoked Potentials in Neurology,- Courjon J., et al., (Eds.), Raven Press, 149-155.
Regan, D. (1972). Evoked Potentials in psychology, sensory physiology and clinical medicine, Chapman and Hall, London.
Ristanović, D. and R. Hajduković (1981). Electroenceph. clin. Neurophysiol., 51, 599-610.
Robinson, K. and P. Rudge (1977). Brain, 100, 19-40.
Robinson, K. and P. Rudge (1978). J. Neurol. Sci., 36, 147-156.
Rossini, P.M., M. Pirchio, M. Treviso, D. Gambi, B. Di Paolo and A. Albertazzi (1981). Electroenceph. clin. Neurophysiol., 52, 435-444.
Roy John, E., D.S. Ruchkin and J.J. Vidal (1978). In: Event-Related Brain Potentials in Man, - Callaway E., et al., (Eds.), Academic Press, 93-138.
Saletu, B. (1977). In: Auditory Evoked Potentials in Man - Psychopharmacology Correlates of Evoked Potentials - Desmedt J.E. (Ed.), Karger, 175-207.
Seppäläinen, A.M., C. Raitta and M.S. Huuskonen (1979). Electroenceph. clin. Neurophysiol., 47, 492-498.
Seppäläinen, A.M., K. Savolainen and T. Kovala (1981). Electroenceph. clin. Neurophysiol., 51, 148-155.
Serra, C., A. Rossi, D. Facciolla, A. Ruocco and A. Biondi (1979). Riv. It. EEG Neurofisiol. Clin., 2, 135-139.
Seyal, M., S. Sato, B.G. White and R.J. Porter (1981). Electroenceph. clin. Neurophysiol., 52, 424-428.
Shagass, C. (1972). Evoked Brain Potentials in Psychiatry, Plenum Press.
Signorino, M., O. Scarpino, L. Provinciali, G.F. Marchesi, M. Valentino and M. Governa (1981). Riv. It. EEG Neurofisiol. Clin., suppl. 1, 167-168.
Sohmer, H. and H. Pratt (1976). Electroenceph. clin. Neurophysiol., 40, 253-260.
Squires, K.C., N.S. Chu and A. Starr (1978). Science, 201, 174-176.
Starr, A. (1977). in: Auditory Evoked Potentials in Man - Psychopharmacology Correlates of Evoked Potentials - Desmedt J.E. (Ed.), Karger, 45-57.
Uhl, R.R., K.C. Squires, D.L. Bruce and A. Starr (1980). Anesthesiology, 53, 273-276.
Yiannikas, C., J.C. Walsh and J.G. Mc Leod (1981). Electroenceph. clin. Neurophysiol., 52, S91.

Regional Cerebral Blood Flow by the 133-Xenon Inhalation Technique

J. Risberg and S. Hagstadius

Laboratory of Neuropsychology, Department of Psychiatry, University Hospital, Lund, Sweden

ABSTRACT

A large body of evidence has shown that regional cerebral blood flow (rCBF) is normally regulated by the regional metabolic activity. Changes in regional cerebral blood flow thus mirror metabolic and functional changes in the brain. Techniques for measurement of rCBF employ radioactive tracers, primarily 133-Xenon, a freely diffusible gamma-emitting radionuclide. Previously it was administered by injection into the internal carotid artery but in recent years this technique has been replaced by the non-traumatic 133-Xe inhalation method. Methodology and applications in psychiatry and neuropsychology are briefly reviewed.

KEYWORDS

Regional cerebral blood flow - 133-Xenon inhalation technique - psychiatric disorders - neuropsychology.

There is a large body of evidence indicating that the regional cerebral blood flow (rCBF) is normally regulated by the regional metabolic activity (Raichle, et al, 1976; and others). These findings have served as the basis for the interpretation of rCBF changes as changes in metabolism and function in the normal and, in most instances, also in the diseased brain. This means that rCBF-measurements are of interest in all cases where an evaluation of the level of cerebral function is of value.

METHODOLOGICAL DEVELOPMENT

Before 1975, the most commonly used method for the measurement of rCBF was based on injection of the inert and diffusible gamma-emitting tracer 133-Xe into the internal carotid artery (Lassen and Ingvar, 1963). The potential of the method for determining cortical patterns of activation during performance of mental tasks was first recognized by Ingvar and Risberg (1967). However, the main drawback was the invasive nature of the method which limited its use to patients undergoing carotid angiography for diagnostic purposes. Furthermore, the technique was unilateral, offering data only from the hemisphere supplied by the injected artery.

The development of the 133-Xenon inhalation technique for the measurement of rCBF has solved some of the problems of the intra-arterial technique. This technique, outlined by Mallet and Veall (1965) and improved by Obrist et al,(1967, 1975), Risberg et al, (1975 a) and Risberg (1980) has obvious advantages. It is non-invasive and non-traumatic and allows simultaneous bilateral measurements. The technique can be used with few limitations in patients and normal subjects and allows repeated measurements in the same patient over short or long periods.

THE MEASUREMENT SYSTEM AND PROCEDURE

Our present system consists of 32 scintillation detectors (0.75 x 0.75-in NaI (Tl) crystals), the lateral part of each hemisphere being covered by 16 detectors placed parallelly. A mixture of 133-Xenon and air (3-5 mCi/ /liter) is inhaled by the subject during one minute, followed by ten minutes of breathing of atmospheric air. The advent and washout of the radioactive tracer is registered by the scintillation detectors and these data form the basis for flow calculation.

For calculation of flow, computer procedures primarily developed by Obrist et al, (1975) and extended and modified by Risberg (1980) are utilized. Essentially, the analysis yields an estimate of the flow of rapidly perfused tissues (grey matter) and a determination of the Initial Slope Index (ISI); (Risberg et al, 1975 a). This flow parameter is dominated by grey matter flow and, because of its greater stability and reliability, it is preferable in most clinical applications of the technique.

Expired air is sampled via a catheter ending in the face mask for determination of 133-Xe and CO_2 concentrations. End-tidal 133-Xe values are used for estimation of arterial tracer concentrations during the period of recording and end-tidal CO_2 values are used to estimate arterial pCO_2. The mask assembly contains a microphone which permits conversation with the subject during the measurements.

APPLICATIONS IN NEUROPSYCHOLOGY

Several rCBF studies in our laboratory have been made in normal healthy male volunteers. The normal flow level and the normal regional variations were established during resting conditions and various types of mental activation. Figure 1 shows the normal flow pattern in resting young, healthy, right-handed male volunteers (Prohovnik et al, 1980 a). The main finding was a pronounced hemispheric symmetry of flow, with values in the frontal lobe which were 5-15% above the hemispheric mean. The central sulcus flow was close to average, and so were the values of parietal and occipital lobes. The temporal lobe had the lowest flow, about 5-15% below the hemispheric mean. The "hyperfrontal" pattern, characteristic of normal resting subjects, has been interpreted as reflecting the mediation of the frontal lobes in such functions as arousal, general control of behaviour, inner speech and anxiety.

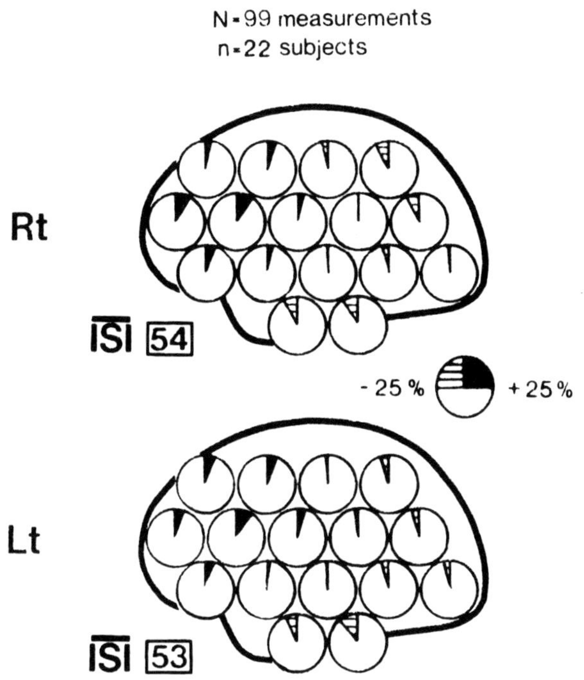

Fig. 1. The bilateral resting landscape in normal volunteers. Hemispheric mean flows (ISI) are indicated in the boxes. Regional distribution flows are shown as deviations from the mean. Black shadowing indicates a value higher than the mean; striped is lower than the mean. A 25% deviation from the mean is equal to 90 degrees in the clock symbols. (Adapted from Prohovnik et al, 1980 a).

Prohovnik et al, (1980 a) also studied interrelations between regional flow values expressed as percentage of the hemispheric average. Correlations between such values for homologous regions revealed the highest correlations in Rolandic, occipital and temporal regions. These regions seem grossly to correspond to the projection areas. Lower correlations were found in regions considered to be engaged in integrative-associative functions. These findings were interpreted as indicating a functional coupling between homologous projection areas and a looser coupling between the association areas in the resting state, that is, in the absence of stimulation or task-oriented activity.

The possibility of studying lateralized mental functions with the inhalation technique was first demonstrated by Risberg et al (1975 b), who showed that activation by a verbal analogies test gave larger flow increase in the left hemisphere, especially over postcentral (Wernicke) areas, while activation by a spatial test (Street) gave a right hemisphere superiority in frontal and parietal regions.

Maximilian et al, (1978) demonstrated the involvement of the left parietal region during learning and recall of visually presented paired associates. There was a greater involvement of the frontal pole during learning as compared to recall.

Risberg et al, (1977) studied habituation to a reasoning test (Raven's Matrices). During a first day of testing, the test gave frontal and parietal increases of rCBF while activation by a parallel version of the test on a second day, resulted in close to no increase of flow in the frontal areas. The parietal activation was essentially unchanged during this repeated activation. These findings indicate the involvement of the frontal region in processes related to habituation to new situations and tasks and to general organization and control of mental processes, as has been suggested by Luria (1973) and others.

Prohovnik et al, (1980 b) studied cortical activity during visual perception. Stimulation was achieved by a Spiral After Effect (SAE) apparatus, and the three conditions were: looking at a stationary spiral, looking at a rotating spiral, and performing the SAE test. The stimulus was thus kept fairly constant with increasing complexity of interpretation. Average flow increases were 5%, 7% and 12% respectively. Both the magnitude and spread of flow increases were related to task complexity. During the stationary spiral condition,regional activation was primarily seen in parieto-occipital (especially right), posterior temporal (especially left) and superior frontal areas. During the rotating spiral condition, there were no obvious changes in the distribution of flow changes.The 3rd condition caused larger rCBF increases in frontal, anterior temporal and right central-parietal regions compared to the stationary and rotating spiral conditions. The activation response was generally slightly larger on the right side. The results are in agreement with several studies of visual pathways showing a sequential processing hierarchy.

CLINICAL APPLICATIONS

Patients with organic presenile and senile dementia were investigated extensively by our group (Gustafson and Risberg, 1974; Gustafson, 1975; Risberg, 1980; Risberg and Gustafson, 1982; and others). A multitude of meaningful relations between rCBF and clinical and psychometric data were found. A clinically very important application of the technique is the differentiation between primary degenerative dementias (Alzheimer's disease), cerebrovascular dementia and pseudo-dementia due to emotional disorders, toxic influence, etc.

The primary degenerative dementias have been found to have fairly distinctive rCBF pathologies (Risberg and Gustafson, 1982). Alzheimer's disease was characterized by a reduction in mean flow and by focal flow decreases in post-central, mainly parieto-occipital and parieto-temporal areas. Pick's disease showed focal decreases in the frontal lobe together with a general flow decrease. Cerebrovascular dementia was characterized by right-left asymmetries and spotty flow abnormalities (Fig. 2.). Patients with pseudo-dementia due to depression showed flow levels and regional distributions within normal variations. An excellent agreement between rCBF diagnosis and neuropathological diagnosis has recently been found (Risberg and Gustafson, 1982). Further significant information can be obtained from studies during mental activation in demented patients. Reduced flow response or abnormal patterns of activation are sometimes seen (Risberg, 1980).

The lack of effective pharmacotherapy in organic dementia is, in spite of many efforts, nearly total. The inhalation technique has been shown to be useful in the evaluation of the effects of drugs on cerebral circulation. We studied the effects of Piracetam (Notropil) in nine patients with presenile dementia (Gustafson et al, 1978) but no improvements could be established. Studying the effects of Vincamine and Bromvincamine on 10 cases of cerebrovascular dementia, Hasgstadius et al, (1982) did, however, demonstrate some improvements in rCBF for both drugs.

A study of patients with depression subjected to ECT indicated that a pre-ECT flow level in the lower normal range, along with further decreases in this level during the ECT series, was associated with less favourable therapeutic outcome and greater chance of negative side effects like memory disturbances (Silfverskiöld et al, 1979). This indicates a potential usefulness of the method for screening patients and also for elucidating the cerebral mechanisms of this form of therapy.

Regional cerebral blood flow in chronic alcoholics has been extensively investigated (Berglund and Risberg, 1981). The first seven weeks of abstinence were followed in a study of 22 chronic alcoholics (Berglund et al, 1980). At the last examination, the group had a mean flow decrease of 17% compared to normal subjects of similar age. A small increase in flow (4%) from the first to the seventh week was observed. Furthermore, the regression between CBF and age, was found to be more pronounced in alcoholics than in normal subjects, indicating that alcohol abuse might speed up aging of the brain.

FINAL REMARKS

The differential diagnosis of dementia seems to be the most important clinical application of the 133-Xe inhalation technique at the moment. However, as has been demonstrated, a multitude of scientific and clinical applications are possible and the meaningfulness and significance of rCBF-data seem very obvious. The measurements elucidate the functional organization of the human brain at rest and at work. Global or regional dysfunction is shown in cases of organic brain disease and toxic influence, providing important information for diagnostic evaluations and for selection and evaluation of therapy. Another paper in this volume will describe findings in subjects exposed to organic solvents (Hagstadius and Risberg, this volume).

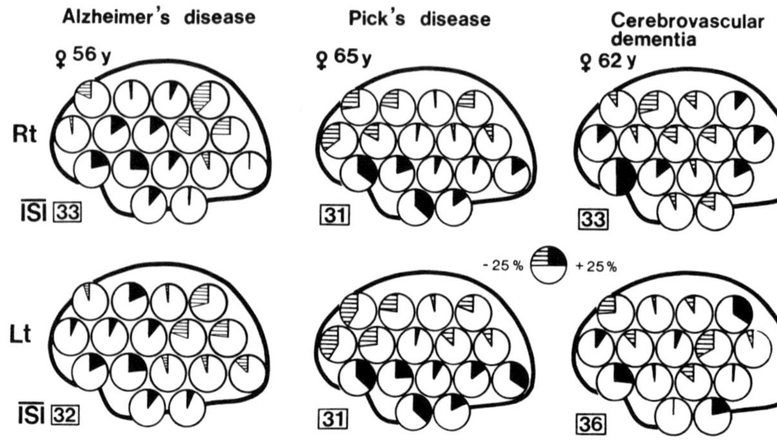

Figure 2. Regional cerebral flow in patients with presenile dementia. (Alzheimer's disease, Pick's disease and cerebrovascular dementia). Symbols as in Fig. 1. Note the postcentral flow decrease in the case of Alzheimer's disease, the frontal flow abnormality in the case of Pick's disease and the spotty flow distribution in the CVD case. The abnormal flow distributions are accompanied by subnormal mean flow levels.

ACKNOWLEDGEMENTS

Supported by grants from the Swedish Work Environment Fund, the Swedish Medical Research Council (contract 04969) and the Swedish Council for Social Science Research.

REFERENCES

Berglund, M., G. Bliding, Å. Bliding and J. Risberg (1980). Acta Psych. Scand. (suppl), 119-127.
Berglund, M. and J. Risberg (1981). Arch. Gen. Psych., 38, 351-355.
Gustafson, L. (1975). Acta Psych. Scand. (suppl) 257, 8-35.
Gustafson, L. and J. Risberg (1974) Acta Psych. Scand. 50, 516-538.
Gustafson, L, J. Risberg, M. Johanson, M. Fransson and V.A. Maximilian (1978). Psychopharmacology, 56, 115-117.
Hagstadius, S., L. Gustafson and J. Risberg (1982). Submitted.
Ingvar, D.H. and J. Risberg (1967). Exp. Brain Res., 3, 195-211.
Lassen. N. and D.H. Ingvar (1963). Arch. Neurol., 9, 615-622.
Luria A.R. (1973). The Working Brain. London: Allen Lane, The Penguin Press.
Mallet, B.L. and N. Veall (1965). Clin. Sci., 29, 179-188.
Maximilian V.A., I. Prohovnik, J. Risberg and K. Håkansson (1978). Brain and Language, 6, 22-31.
Obrist, W. D., H.K. Thompson, C.D. King and H.S. Wang (1967). Circ. Res. 20, 124-135.
Obrist, W .D., H.K. Thompson, H.S. Wang and W.E. Wilkinson (1975). Stroke 6, 245-256.
Prohovnik, I., K. Håkansson and J. Risberg (1980 a) Neuropsychologia, 18, 203-217.

Prohovnik, I., J. Risberg, S. Hagstadius and V.A. Maximilian (1980 b) in: Prohovnik, I. (1980). Mapping Brainwork. Theoretical and Methodological Considerations in Applying the Regional Cerebral Blood Flow Method to Neuropsychology. Thesis, Lund: Liber, CWK Gleerup.

Raichle, M.E., R.L. Grubb, M.H. Gado, J.O. Eichling and M.T. Ter-Pogossian (1976). Arch. Neurol. 8, 523-526.

Risberg, J. (1980). Brain and Language, 9, 9-34.

Risberg, J., Z. Ali, E.M. Wilson, E.L. Wills and J.H. Halsey (1975 a). Stroke, 6, 142-148.

Risberg, J. and L. Gustafson (1982). In: Proceedings of symposium on Radionuclides and Brain Disease. Crans sur sierre. Valais, Switzerland, Sept. 7-10, 1982. Serono-symposia, Raven Press. In press.

Risberg, J., J.H. Halsey, E.L. Wills and E.M. Wilson (1975 b). Brain, 98, 511-524.

Risberg, J., V.A. Maximilian and I. Prohovnik (1977). Neuropsychologia, 15, 793-798.

Silfverskiöld, P., L. Gustafson, M. Johanson and J. Risberg (1979). In: Obiols, J., C. Ballũs, E. Gonzãles Maclũs and J. Pujol (eds). Biological Psychiatry Today. Amsterdam: Elzevier/North Holland. 1178-1183.

Neurobehavioral Methods in Occupational Health: Computerized Tomography

G. Scotti, D. Perani, L. Caverni, N. Colombo and C. Tonon

Neuroradiological Service, Ospedale Maggiore Niguarda, Milan, Italy

Clinical manifestations of Central Nervous System (CNS) disease related to acute or chronic intoxication by organic or inorganic chemical agents are well known and have been extensively described (Gilioli et al, 1979).

The pathophysiological mechanisms involved can be different for the different agents. Anoxia is the most common event involved in the majority of cases. Carbon monoxide intoxication classically results in the formation of carboxyhemoglobin, since CO has an affinity for hemoglobin 200 or 300 times that of oxygen.
Cyanide is also responsible for cytotoxic hypoxia blocking tissue utilisation of oxygen by inhibition of the cytochrome oxidase system. Lead encephalopathy is mainly mediated by vascular changes involving both arteries and veins. Intoxication by organic solvents is usually chronic, most likely due to accumulation of the toxic agent in fat.

The anatomopathological changes reflect the different pathophysiological mechanisms and the sensitivity of the different areas of the brain to the toxic agents. Anoxic changes result in demyelination of the white matter, usually in the so-called watershead areas more sensitive to lack of oxygen. Basal ganglia are also frequently involved. Bilateral pallidal lesions with necrosis can be the result of acute intoxication by carbon monoxide, cyanides, carbon disulfide, methylchloride, phosphorus, or of chronic intoxication by carbon monoxide, cyanides, manganese, lead, methylchloride, carbon disulfide. Symmetrical areas of necrosis in each putamen is a typical neuropathological finding in lethal methanol intoxication (Aquilonius et al, 1980). Extensive hemorrhages in the basal ganglia can also be found in the severe acute cases. In lead encephalopathy the pathologic changes are mainly in the capillaries, venules and larger veins and the parenchymal changes are thought to be largely secondary. Patients exposed to industrial solvents or solvent-exposed painters have been found to have larger cerebral and cerebellar sulci and ventricles, suggesting brain atrophy (Arlien et al, 1979; Juntunen et al, 1980).

THE ROLE OF NEURORADIOLOGICAL INVESTIGATIONS

When the acute or chronic intoxication of the CNS by neurotoxic agents results in morphological changes at a macroscopic level, these can be detected by neuroradiological investigations.

In the past, angiography and pneumoencephalography were the only available neuroradiological techniques. Only PEG however can indirectly reflect atrophic changes of the brain by showing enlarged ventricles and sulci. In fact PEG has been employed as a method to detect cerebral atrophy in subjects exposed to organic solvents (Juntunen et al, 1980). The disadvantages of PEG are its invasiveness and discomfort to the patient. Air must be introduced in the subarachnoid spaces and CSF removed in variable amounts from 15 to 35/40 ccs and this can modify the condition of the subarachnoid spaces and ventricular size.

COMPUTERIZED TOMOGRAPHY

The advent of Computerized Tomography (CT) has introduced a totally new dimension in the in vivo evaluation of the brain. It is now possible to clearly detect not only the CSF spaces, their morphology and size, but also the brain parenchyma. Distinction between white and grey matter and basal ganglia is now possible. The density resolution is very high and very fine structures or lesions, down to a few millimeters in diameter, can be detected. Electronic measurements of size and density, sagittal and coronal computer reconstructions increase the accuracy of the technique.

Characteristic pathological changes are well known. Ischaemic or anoxic lesions appear as areas of diminished density with respect to the surrounding brain parenchyma. Haemorrhagic lesions are hyperdense areas in the first two or three weeks and become hypodense when transformed into necrotic or cystic cavities. Intravenous injection of an iodinated contrast medium can result in contrast enhancement of ischaemic areas or where the blood-brain barrier is disrupted. Being a non-invasive and safe method, CT can be repeated at short intervals, providing an accurate follow-up evaluation of the evolution of brain lesions.

Only a very limited number of papers has been published up to now on the role of CT in occupational toxic encephalopathies. Aquilonius et al (1980), reported necrotic changes in the area of the putamina, shown by CT, in four out of six patients with methanol intoxication. Hemorrhages were also seen in some cases. CT is considered potentially very useful for prognosis and possible therapy guidance since there is a relationship between the severity of the clinical symptoms and the CT findings. Symmetric globus pallidus infarctions were seen on computed tomography in a case of cyanide poisoning (Finelli et al, 1981). In carbon monoxide poisoning CT shows decreased density in the basal ganglia, most likely due to anoxia (Spinnler et al, 1980).
The baseline CT scan can be normal but intravenous contrast injection can produce enhancement in the basal ganglia and dentate nucleus (Nardizzi et al, 1979).

CONCLUSIONS

Computerized Tomography has become the most valuable neuroradiological modality for the evaluation of the CNS. It can be foreseen that it will be increasingly used in the evaluation of brain lesions due to neurotoxic agents.

In acute cases, however, the CT pictures can be negative for a while since the early changes occur at a biochemical level without morphological changes. Contrast enhancement could help in better defining anoxic areas still isodense with the brain parenchyma. Early brain oedema will be detected as a possible diffuse decrease of brain density and reduction in size of ventricles and sulci. CT may provide essential information for differential diagnosis in severely ill patients when exposure to neurotoxic agents is not known since the anoxic changes in the basal ganglia can be quite characteristic although non-specific.

The use of CT for the evaluation of chronic effects of neurotoxic agents, i.e., mainly brain atrophy, requires large series of patients and control groups. The size of ventricles and sulci in fact varies widely from individual to individual and increases with ageing. It is hence very difficult to attribute clinical significance to slightly enlarged ventricles and sulci in a single patient although exposed to neurotoxic agents.

New techniques, like Positron Emission Tomography (PET) and Nuclear Magnetic Resonance (NMR) will introduce different parameters in the evaluation of brain. The metabolism of the brain and its physical properties will be better analysed and understood with minimal invasiveness for the patient.

REFERENCES

Aquilonius, S.M., K. Bergstom, P. Enoksson, V. Hedstrand, P.O. Lundberg, U. Mostrom and Olsson (1980). J. Comput. Assist. Tomogr., 4, 425-428.

Arlien-Loberg, P., P. Bruhn, C. Gyldensted and B. Melgaard (1979). Acta. Neurol. Scand., 60, 149-156.

Finelli, P.F. (1981). J. Comput. Assist. Tomogr., 5, 755-756.

Gilioli, R., G. Filippini, M.G. Cassitto and C. Bulgheroni (1979). In Trattato di Medicina del Lavoro, ed. Piccin, Padova.

Nardizzi, L. (1979). Arch. Neurol., 36, 38-39.

Juntunen, J., S. Hernberg, P. Eistola and V. Hupli (1980). Eur. Neurol., 19, 366-375.

Spinnler, H., R. Sterzi and G. Vallar (1980). Arch. Suis. Neurol. Neurochir. Psychiat., 127, 79-88.

Lymphocytic Neurotoxic Esterase

A biochemical monitor of organophosphorus-induced delayed neurotoxicity in man

M. Lotti

Division of Occupational Medicine and Toxicology, Medical Service, and Northern California Occupational Health Center, San Francisco General Hospital Medical Center, and the Department of Medicine, University of California, San Francisco, California, USA

KEYWORDS

Organophosphorus esters; delayed neurotoxicity; human peripheral neuropathy; human neurotoxic esterase; lymphocyte NTE: neurotoxicity biomonitor.

INTRODUCTION

Organophosphorus esters (OP) have different toxic effects on the nervous system. Most extensively studied is the acute toxicity due to inhibition of acetylcholinesterase (AChE) at nerve endings that is characterized by muscarinic, nicotinic, and central nervous system symptoms. Another form of toxicity caused by OP affects certain animals including humans. It is not related to AChE inhibition and is called OP-induced delayed neurotoxicity (OPIDN) (reviewed recently by Davis and Richardson, 1980).

In humans with OPIDN the main clinical signs are pain and sensory disturbances in the calf muscles followed by paralysis of limb muscles. In a single dose situation, the delay between poisoning and onset of symptoms is 2 to 4 weeks, whereas with chronic exposure the time of onset of symptoms might be unpredictable. Neurophysiologic findings are widespread neuropathic changes in the distal limb muscles and prolongation of distal latencies with relatively mild slowing, if any, of motor conduction velocities (Senanayake, 1981). These physiologic changes are consistent with the pathologic changes, which are characterized by distal axonopathy with multifocal origin and only a secondary loss of myelin (Bouldin and Cavanagh, 1979a and b). Young persons are apparently more resistant to OPIDN and may recover completely (Senanayake, 1981), whereas recovery in adults, over a 47-year follow-up study, is poor (Morgan and Penovich, 1978).

Several OPs have been associated with neurotoxicity in humans including tricresylphosphates (Senanayake, 1981), mipafox (Bidstrup, *et al*, 1953), leptophos (Xintaras, *et al*, 1978), trichlorphon (Hierons and Johnson, 1978), trichlornate (Jedrezejowska, *et al*, 1980), EPN (Xintaras and Burg, 1980), and methamidophos (Senanayake and Johnson, 1982).

NEUROTOXIC ESTERASE (NTE): THE MOLECULAR TARGET FOR OPIDN

For a better understanding of the molecular mechanism of initiation of OPIDN, a simplified scheme of the biochemistry of OP is shown below (for a more extensive review, see Aldridge and Reiner, 1972):

Biochemistry of Esterases and Organophosphorus Esters

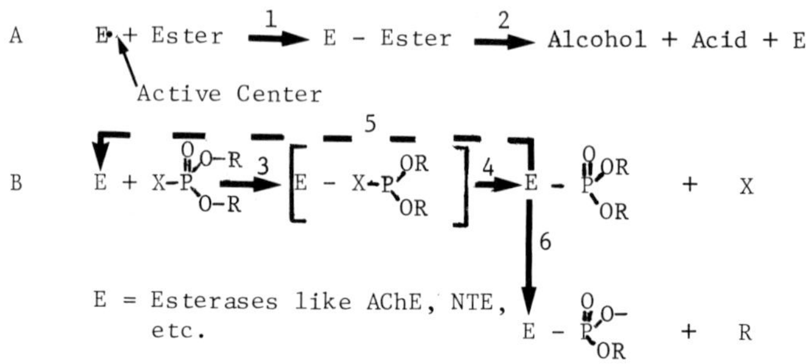

A *Enzyme-Substrate Interactions:* Reaction 1 is the formation of the Michaelis complex and 2 is the hydrolysis of the ester with immediate reappearance of the free enzyme.

B *Enzyme-Organophosphorus Ester Interactions:* Reactions 3 and 4 are analogous to 1 and 2. X is the acidic group that leaves after phosphorylation of the enzyme. Reaction 5 (restoring free enzyme) is very slow: the speed of this reaction differentiates inhibitors (OP) from substrates. Reaction 6 is called "aging" because it is time-dependent. The nature of this reaction is not enzymatic but depends on the chemical bond of groups R. The result is the loss of an R group, which leaves a negatively charged phosphoryl-enzyme complex.

The mechanism of OPIDN was recently reviewed extensively by Johnson (1982). In brief, in the early 1970s, Johnson demonstrated that OPIDN is initiated by the phosphorylation of a protein in the nervous tissue designated NTE. A second step is further required to produce the delayed neuropathy: the "aging" of the phosphoryl-enzyme complex. It is usually a rapid reaction and involves the loss of a group attached to the phosphorus, leaving a negatively charged phosphorus group attached to the protein; groups lost during aging might have a wide range of reactivity. The aging depends only on the chemistry of the OP. OP-NTE complexes that are able to age are those formed by phosphates, phosphonates, and phosphoroamidates: those formed by phosphinates, carbamates, and sulphonates are not able to age. The former compounds might cause delayed neuropathy when the threshold of inhibition of NTE *in vivo* is reached, whereas the latter will never cause OPIDN. A high level (70 to 80%) of inhibition of NTE in the nervous tissue of the experimental animal, soon after dosing with an OP capable of causing OPIDN, predicts the development of ataxia 2 weeks later in those that survive.

The phosphorylation and subsequent aging of NTE represents the two-step mechanism of initiation of OPIDN; the most convincing experimental evidence available is based on "protection" experiments. When high inhibition of NTE is achieved in animals dosed with phosphinates, carbamates, or sulphonates, no neuropathy occurs 2 weeks later because the inhibitor-NTE complex cannot age. If similarly treated animals are further dosed with a high dose of a potent neurotoxic OP (i.e., a phosphate), neuropathy will also not develop because the neurotoxic OP will find the catalytic site of NTE already phosphorylated. Therefore, there is no possibility of forming an aged NTE. The mechanism is summarized below.

The Mechanism of Initiation of OPIDN

Type of OP	NTE inhibition [% inhibition required to produce the described effect]	Aging of inhibited NTE	Clinical effect
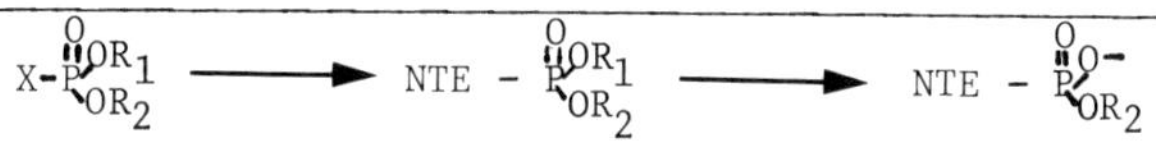			NEUROPATHY
PHOSPHATE	[70-80%]		
		NO AGING	NO NEUROPATHY PROTECTION from a subsequent challenging dose of a phosphate
PHOSPHINATE	[>30-40%]		

It is therefore obvious that the catalytic activity of NTE is not vital *per se* in maintaining the health of the neuron. It is not known yet if the negative charge left after aging or the R group itself is responsible for the development of the axonal lesion. The cascade of biochemical reactions leading to the onset of clinical symptoms 2 weeks later is also unknown. To summarize the actual level of biologic understanding about the role of NTE in the development of OPIDN, *it is not known which physiologic functions are associated with this protein that has an experimentally useful but not physiologically relevant NTE catalytic activity.* The mechanistic difference from the other toxicity caused by OPs in which the lack of catalytic activity of inhibited AChE is experimentally useful and also the causes of the toxicity should be evident.

Two possible pathophysiologic mechanisms (among others) have been suggested for OPIDN: a defective protein synthesis (Cavanagh, 1973) and inhibition of fast axoplasmic transport (Reichert and Abou-Donia, 1980). A study of NTE in peripheral nerves indicated that the rate of return to NTE in peripheral nerves after inhibition with an effective dose of neurotoxic OPs was similar to that in other nervous tissues, suggesting that the rate of axonal transport of NTE or of its protein components is not affected (Caroldi and Lotti, 1982). Furthermore, it seems that the axon, and not the cell body (where protein synthesis takes place), plays a major role as a target site in this kind of peripheral neuropathy. This is supported by the evidence that intra-arterial injection into the leg of hens (the animal of choice for OPIDN experiments) of diisopropylphosphorofluoridate (DFP) or phenylmethane sulfonylfluoride (PMSF) produces, respectively, unilateral neuropathy and protection; both toxic and protective effects are related to the level of inhibition in the axon and not to that in other parts of the nervous system or the cell (Caroldi, *et al*, 1982).

FROM EXPERIMENTAL ANIMALS (HENS) TO MAN: EXTRAPOLATION OR VALIDATION?

Studies of the sensitivity of NTE and related esterases to OPs in human central and peripheral nervous tissue samples (biopsied and autopsied) have been performed and the results have been compared with those obtained in hens. The pattern of response to inhibitors indicates that human and hen nervous tissues contain similar amounts of NTE. This enzyme is defined as mipafox-sensitive phenyl valerate esterase activity among the diethyl *p*-nitrophenyl phosphate (paraoxon) resistant ones (Lotti and Johnson, 1980a).

Analysis of the response of human NTE *in vitro* to a large number of compounds known to affect hen NTE showed that I_{50}s for the enzyme from both species were usually similar within a threefold range (Lotti and Johnson, 1978). We proposed that the relative potency of an OP to react with AChE or NTE *in vitro* should be correlated to the corresponding potency *in vivo* to kill the animal or to produce delayed neurotoxicity, respectively. This was the case for several OPs: the *in vitro* ratios AChE I_{50}/NTE I_{50} correlated with the corresponding *in vivo* ratios LD_{50}/neurotoxic dose in hens. Similar *in vitro* ratios have been obtained using human enzymes. Although I_{50}s for both AChE and NTE differed between the 2 species by no more than a factor of 4, the ratios differed in some cases up to eightfold. *In vivo* ratios for humans are obviously not available. The extrapolation to man is based, therefore, on comparisons of *in vitro* ratio: compounds with a lower ratio for humans than for hens suggest that it will be more difficult to produce neuropathy in the former. However, an *in vitro* ratio that is higher for humans than hens might reflect a greater relative sensitivity of humans for delayed neurotoxicity. It is obvious that this type of study does not say anything about the dose that might be needed to produce such effect in humans but suggests only which toxic effect might predominate. It has to be stressed that such *in vivo* and *in vitro* correlations are based on several assumptions: the ultimate inhibitor is assumed to have equal access to both targets, the NTE and AChE rates of resynthesis are sufficiently slow, the rate of aging of NTE is rapid compared with the rates of synthesis of new enzyme, and the spontaneous reactivation of the enzymes *in vivo* is small. The evidence indicates that these assumptions are correct. Indeed, the evidence so far suggests that OPs that are neurotoxic to humans are also neurotoxic to hens.

It is not known whether or not the percentage of inhibition of NTE necessary to initiate neuropathy in humans is the same as that in hens (70 to 80% inhibition). Therefore, extrapolation to humans by comparing ratios is based on the further assumption that the same amount of inhibition of NTE is necessary. Trichlorfon (dichlorvos is the active inhibitory metabolite) can produce neuropathy in humans after a single massive dose (Hierons and Johnson, 1978). Initially, dichlorvos was thought to be neurotoxic in adult hens after repeated high doses but not after a single dose (Johnson, 1975). However, a single massive dose of dichlorvos caused neuropathy in the hens (Caroldi and Lotti, 1981). The 2 species are therefore apparently equally susceptible to dichlorvos. To explain the higher ratio in humans, we suggest that a smaller proportion of NTE in the human nervous system is required to be phosphorylated and aged to reach the initiation threshold. This might also have occurred in the several reported cases of neuropathy due to methamidophos in humans (Senanayake and Johnson,

1982); a dose of about twice the unprotected median lethal dose causes 50% inhibition of NTE and no neuropathy in hens (Senanayake and Johnson, 1982).

The question about the threshold for humans is a key one and might be answered by postmortem assay of both AChE and NTE in tissue after OP intoxication. This would help to build up a data bank for comparisons with observed clinical responses in nonfatal cases; as more data are published, it may be possible to assess the maximum inhibition of NTE that can be tolerated by humans without expectation of a clinical neuropathic response. Such data have been collected for omethoate, an OP with negligible neurotoxic potential in both hens and humans (Lotti, *et al*, 1981). In an autopsy sample of tissue after omethoate poisoning, NTE activity was 100% and AChE was highly inhibited. These results were expected based on data obtained in hens: omethoate preferentially inhibited AChE rather than NTE and did not cause OPIDN.

The clinical evidence validated the assumption that toxicologic and enzymatic tests in hens can identify those OPs that may or may not induce OPIDN in humans. Nevertheless, hen and human neurons may differ in their tolerance to inhibition of NTE: in hens an OP dose that effects 70% loss of NTE activity also produces OPIDN (if the OP-NTE complex can age). Whether or not this is also true in humans obviously cannot be answered by controlled experimentation, but it may be addressed indirectly, as will be explained in the next section.

NTE ACTIVITY IN LYMPHOCYTES: A CLINICAL TEST?

In a recent comprehensive study of the occurrence of NTE in tissues of the adult hen by Dudek and Richardson (1982), NTE had limited distribution among the tissues and was present in lymphatic as well as neural tissue (Dudek and Richardson, 1982). These investigators suggested that if inhibition of NTE in blood lymphocytes correlates with inhibition of NTE in nervous tissue, it might be possible to monitor and assess the potential neurotoxic hazard after an exposure to OP by measuring NTE activity in peripheral lymphocytes. In their study, performed in 1978, inhibition of NTE in lymphocytes correlated highly with inhibition of NTE in brain tissue in hens 4 hours after a single exposure to an OP. However, this correlation was not evident at any other time in hens after an acute exposure or during a chronic exposure to OP in an earlier study (Lotti and Johnson, 1980b). Furthermore, we subsequently measured lymphocyte NTE activity in humans after both single and repeated exposures to different OPs.

CASE REPORTS

A 53-year-old woman accidentally ingested an estimated dose of 35 mg of dimethyl 2,2,2-trichloro-1-hydroxyethylphosphonate (trichlorphon) per kilogram and was admitted to Ospedale Civile di Cittadella, Padua, Italy in November, 1980. Acute cholinergic symptoms correlated with high inhibition of cholinesterases. Lymphocyte NTE was measured several times and electromyography and nerve conduction studies were performed. No inhibition of NTE or electrophysiologic changes in the peripheral nerve were detected because of the low dose ingested (Caroldi *et al*, 1981).

A 26-year-old man intentionally ingested an estimated dose of 20 g of chloropyrifos together with about 40 g of 2,4-dichlorophenoxyacetic acid (2,4-D) and 40 g of 2-methyl-4-chlorophenoxy-proprionic acid (MCPP). Lymphocyte NTE was inhibited (about 50%) in the blood sample obtained 13 hours after the ingestion. NTE measurements in a previous blood sample obtained at the time of admission to San Francisco General Hospital Medical Center in December 1981 and 3 other samples obtained between 13 and 30 hours after the ingestion, when the patient died, were within the normal range. The patient had few cholinergic symptoms. Inhibition of NTE in autopsied peripheral nerve was about 40% of a normal control value (Pond *et al*, 1982).

From these acute poisoning cases due to potentially neurotoxic OPs, we learned that there is no likelihood of development of OPIDN if lymphocyte NTE activity is not affected. Furthermore, lymphocyte NTE activity correlates with that in peripheral nerves. However, this correlation is evident only over a limited period of time after the poisoning, confirming the same intriguing observation in the animal studies (Dudek, 1979).

Lymphocyte NTE activity was also studied in workers occupationally exposed to S-S-S,tributylphosphorotrithioate (DEF), a cotton defoliant widely used in California that is capable of causing OPIDN in the hen (Lotti *et al*, 1982). Its relative potency as an AChE inhibitor *in vivo* in the hen is low if compared with its ability to inhibit NTE. *In vitro* tests challenging human enzymes confirmed this toxicologic characteristic of DEF (Lotti, unpublished observations). The exposure of 7 workers was monitored during aerial and ground application of DEF by measuring air and dermal concentrations. Whole blood AChE, plasma butyrylcholinesterase (BuChE), and lymphocyte NTE together with electromyography and nerve conduction velocity studies were performed before and after the exposure. Whole blood AChE and plasma BuChE did not change significantly from the preexposure values when measured during the course of the exposure (the length of exposure was from 10 to 35 days). Electromyographic and nerve conduction velocities did not change from the preexposure values when measured 3 weeks after the end of exposure. NTE activity in lymphocytes increased from the preexposure values on day 1 of exposure (+ 40 to 80%) but was inhibited (- 40 to 60%) after 25 to 30 days of exposure. Three weeks after the end of exposure, the values of NTE approached preexposure values. The initial increase in activity, which has been described in animals as well, cannot be explained at this time. Although producing a decline in lymphocyte NTE activity, a prolonged, mainly dermal exposure to DEF caused no signs whatsoever of peripheral nerve involvement. DEF is not a direct inhibitor of esterase and it needs metabolic activation: it is likely that the active ultimate inhibitor would be extremely unstable. It is therefore possible that during this exposure to DEF more inhibitor was offered to the peripheral lymphocytes than to the nervous system. In this case the NTE in lymphocytes might not represent the actual level of the enzyme in the nervous system; further studies during exposures to DEF and other OPs are necessary to confirm this hypothesis.

CONCLUSIONS

The hen is a good model for studying human OPIDN. The target protein NTE is similar in both species and the sensitivity to several inhibitors is comparable. There is evidence that *in vivo* effects of OPs on the target enzymes NTE and AChE are similar in both species.

There is no direct evidence that inhibition of NTE in nervous tissue soon after the poisoning correlates with the development of clinical signs of peripheral neuropathy 2 weeks later in humans. If, whenever a death due to OP poisoning occurs, NTE and AChE activity are measured soon after the poisoning, it will be possible to compare the levels of inhibition with the clinical responses in other nonfatal cases; identification of the OP and measurement of tissue levels of OP are also absolutely necessary. It is not known what level of NTE inhibition in the nervous tissue is required in humans to initiate OPIDN; some evidence indicates that it might be lower than the level in hens.

Lymphocyte NTE activity is affected in humans after both single and repeated exposures. Lymphocyte NTE activity represents, over a limited period of time after a single exposure, a good mirror of the amount of NTE in the target organ (peripheral nerve). After subacute exposure to DEF, a potential neurotoxic OP, lymphocytic NTE activity was significantly affected (but less than the 70% threshold required in the hen), but there was no correlation with neuropathy; no neurologic abnormalities were detected. The altered lymphocyte NTE activity might represent a false-positive case. Some answers will be available through the data obtained from future fatal cases, from further experiments with the hen model, and from observations and experiments in occupational medicine.

All of the available evidence suggests that it will be possible to determine the level of lymphocyte NTE inhibition in humans after an acute exposure or during a chronic exposure and to rationally predict in clinical conditions the likelihood of development of OPIDN.

REFERENCES

Aldridge, W.N., and E. Reiner (1972). Enzyme Inhibitors as Substrates, North Holland Publishing Co., Amsterdam.

Bidstrup, P.L., Bonnell, J.A., and A.G. Beckett (1953). Br. Med. J., 1, 1068-1072.

Bouldin, T.W., and J.B. Cavanagh (1979a). Am. J. Pathol., 94, 241-252.

Bouldin, T.W., and J.B. Cavanagh (1979b). Am. J. Pathol., 94, 253-270.

Caroldi, S., and M. Lotti (1981). Toxicology Lett., 9, 157-159.

Caroldi, S., and M. Lotti (1982). Toxicol. Appl. Pharmacol., 62, 498-501.

Caroldi, S., Lotti, M., and M. Pegoraro (1981). Proceedings: 44 Congresso Nazionale della Società Italiana di Medicina del Lavoro e di Igiene Industriale, CLEUP-Padova, pp. 359-366.

Caroldi, S., Lotti, M., and A. Masutti (1982). Vet. Hum. Toxicol., 24, 218 (Abstract).

Cavanagh, J.B. (1973). CRC Crit. Rev. Toxicol., 2, 365-417.

Davis, C.S., and R. J. Richardson (1980). Experimental and Clinical Neurotoxicology, P.S. Spencer and H.H. Schaumburg, Eds., Williams & Wilkins, Baltimore, pp. 527-544.

Dudek, B.R. (1979). Brain and Leucocyte Neurotoxic Esterase as Biomonitors of Organophosphorus Delayed Neurotoxicity, Ph.D. Thesis, the University of Michigan, Ann Arbor, MI.

Dudek, B.R., and R.J. Richardson (1982). Biochem. Pharmacol. 31, 1117-1121.

Hierons, R., and M.K. Johnson (1978). Arch. Toxicol., 40, 279-284.

Jedrezejowska, H., Rowinska-Marciuska K., and B. Hoppe (1980). Acta Neuropathol., 49, 163-168.

Johnson, M.K. (1975). Arch. Toxicol., 34, 259-288.

Johnson, M.K. (1982). Rev. Biochem. Toxicol., 4, 141-212.

Lotti, M., and M.K. Johnson (1978). Arch. Toxicol., 41, 215-221.

Lotti, M., and M.K. Johnson (1980a). J. Neurochem., 34, 747-749.

Lotti, M., and M.K. Johnson (1980b). Arch. Toxicol., 45, 263-271.

Lotti, M., Ferrara, S.D., Caroldi, S., and F. Sinigaglia (1981). Arch. Toxicol., 48, 265-270.

Lotti, M., Becker, C.E., Aminoff, M.J., Woodrow, J.E., Seiber, J.N., Talcott, R.E., and R.J. Richardson (1982). Vet. Hum. Toxicol., 24, 220-221 (Abstract).

Morgan, J.P., and P. Penovich (1978). Arch. Neurol., 35, 530-532.
Pond, S., Donohue, B., Stevens, B., Lotti, M., Osterloh, J., and C.E. Becker (1982). Program Intern. Congr. Clin. Toxicol. (in press).
Reichert, B.L., and M.B. Abou-Donia (1980). Mol. Pharmacol., 17, 56-60.
Senanayake, N. (1981). J. Neurol. Neurosurg. Psychiatry, 44, 775-780.
Senanayake, N., and M.K. Johnson (1982). N. Engl. J. Med., 306, 155-157.
Xintaras, C., and J.R. Burg (1980). Experimental and Clinical Neurotoxicology, P.S. Spencer and H.H. Schaumburg, Eds., Williams & Wilkins, Baltimore, pp. 527-544.
Xintaras, C., Burg, J.R., Tanaka, S., Lee, S.T., Johnson, B.L., Cottrill, C.A., and J. Bender (1978). Occupational Exposure to Leptophos and Other Chemicals, US Government Printing Office, Washington, DC, DHEW (NIOSH) Publication No. 78-136.

ACKNOWLEDGMENT

I am grateful to Drs. Charles E. Becker and Ronald E. Talcott for reviewing this manuscript.

Recent Behavioral Developments

Chairmen: K. Battig (*Switzerland*)
C. Xintaras (*W.H.O. - Geneva*)
G. Scarlato (*Italy*)
E. A. Emmett (*USA*) (unable to attend)

Behavioral Performance Assessment as a Biological Control of Occupational Exposure to Neurotoxic Substances

F. Gamberale and A. Kjellberg

National Board of Occupational Safety and Health, Stockholm, Sweden

ABSTRACT

Assessment of the behavioral performance of humans is used with increasing frequency as a tool in the investigation of the effects on the nervous system of exposure to neurotoxic substances. Thus, psychometric tests or batteries of tests have been used in experimental investigations as well as in epidemiological or clinical studies. However, psychometric techniques can also be usefully applied in the monitoring of hazards to health and safety, which can be carried out directly in the work environment. This field application of psychometric technique is illustrated and discussed in the present paper on the basis of 1) a description of a 10-min vigilance test, 2) a review of a series of field studies on the acute effect of exposure to industrial solvents, in which the above vigilance test was used and 3) an analysis of the combined results of the vigilance test from a large population of industrial workers tested in the field.

KEYWORDS

Behavioral performance; reaction time; field studies; industrial solvents; biological control.

INTRODUCTION

There is by now a large body of evidence showing that effects on the nervous system due to occupational exposure to neurotoxic substances can be assessed with the aid of psychometric techniques. Thus, deterioration of performance on tests of various psychological functions has several times been observed prior to the occurrence of manifest medical symptoms of any toxic action of the substance (Hänninen, 1971; Gamberale, 1975).

Psychometric tests or batteries of tests have mainly been used in experimental or in epidemiological investigations. These methods, however, are now also being used by several clinical laboratories for diagnostic purposes. Undoubtedly the application of performance tests in toxicological investigations has increased our knowledge of the potential noxious action of certain substances on the nervous system. The impairments of the behavioral performance of exposed workers, which have been observed in several epidemiological investigations, have clearly shown that occupational exposure to neurotoxic substances constitutes a hazard to health and safety. Consequently, these findings have stressed the need for prevention. However, the behavioral effects observed in epidemiological or clinical investigations are generally of very little use when the issue is to decide whether, or to what extent, preventive measures are needed in a specific work environment. Thus, in these types of studies, too little is usually known about the specific work environment conditions which have caused the adverse health effects observed. Experimental studies do not suffer from this shortcoming and they have probably influenced the setting of TLV's for several substances. However, the basic limitation of the experimental approach is that the exposure conditions which can be simulated in an exposure chamber are rarely if ever very representative of the condition existing in the real work environment. In spite of these problems there is some evidence that the assessment of behavioral changes caused by toxic substances could be of practical importance with regard to the monitoring and prevention of hazards to health. This evidence comes from a few investigations carried out directly at the working site (Götell, et al, 1972; Gamberale, et al, 1974; Gamberale, et al, 1976; Binaschi, et al, 1976; Kjellberg and Strandberg, 1979; Kjellberg, et al, 1979; Cherry, et al, 1980; Cherry, et al, 1981; Anshelm Olson, et al, 1981; Anshelm Olson, 1982). A brief review of the main results and characteristics of these investigations is given below to give some idea of the potentialities of this type of field study. In some of these investigations (Gamberale, et al, 1975; Kjellberg and Strandberg, 1979; Kjellberg, et al, 1979; Anshelm Olson, et al, 1981; Anshelm Olson, 1982) the same reaction time test was employed. Some data from these studies will be described more fully later in the paper, following a presentation of the background leading to the choice of this test and a description of the test itself.

SOME FIELD INVESTIGATIONS OF PERFORMANCE CHANGES

In the study of Götell and coworkers (1972) the performance on a reaction time test for a group of workers exposed to an average of 150 ppm styrene was compared to that of a reference group of non exposed workers of the same age. The mean reaction time of the exposed workers appeared to be affected both before and after a work day.

One of the studies of Gamberale and coworkers (1974) dealt with a possible acute effect on vigilance due to the low-dose exposure to anesthetic gases to which anesthetics staff may be exposed during their work. It was found that anesthetics nurses showed a greater intraindividual variability in a reaction time test after work than did age-matched nurses with intensive care duties. No such performance difference could be observed when measurements were carried out before the working day.

In the other study by Gamberale and coworkers (1976) workers from four factories building glass-fibre boats, who were exposed to styrene in concentrations varying between 20 and 120 ppm, were compared with regard to reaction time to age-matched workers from two mechanical industries. The workers exposed to styrene had a longer reaction time, a greater deterioration of reaction time over time and a greater irregularity of performance in the test than the non exposed workers both before and after a work day. However, these differences between the groups were more pronounced at the end of the work day.

The performance of a group of workers exposed to solvent mixtures in concentrations clearly exceeding the TLV's was studied by Binaschi and coworkers (1976) using a battery of several tests. No control group was studied in this investigation. However, the effect of exposure manifested itself in a pronounced deterioration of performance after a work day. This deterioration of performance was observed in spite of the use of tests which in control conditionswould have revealed a learning effect.

Kjellberg and coworkers (1979) were able to follow up a group of styrene exposed workers after the closing down of a glass-fibre boat factory. They found that the reaction time of the exposed workers was prolonged compared to that of a control group of non-exposed workers. This effect could still be observed four days after the cessation of exposure. No effect was noticeable 35 days after cessation of exposure.

Kjellberg and Strandberg (1979) studied the performance of anesthetics nurses in a reaction time task before and after surgical operations. They were not able to find any performance differences between this group and a group of nurses from an intensive care unit.

In the investigations of Cherry and coworkers (1980, 1981). which are described in detail elsewhere in the proceedings. reaction time was found to be prolonged among styrene exposed workers before the beginning of a work day. They also found that early morning urinary mandelic concentrations after two days without exposure correlated with reaction time measured on arrival at work.

In a longitudinal investigation (Anshelm Olson, et al., 1981) of workers handling organic solvents and solvent based paints in the production of plastic-coated sheets, reaction time was measured several times during a period of more than two years. The performance of the workers was found to improve over time as a result of changes in the ventilation system and in the working process, which had brought about a considerable reduction in solvent vapor concentration at the work-place.

The effects of exposure to a mixture of organic solvents was also studied by Anshelm Olson (1982) among workers in the paint industry. In this study the exposed group performed less well than the control group in most of the behavioral tests used. The differences were evident for both morning and afternoon measurements. Particularly noticeable was a marked decrease in performance in a reaction time test over the course of the day for the workers exposed to the highest solvent levels.

A common characteristic of the designs adopted in all the investigations referred to above is that performance measures were collected both at the beginning and at the end of a work day. The rationale of this design is based on the underlying assumption that the effect of exposure during the day may manifest itself in an impairment of performance at the end of a work day, with a return to normal levels of performance by the beginning of the following working day, i.e., after approximately 16 hours of rest from exposure. If the action of exposure lasts long enough to bridge the exposure-free period between two consecutive work days, performance should be negatively affected also at the start of a work day. Since this type of field study is based on repeated measurements performed on the same subjects, the behavioral tests used must be as insensitive to learning or training effects as possible. It is therefore not surprising that, with only one exception (Binaschi, et al, 1976), all the researchers have adopted tests of simple reaction time which satisfy this requirement better than other types of behavioral tests.

When measurements of performance have to be carried out at different times during the day, the circadian rhythms may constitute a confounding factor with regard to performance changes which is as important or even more important than learning or motivation. However, it does seem possible to minimize or control the effects of these confounding factors using operations of experimental control such as the use of an age-matched reference group and a balanced design. This was accomplished in some of the investigations reviewed above using a quasi-experimental study design of which a schematic representation is given in Table 1.

Table 1. Experimental design in field studies of acute effects of occupational exposure to neurotoxic substances.

MEASUREMENT OCCASIONS

BEFORE WORK	AFTER WORK	BEFORE WORK
Eg 1	Eg 1	
	Eg 2	Eg 2
Cg 1	Cg 1	
	Cg 2	Cg 2

Eg 1, Eg 2 Exposed groups

Cg 1, Cg 2 Control groups of age-matched workers

As shown in the table, half of the workers in both the exposed and the control group perform the tests before work the first time and after work the second time, whereas the other half perform the tests after work the first time and before work the second time. Balancing the study in this way, a single analysis of variance for each performance variable is sufficient to test the effects of exposure, of learning and of the circadian rhythm independently of each other and to test the possible effect of the interactions between these factors.

ENVIRONMENTAL CONTROL BY MEANS OF REACTION TIME TESTING

Considering the results obtained in the investigations reviewed above, it seems reasonable to regard the assessment of behavioral performance as constituting a potential means for the monitoring and control of exposure to neurotoxic substances in a specific work environment. The detection of,e.g. working conditions leading to CNS depression would clearly indicate the need for improvements in the hygienic quality of the environment under investigation and would therefore be a very valuable complement to traditional environmental monitoring. The use of the study design outlined in this paper or the application of a similar strategy to collect measurements of performance directly at the working site appears to be practicable. Some behavioral tests suitable for this purpose are already available and others can be developed. To be suitable for this use in the field a performance test must satisfy a number of requirements: the test must be sensitive, short, reliable, simple to administer, resistant to learning or training effects and, most important, the performance measures obtained must be related to early signs of functional disorders,

A performance test which in the opinion of the authors has been shown to meet these requirements better than any other test known to them is a 10-min simple reaction time test developed by Lisper and Kjellberg (1972) and used extensively in behavioral toxicology research by our organization.

A point of departure for the development and use of this test was the work by Buck (1966) in which reaction time was shown to be highly correlated with the probability for a stimulus to be detected. This finding implied that reaction time could be used as a measure of perceptual vigilance, avoiding the time-consuming procedure of traditional vigilance tests. The present test was also strongly influenced by the investigations by Lisper (1969), who demonstrated that the shorter the inter-trial interval a reaction time test has, the greater will be the deterioration of performance over time. Thus a very high signal density (16 signals/min) was chosen for the test with inter-trial intervals varying between 2.5 and 5.0 seconds in steps of 0.25 seconds. A random series of 32 inter-trial intervals with an approximately quadratic distribution is repeated five times. The testing is preceded by a one minute training period.

From the beginning the test was used with acoustic stimuli. However, it was subsequently shown that the test could be used with optic stimuli without its characteristics being changed. Using this test. Lisper and Kjellberg (1972) could demonstrate an adverse effect of only one night of sleep deprivation. Prior to their study it had not been possible to observe any adverse effect of a period of lost sleep as short as this without the use of time-consuming vigilance tests (Naitoh, 1970). The 10-min SRT test has later been found sensitive to the effects of e.g., physical work (Lisper, 1977), alcohol (Blomquist and Hedberg-Sowa, 1976) and of caffeine and diazepam (Lisper, et al, 1981). The most comprehensive use of the test, however, has been in the field of behavioral toxicology. Besides its use in the field studies referred to earlier, the test has been adopted in experimental investigations (Anshelm Olson, et al, 1982; Gamberale, et al, 1975; Gamberale, et al, 1978;) as well as in epidemiological studies (Elofsson, et al, 1980; Iregren, 1982; Knave, et al, 1978; Knave, et al, 1979;).

ANALYSIS OF EMPIRICAL DATA ON THE SRT TEST

Six of the field and epidemiological investigations (Anshelm Olson, 1982; Elofsson et al, 1980; Gamberale et al, 1975; Iregren, 1982; Knave et al, 1978; Knave et al, 1979;) in which the SRT test has been used have been subjected to detailed statistical analyses by Söderman et al.(1982). In total the analyses comprised data from 292 workers occupationally exposed to solvents and 424 non-exposed control cases. This large data base permitted a detailed description of various test characteristics and of the performance changes caused by solvent exposure and increasing age. A further purpose of the analyses was to arrive at a standardized treatment of data.

Figs.1 and 2 illustrate two fundamental characteristics of the RT performance: The distribution of RT's is slightly positively skewed (Fig. 1) and the RT is gradually prolonged during the ten minutes of the test (Fig. 2). Both these figures are based on data from the non-exposed subjects.

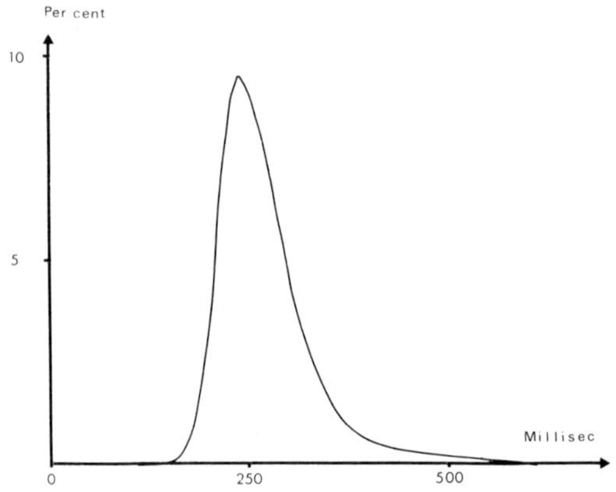

Fig. 1 Distribution of reaction times in the 10-min SRT test. N = 67840 (424 subjects x 160 stimuli).

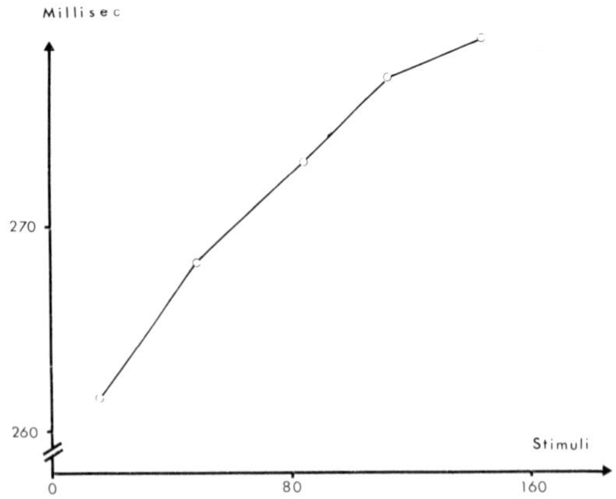

Fig. 2 Reaction time changes over test time. The points indicate the means of the reaction times during successive two-min time blocks (32 stimuli) for 424 subjects.

Fig. 3 shows the difference between the exposed and non-exposed workers at different points of the distribution. The figure has been constructed by ordering the 160 RT's of each individual from the fastest to the slowest RT.

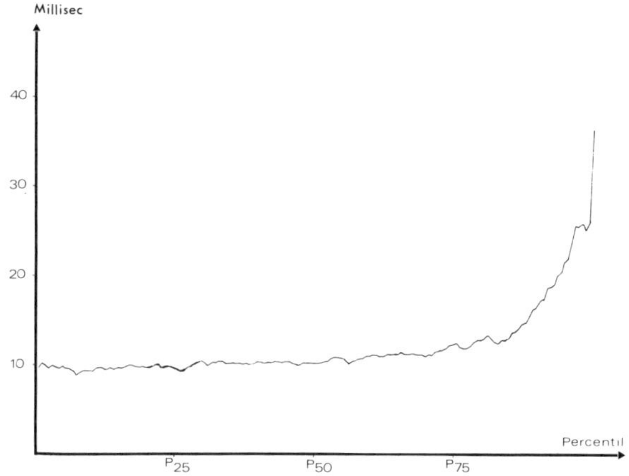

Fig. 3 Mean reaction time difference between solvent exposed (N = 292) and non exposed (N = 424) industrial workers at different points of the reaction time distribution.

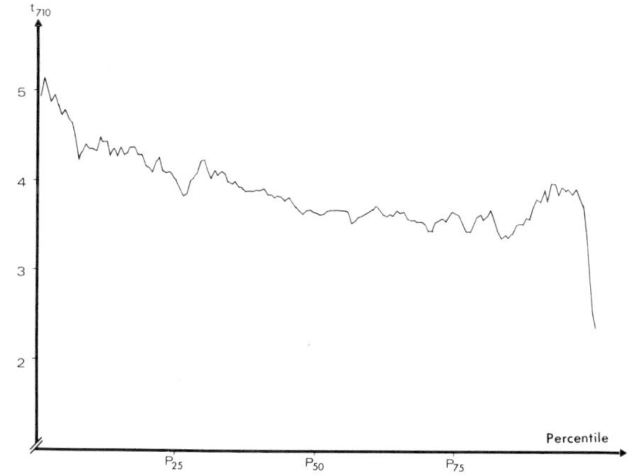

Fig. 4 T-values for the difference between solvent exposed (N = 292) and non-exposed (N = 424) industrial workers at different points of the reaction time distribution.

The mean difference between the two groups at each point of this order was then computed. It is evident from the figure that solvent exposure leads to a general prolonging of the RT's but that this effect is most pronounced

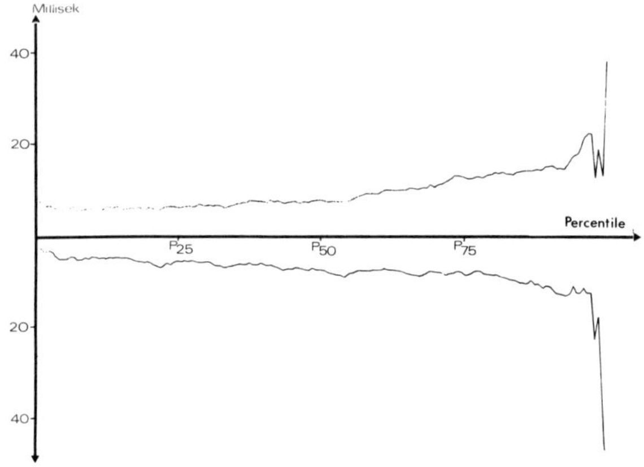

Fig. 5 Mean reaction time difference between three age groups of industrial workers at different points of the reaction time distribution. (N = 129 under 35 years, N = 97 between 36 and 45 years and N = 196 over 45 years).

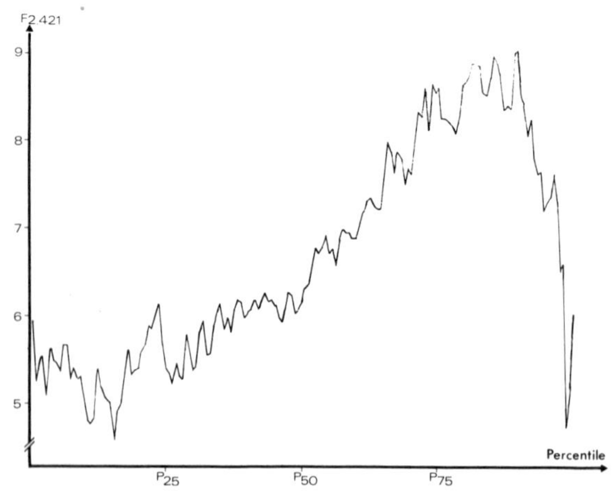

Fig. 6 F-values for the difference in mean reaction time between three age groups of industrial workers at different points of the reaction time distribution. (N = 129 under 35 years, N = 97 between 36 and 45 years and N = 196 over 45 years).

at the upper end of the distribution. However, since the RT variance of the longest RT'S is much larger than that of the faster ones, the point of maximum difference does not coincide with the point of maximum sensitivity to the effects of solvents. This is shown by Fig. 4 which depicts the result of a series of t-tests performed on the ordered RT-data. It is clear that the sensitivity is about the same over the whole distribution except for the highest percentiles where it shows a sharp decline. Figs.5 and 6 show the results of similar analyses of the age effects on the RT performance. From these and other types of statistical analyses Söderman et al. (1982) concluded that the most adequate measures of the RT performance were the mean and standard deviation of the 160 RT's Table 2 presents the means of these parameters in the exposed and non-exposed groups and Table 3 gives the corresponding values for the three age groups. One more parameter is accounted for in the tables, namely the increase in the RT during the ten minutes of the test. It is evident from the tables that no effect was observed in this parameter of either solvent exposure or age when the total material was analyzed. However, when more homogeneous groups are studied, particularly in repeated measurements designs, this has proven to be an informative variable (Anshelm Olson, 1982; Gamberale et al., 1975).

The balanced design used in the studies where measurements have been made before and after work makes it possible to evaluate learning effects or other effects of the repetition of measurement. There was not the slightest tendency in any of these cases of the first testing affecting the second one (Anshelm Olson, 1982; Gamberale et al., 1975). However, in an experiment where three testings were carried out during four hours a very slight but progressive slowing of the SRT was observed (Anshelm Olson et al., 1982).

The stability of the mean RT is also supported by high test-retest correlations. Such correlations have been computed both for field and laboratory data and for testings separated by intervals of varying lengths. With a two hours interval in the laboratory setting the correlations varied between .89 and .95. For measurements occurring six weeks apart it was .75 even though exposure conditions were not the same during the two testings (Anshelm Olson et al., 1982). In the field studies the testings before

Table 2. Mean, standard deviation and the increase in RT (in ms) during the 10-min reaction time test for a solvent-exposed group and a reference group of industrial workers, and p-values from the analyses of variance of the differences between the groups.

GROUP	X	s	INCREASE	N
EXPOSED	282	54	24	292
REFERENCE	270	50	22	424
p-VALUE	.000	.020	.469	

Table 3. Mean, standard deviation and increase in RT (in ms) during the 10-min reaction time test for three age groups of industrial workers, and p-values from the analyses of variance of the differences between the groups.

GROUP	$\overline{X}$	s	INCREASE	N
- 35 years	263	45	23	129
36 - 45 years	270	50	20	97
46 - 65 years	279	54	23	197
p-VALUE	.001	.001	.750	

and after work have given a correlation of .72. This was lowered to .56 when a 27 month interval separated the measurements (Söderman *et al.*, 1982).

CONCLUDING REMARKS

The analyses of data collected with the 10 min simple SRT test have shown that it fulfils the requirements stated above of an instrument for field studies of neurotoxic effects. The task is short and simple to administer. The measurements have high reliability and the training effects are negligible. Most important, a series of studies on workers exposed to industrial solvents have demonstrated the sensitivity of the test as an indicator of neurotoxic effects. It is true that the differences between the exposed and non-exposed groups shown in the tables and figures based upon the combined material are rather small. However, this group includes many persons exposed to very low levels of solvents. Subgroups exposed to higher levels show a much larger effect, which thus makes it possible to detect the effects also in quite small groups (e.g. Anshelm Olson, 1982). Furthermore, in an epidemiological study in which the SRT test was included in a very large battery of psychometric tests (Elofsson *et al.*, 1980) it proved to be the most sensitive of all tests used. Of course, both the research strategy and the behavioral test described in this paper have many limitations. However, they constitute a promising point of departure for the development of methods to be used in preventive programs.

REFERENCES

Anshelm Olson, B. (1982). *Neurobehav. Toxicol. Teratol.* (In print).

Anshelm Olson. B., F. Gamberale and B. Grönqvist (1981). *Int. Arch. Occup. Environ. Health, 48*, 211-218.

Anshelm Olson, B., F. Gamberale and A. Iregren (1982). *Arbete och Hälsa.* National Board of Occupational Safety and Health (Solna, Sweden) (to be published).

Binaschi. S., G. Gazzaniga and E. Crovato (1976). In: M. Horváth (Ed.) *Adverse Effects of Environmental Chemicals and Psychotropic Drugs. Vol. 2*, Elsevier Scient. Publ. Co. Amsterdam. 91-98.

Blomquist. J., and J. Hedberg-Sowa (1976). Department of Psychology, University of Uppsala. Sweden.
Buck, L. (1966). Psychol. Bull., 65, 291-304.
Cherry, N, B. Rodgers, H. Venables. H.A. Waldron and G.G. Wells (1981). Br. J. Ind. Med., 38, 346-350.
Cherry, N., H.A. Waldron, G.G. Wells, R.T. Wilkinson, H.K. Wilson and S. Jones (1980). Br. J. Ind. Med., 37, 234-240.
Elofsson, S.A., F. Gamberale, T. Hindmarsh, A. Iregren, A. Isaksson, I. Johnsson, B. Knave, E. Lydahl, P. Mindus, H.E. Persson, B. Philipson, M. Steby, G. Struwe, E. S.derman, A. Wennberg and L. Widén (1980). Scand. J. Work Environ. Health, 6, 239-273.
Gamberale, F. (1975). AMBIO, 4, 43-46.
Gamberale, F., G. Annwall and B. Anshelm Olson (1976). Scand. J. Work Environ. Health, 4, 220-224.
Gamberale, F., G. Annwall and M. Hultengren (1978). Scand. J. Work Environ. Health, 4, 204-211.
Gamberale, F-, H.O. Lisper and B. Anshelm Olson (1976). In: M. Horváth (Ed.) Adverse Effects of Environmental Chemicals and Psychotropic Drugs. Vol. 2, Elsevier Scient. Publ. Co. 135-148.
Gamberale. F. and G. Svensson (1974). Work Environm. Hlth, 11, 108-113.
Götell. P., O. Axelson and B. Lindelöf (1972). Work Environm. Hlth, 9, 76-83.
Hänninen, H. (1971). Br. J. Ind. Med., 28, 374-381.
Iregren, A. (1982). Neurobehav. Toxicol. Teratol. (In print)
Kjellberg. A. and M. Strandberg (1979). Research reports from National Board of Occupational Safety and Health. Sweden. 11, 1-12.
Kjellberg. A., E. Wigaeus, J. Engström, I. Åstrand and E.Ljungqvist (1979). Arbete och Hälsa. National Board of Occupational Safety and Health (Solna, Sweden) 18, 1-25.
Knave, B., B. Anshelm Olson, S. Elofsson, F. Gamberale, A. Isaksson, P. Mindus, H.E. Persson, G. Struwe. A. Wennberg. and P. Westerholm (1978). Scand. J. Work Environ. Health, 4, 19-45.
Knave. B., F. Gamberale. S. Bergström. E. Birke, A. Iregren, B. Kolmodin Hedman and A. Wennberg (1979). Scand. J. Work Environ. Health, 5, 115-125.
Lisper, H.O. (1969). Psychon. Sci., 17, 217-218.
Lisper, H.O. (1977). Fatigue in Traffic [in Swedish]. Bilaga 1, SOU 1977:2.
Lisper, H.O. and A. Kjellberg (1972). J. Exp. Psychol. 96, 287-290.
Lisper, J.O., J. Törnros and J. vanLoon (1981). In: L. Goldberg. (Ed.) Alcohol, Drugs and Traffic Safety, Vol. 3.
Naitoh, P. (1970). Human Factors, 12, 575-585.
Söderman, E., A. Kjellberg, B. Anshelm Olson and A. Iregren (1982). Research reports from National Board of Occupational Safety and Health, Sweden, 27, 1-40.

Psychological Test Batteries: New Trends and Developments

H. Hänninen

Institute of Occupational Health, Helsinki, Haartmaninkatu 1, SF-00290 Helsinki 29, Finland

First something about the background of this paper: At the Institute of Occupational Health in Finland, I have been involved in behavioral toxicology since the early 1960's. In 1981 I had the opportunity of working at the National Institute for Occupational Safety and Health (NIOSH) in the United States.
I was there to review the literature on behavioral toxicity of chemicals at the workplace and to evaluate the methods applied in such research. During my stay I also had the possibility of learning not only about the research currently underway at NIOSH but also about research that was going on or being planned by some research groups outside NIOSH. And finally, at the beginning of June 1982 I participated in the Nordic Conference on Behavioral Toxicology in Örebro, Sweden. The conference dealt with the sensitivity and specificity of instruments used in behavioral toxicology and with the need for developing methods for clinical practice and research. My paper will be a résumé both of the observations and conclusions that I made when reviewing the literature and also of the thoughts and ideas presented in several methodological discussions at NIOSH and elsewhere in the United States, at the Nordic Conference, and at the Institute of Occupational Health, Helsinki.

The objective measurements of the psychological changes which follow long-term exposure to neurotoxic chemicals began in the 1960's and 1970's rather independently at several institutes. One initiative for such research came from observations made in connection with clinical psychological assessments of workers with diagnosed or suspected occupational poisonings, and partly the same methods were applied in research as were applied in the clinical assessments. But because not all of the neuropsychological assessment methods are good research tools, only the most promising ones were selected. Additional methods were provided by diverse aptitude test batteries. Later on, methods developed within the field of experimental psychology were also included.

The different professional backgrounds and psychological traditions of each country led the various psychological research groups to select different test batteries or sets of tests. Without doubt methodological diversity is one of the problems of behavioral toxicology today. It complicates the comparison of the results obtained by different research groups, and it slows down the accumulation of a common body of knowledge in this field. However, the magnitude of this problem should not be overestimated.

When using the term "test battery" I refer to a test series constructed within a certain theoretical frame, and intended more or less for continuous use for a certain purpose. The test battery is generally designed to cover a wide range of different capabilities in a reasonably structured and balanced manner. The optimal situation is to standardize all of the tests in the battery with the same population, and to calculate the internal correlations between the tests. One advantage of a balanced and standardized test battery is that it allows a structured description of the findings to be made.

The term "set of tests" is better suited to series of tests that cover a narrower area of behavioral functions, and to looser combinations of diverse test methods which may be based on different behavioral theories.

Test series that fulfill at least most of the criteria of a test battery have been constructed for the purposes of behavioral toxicology in Finland (Hänninen and Lindström, 1979), Sweden (Hogstedt et al, 1980) and the German Democratic Republic (Schneider and Seeber, 1979). The Italian methods also approximate the criteria (e.g. Cassitto et al, 1978; Angotzi et al, 1980). These test batteries have a rather similar structure: they are designed to assess separately the domains of intellectual performances, memory, perceptual motor speed and accuracy, and psychomotor abilities. Their similar structures facilitate the comparison of results even when the single tests are not the same.

Moreover, some single tests are used as elements of several test series or test batteries applied in behavioral toxicology. For instance, the use of some of the subtasks of the Wechsler Adult Intelligence Scale (WAIS) has been widespread. As the tests for perceptual motor speed are often found to be rather equivalent measures of the same ability, their results are comparable, too. The same applies at least partly to those eye-hand coordination tests, the factor structure of which has been stated. On the other hand, the different variants of the Choice Reaction Time test are not necessarily equivalent or comparable in spite of the same name: depending on the design of the task, they may assess quite different behavioral functions.

When searching through the literature, I found about 50 studies where the study design involved the application of psychological test methods to exposed workers, whereafter the test results were either compared with those yielded by a reference group, or correlated with the measures of exposure, or both. Table 1 is an overview of the most commonly used tests and their sensitivity in 32 studies, all of which dealt with the effects of long-term exposures. The first column of numbers gives the number of studies where the test was applied. The second column gives the number of studies where the test showed or suggested some effects, and the third column gives the number of studies where the method in question could be considered the best detector or was one of the two or three best detectors.

The main conclusions drawn from this table and from some other corresponding scrutinies were as follow:

1. Some of the tests that were most commonly used have shown fairly consistent sensitivity to toxic effects (Benton visual memory test, the tests for

perceptual speed and accuracy, the Santa Ana dexterity test).

2. Some other tests yielded statistically significant results less consistently but still so often that it seems unlikely that the results were due to chance. Two kinds of tests can be distinguished in this respect: A) Some tests most often yielded results near, a bit above, or a little below the threshold of statistical significance (Digit Span, the Visual Memory test of the Wechsler Memory Scale). It is probable that these tests assess functions indeed sensitive to neurotoxic agents, but that the tests themselves are weak tools of measurement. B) Some other tests were among the best detectors in some studies but yielded clearly negative results in some others. These inconsistencies possibly reflect qualitative differences in the effects of different chemicals, i.e., such tests may detect impairments that are specific to certain kinds of exposure (examples are Similarities and the verbal memory tests).

3. Tests of vocabulary never showed any relationship to exposure. When applied they were used to control for the primary intelligence level as a possible confounder. They do in fact seem to be suitable for this purpose.

The need for improved test methods is keenly felt by psychologists active within this area of research. This need is not so urgent because of failures in previous studies as, with a few exceptions, the applied methods were able to detect behavioral effects in the groups studied, sometimes even when the level of exposure was rather low (e.g., Elofsson _et al._, 1980; Hänninen _et al_, 1976; Lindström _et al_, 1976), or when the groups under study were small (e.g., Angotzi _et al_, 1980; Hänninen _et al_, 1978).

But neurobehavioral research is now faced with new challenges which add new and more strict requirements to the methods. Improved sensitivity is required both when the more subtle effects of low exposures are investigated and also when the health hazards of small worker groups are evaluated. Descriptions of exposure-effect relationships call for tests which allow the accurate scaling of the effects. And if the tests are used in health surveillance, more knowledge about the specificity of the test findings is needed. Another aim for method development is to increase the economy of the test batteries so that the maximum amount of information can be obtained at minimum costs in terms of time, expertise and equipment.

There are different possibilities open to the process of method development. The conservative way is to rely on the present test batteries, improving their most obvious weaknessess A) by substituting weak tests with better ones which measure the same functions and B) by improving the psychometric qualities of tests that seldom yield clear results. Substituting poor items with better ones, increasing the items, and modifying the difficulty of the task are methods by which the reliability and/or sensitivity of psychological tests can be improved. Sometimes it is also possible to increase the information obtained by a test or to improve its reliability by modifying the scoring system.

A few examples of the ways single tests can be improved: the sensitivity of the Simple Reaction Time test has been improved by increasing the number of stimuli and by using the individual variability of reaction times and the performance curve as additional measures (Gamberale, 1976). A modified scoring system for Block Design, one that yields separate scores for time and erroneous reproductions, has proved to be superior to the traditional scoring (Eskelinen, L., Comparison of solvent intoxication patients with a group of neurological patients: unpublished manuscript). The perceptual motor speed tests might be improved by applying alternative scoring methods for better treating the speed-accuracy trade-off. It might also be useful to apply different modifications of the memory tests so that the most sensitive versions could be found. The rate of presentation, delay for the recall or recognition, and the sense modality used in the task are some of the task characteristics that could be varied.

Modern technology can be applied and has been applied to improve the methods. The application of computerized registration increases the objectivity and accuracy of the method and often makes it possible to analyze performance more thoroughly. The application of technical means - such as a visual display or a recording tape - to present the task and for the performance itself similarly increases the objectivity and accuracy of the measurement and is particularly well suited to the assessment of relatively simple behavioral functions. It is, however, more problematic when they are applied to tasks intended to assess more complex behavior. This particuarly decreases the possibilities of controlling the extent to which the subject follows the given instructions. Another limitation of computerized instruments is that they usually require standard conditions and are therefore more suitable for use in laboratories than in field studies. Good technically reliable equipment is also expensive.

Successful methodological development is not only a matter of how to design good tests and how to improve the old ones. Rather, it also presupposes a clear theoretical frame of reference and a good understanding of the functions being measured and the behavioral effects being sought. One step of my work at NIOSH was an attempt to organize and classify the methods used to date into the frame of a common taxonomy of psychological functions. This procedure is useful when attempts are made to evaluate the sensitivity of various functional domains to neurotoxic effects in the light of present empirical results. A good taxonomy is also helpful when attempts are made to evaluate the coverage of the applied test batteries and to identify those potentially sensitive behavioral domains that have not been studied sufficiently. Moreover, it is helpful for evaluating the theoretical constructs themselves on which the descriptions of the tests and the interpretations of the test findings are based (Fleishman, 1975).

The classical psychological test batteries are based on the factor analytical theories of abilities. The European test batteries constructed for use in behavioral toxicology follow the same lines but have an additional neuropsychological aspect. The frame of the ability theory provides the batteries with a clear empirically based structure but leads to a somewhat crude and perhaps rather inadequate description system. In behavioral toxicology the interest lies in the functions and processes that may be sensitive to toxic effects on the central nervous system, not so much in the abilities as they are understood in the classical factor theories.

Since most of the toxicopsychological test batteries were designed, both theoretical and methodological progress has been made in different branches of psychology. The advances within two areas, neuropsychology and cognitive psychology, are particularly valuable to the development of behavioral toxicology and its methods. Both neuropsychology and cognitive psychology represent a dynamic view of man's capabilities, functions, and processes. These disciplines can provide us both with new methods and with better description systems. They can also provide new insights to the mental functions and processes accounting for the performances in the "old" tests.

As to the new methods, these two areas can be considered especially profitable for the design of tests to assess memory and attention. According to the descriptions of the exposed subjects themselves, impairments of these functions are the central effects of chemical exposure, yet memory and attention are only poorly represented in most of the current test batteries.

Scientists working in the field of behavioral toxicology are increasingly inclined to see their research as a distinct area of neuropsychological research. Behavioral toxicology is distinct because it deals with more subtle central nervous system dysfunctions than clinical neuropsychology in general, and maybe also with partly different dysfunctions. But the relationship between the brain and behavior is the subject matter of both behavioral toxicology and of neuropsychology in general.

There is also increasing interest in cognitive psychology, which has been one of the mainstreams of psychology throughout the last decade. Some psychologists working in the field of behavioral toxicology are now adopting single methods developed by the cognitive psychologists, whereas other psychologists are applying the theories and methodology of cognitive psychology to design a totally new system of measurement in behavioral toxicology.

It should be noted that cognitive functions are measured by many of the "old" methods, too, particularly by tests of intelligence and memory. Thus the term "cognitive test" when used today may refer either to the classical intelligence and memory tests or to tasks derived from the theories of cognitive psychology. The discipline of cognitive psychology investigates human behavior as an information processing system and methods have been developed to study separately several distinct elements of information processing. Currently there is also growing interest in the individual differences revealed by the tasks of cognitive psychology. The application of these methods to behavioral toxicology may in the future lead to marked advances in this field, yielding a better understanding of the toxic effects on behavior and perhaps also providing better means for their early detection.

Thus it seems probable that the diversity of methods used in behavioral toxicology will persist. Increasing both the coherence and the comparability of research results in the field of behavioral toxicology is a goal that is not easy to reach in the present circumstances. Increased communication between the psychologists who carry out this research might be the first necessary step in that direction.

Table 1. The most frequently used tests in studies on work-place toxicity: 32 studies (*)

		N OF STUDIES	WITH RESULTS	"BEST DETECTOR"
INTELLECTUAL BEHAVIOR	VERBAL			
	Vocabulary Tests	6	0	
	Similarities	9	5	2
	VISUAL			
	Block Design	15	9	3
	Picture Completion	8	3	
	Raven	3	3	
MEMORY	VERBAL			
	Digit Span	16	10	
	Story Recall	8	3	2
	Associative Learning	3	2	1
	VISUAL			
	Benton	7	6	3
	Wechsler	5	3	
PERCEPTUAL SPEED AND ACCURACY	Digit Symbol	14	12	
	Bourdon-Wiersma/Dots	5	3	
	Neisser/"D-2"	5	5	3
PSYCHOMOTOR SPEED	EYE-HAND			
	Santa Ana	11	9	4
	Michigan	5	3	
	FINGER-DEXTERITY			
	Tests	3	3	3
	REACTION TIMES			
	Simple	12	5	3
	Choice	12	7	2
PSYCHOMOTOR CONTROL	Mira	6	5	1
	Tremor	3	2	1

(*) The reference list for this table can be obtained from the author.

REFERENCES

Angotzi, G., M.G. Cassitto, D. Camerino, R. Cioni et al. (1980). Med. Lav. 6, 463-480.

Cassitto, M.G., P.A. Bertazzi, D. Camerino, C. Bulgheroni, A.M. Cirla, R. Gilioli, C. Graziano and M. Tomasini (1978). Med. Lav., 69, 144-150.

Elofsson, S.A., F. Gamberale, T. Hindmarsch, A. Iregren, A. Isaksson, I. Johnsson, B. Knave et al. (1980). Scand. J. Work Environ. Health, 6, 239-273.

Fleishman, E.A. (1975). American Psychologist, 30, 1127-1149.

Gamberale, F., H.O. Lisper and B. Anshelm Olson (1976). The effect of styrene vapour on the reaction time of workers in the plastic boat industry. In: M. Horvath (ed.), Adverse effects of environmental chemicals and psychotropic drugs. Vol. 2. Elsevier, Amsterdam, pp. 135-148.

Hänninen, H., L. Eskelinen, K. Husman and M. Nurminen (1976). Scand. J. Work Environ. Health, 2, 240-255.

Hänninen, H., S. Hernberg, P. Mantere, R. Vesanto and M. Jalkanen (1978). J. Occup. Med., 20, 683-689.

Hänninen, H. and K. Lindström (1979). Behavioral test battery for toxicopsychological studies: Used at the Institute of Occupational Health in Helsinki. Second revised edition. Institute of Occupational Health, Helsinki. Reviews 1.

Hogstedt, C., M. Hane and O. Axelson (1980). Diagnostic and health care aspects of workers exposed to solvents. In: C. Zenz, Developements in Occupational Medicine. Year Book Medical Publishers, Inc., Chicago.

Lindström, K., H. Härkönen and S. Hernberg (1976). Scand. J. Work Environ. Health, 3, 129-139.

Schneider, H. and A. Seeber (1979). Z. Psychol., 187 H.2., 178-205.

Reaction Times

M. G. Cassitto

Institute of Occupational Health, Department of Preventive Medicine, University of Milan, Milan, Italy

ABSTRACT

The use of reaction times as a diagnostic tool in neurobehavioral toxicology is examined in the light of the encouraging results of the investigations carried out mainly in the field of solvent intoxication. Being a quick, simple, reliable and non-intrusive method, with low if any susceptibility to compensatory mechanisms, scarce attention has been given to the many factors which can influence the results. These factors can be distinguished into physical factors originated by the test structure, psychophysiological and psychological factors issuing from the subject, environmental factors. An administration schedule is suggested indicating the information necessary to make the results entirely replicable.

KEYWORDS

Reaction times; Arousal; Neuropsychology; Testing paradigm.

Reaction times and especially simple reaction times is one of the tests most frequently used in neurobehavioral toxicology. The results obtained, mainly in solvent investigations, seem to confirm the interest in this choice. So a discussion and evaluation of its characteristics, laws, administration procedures, confounding factors and analysis of results would be worthwhile to see whether more information can be obtained by means of this simple, rapid and non-intrusive method. The test itself does not need any introduction. It supplies the latency of the response measured from the onset of the stimulus.

The stimulus can be visual (a light or a shape, white or coloured) or auditory, (low, medium, high and bass, medium, acute sounds). The test is generally well accepted by the subjects who have no difficulty in establishing a direct relationship between their rapidity of response and the integrity of their nervous system. It can be administered in various experimental paradigms.

Figures 1 and 2 illustrate a typical RT equipment. Some others are more compact and portable but certain characteristics are constant: all administer visual and/or auditory stimuli according to selected procedures in terms of number of stimuli/minute, intertrial interval, total length, stimuli intensity.

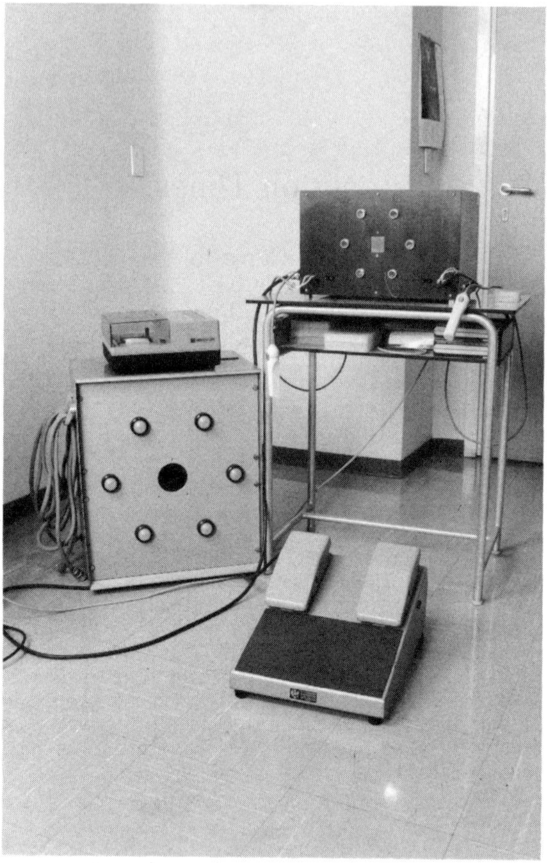

Fig. 1.

The Figure shows the equipment used in the Neuropsychological Unit of the Institute of Occupational Health of Milan University. It is French-made equipment displaying a series of possible programs, from one simple stimulus to several stimuli which can be combined to form a wide variety of testing paradigms. These programs deal with simple visual and auditory reaction times, choice reaction times with multiple variables, complex reactions, diffuse attention, concentrated attention, and vigilance. It utilizes 24 colour stimuli, 6 shapes, and 3 auditory stimuli. Analyzing the literature on reaction time applications in behavioral toxicology, it appears that in general the results have been analysed comparing the means or median and standard deviation of the single trial for each subject and for the groups examined compared to themselves or to control groups. Analysis of variance, regression analysis, and distribution of frequencies have also been used. From this point of view, administration schedules for reaction times are sufficiently clear to be replicated by other groups and in other countries. But many other factors can influence the results and the way they have been handled in the research schedule should also be indicated in order to make

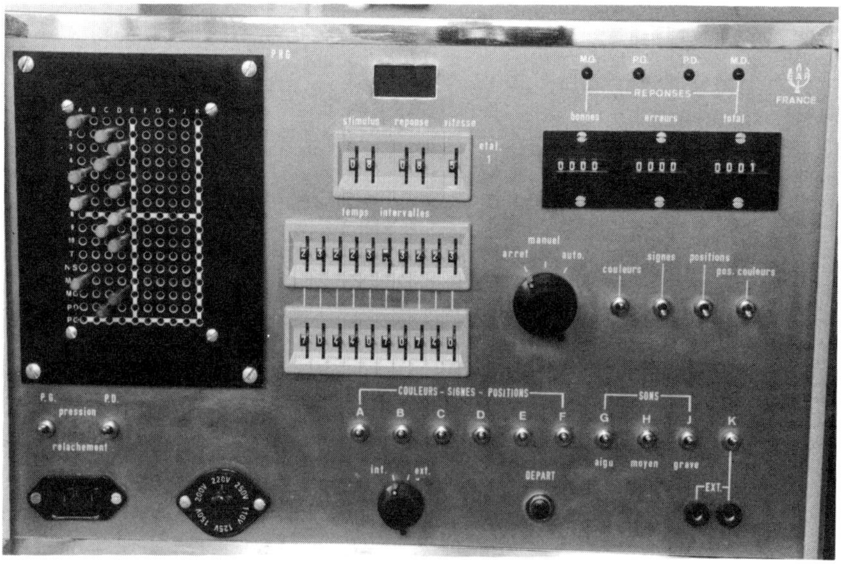

Fig. 2.

the results totally replicable. Three groups of factors seem to consistently influence the results: a group of physical factors originated by the test structure, a group of psychophysiological and psychological factors issuing from the subject, and environmental confounding factors. No mention will be made in this list of those individual or external factors which can be equally present and active in all test situations. The factors mentioned belonging to the first group, namely the physical characteristics of the RT paradigm, include: stimulus intensity, intertrial interval and foreperiod duration and the characteristics of the warning signal. It is an established fact that reaction times are inversely related to the intensity of the stimulus both in visual and auditory stimulation. This finding was first reported by Cattell in 1886, replicated by Pieron in 1920 and by many other investigators afterwards. In the 1970's, Teichner, who has contributed much to defining the laws regulating reaction times, confirmed the relationship between intensity and time of response. More recently, two Finnish investigators not only confirmed the above findings, but stressed that not only the intensity of the stimulus is relevant but also the total psychophysical intensity of the stimulation situation.

In fact, while increasing the intensity of small sources of light did not significantly influence the time of response, the times were consistently reduced when larger lights were used. Intensity in auditory stimuli is less relevant because the responses to auditory stimuli usually give values that are already near the lower physiological limits even at normal intensity. Not only intensity but intertrial intervals and interval foreperiod duration interfere with reaction times values. The intertrial (ITI) is the empty time between two stimuli; sometimes a warning signal is administered; in this case, the preparatory period between the warning signal (WS) and reaction times is known as foreperiod (FP). When no warning signal is given, as in most administration paradigms used in behavioral toxicology, the ITI func-

tions as a warning signal since the end of the stimulus is in itself the signal that a new stimulus is approaching. When the duration of the foreperiod and/or the intertrial interval is constant, there is a direct relationship with the reaction times: the longer the foreperiod duration, the longer the reaction times. Systematic variation of the foreperiod interval leads to the conclusion that about two seconds is the optimum value.
When the interval is below that value, the reaction time increases again, perhaps because the time is not sufficient to fully mobilize the response. The interval variation through the different trials also influences the reaction time. Up to a certain value these variations have a stimulating effect, above that value they introduce the variable of uncertainty in the experimental schedule which, accordingly, slows down the response speed. The nature of the warning signal is largely undefined. A warning signal is both a separate signal intended to inform the subject of the forthcoming stimulus but the end of the preceding stimulus can have exactly the same function for the subject. In such case, ITI and FP coincide. Whatever the nature of the warning signal, it usually yields two kinds of effects: first, it exerts a direct influence on the subject's general alertness, second, it starts the expectation processes. The expectation process consists of a response initiation which starts activating the two basic mechanisms of the final reaction: the perceptual action and the motor action. Perceptual preparation consists of a mental image rehearsal which facilitates its immediate detection, while the motor preparation, like a prepared reflex, leads to the final action declenchement. The other parameter, general alertness, refers obviously to the arousal level which can vary according to parameters of immediate arousal efficacy, and maintenance of optimal activation levels. The level of arousal is known to be related to individual characteristics, namely age, sex, cultural background, previous psychological and/or neurological alterations as well as personality dimensions such as introversion/ /extraversion. According to Eysenck, these dimensions seem to consistently influence the individual level of arousal with important effects in a task, like reaction times, where the uncertainty of input and the requested output promptness require high activation levels. The second parameter, maintenance of an optimal activation level can be untenable owing to short-term exhaustion and fatigue. Bevan has shown that these effects are even more likely to occur when ITI and/or FP time is below 2 seconds. A tentative explanation is that subjects are not able to maintain such a prolonged and elevated level of expectancy.
We shall now examine the second group of factors interfering with the response time: namely the psychophysiological and psychological factors.
In simple visual and auditory reaction times, errors are rare and of two kinds: response without signals or signals not followed by response; in choice reaction times, the error possibility is much higher.
In choice reaction times, subjects must choose between high accuracy and lower speed or high speed and higher error possibility. According to their choice, they will or will not accept a certain number of errors to achieve a higher rate of response. Of course, the response strategy, does not necessarily remain unmodified by the success or failure feedback through the trials.
A third function is then activated: decision process during which all the parameters discussed become cues in the information processing mechanism which leads to the adoption of a behavior strategy. We have seen that in simple visual and auditory reaction times, the role of the physical properties of the test elements and the psychophysical mechanisms of the subjects were prominent. In choice reaction times, other individual factors take on decisive importance in response validity: namely, motivation, learning and memory. Motivation is the energy releasing factor. There is a direct relationship between highly motivated subjects and efficacy of the strate-

gies adopted, and between motivation and optimal utilization of learning and mnestic abilities. Motivation selectively strengthens the capacity of the organism to attain its goal and, in this sense, becomes directly related to the time of response and the correctness of the choice.

Therefore all these parameters should be controlled if possible and in any case considered in the evaluation of reaction time results. This is true especially for monitoring programs when retesting of the subjects is programmed at fixed intervals and fluctuations of performance must be evaluated.

The last group of factors interfering with the response time is the group of environmental confounding factors. Most of them (namely noise, time of the day, examiner's personal influence, etc.) are the usual factors present in all testing situations and will not therefore be discussed in this paper. For these testing situations, noise is certainly the most important environmental factor interfering with the results. It can be of two kinds: occasional and permanent. When registering the reaction times of subjects, I think many of us have observed that when a sudden noise is produced, the response latency is significantly influenced and is lengthened,usually largely exceeding the individual standard deviation. When noise is not an accidental occurence but a permanent experimental or environmental condition, its influence can be twofold and must be interpreted according to the arousal laws and to Eysenck personality factors of extraversion and introversion. As we know, performance is expected to increase with increasing arousal up to an optimum level, after which it declines following an inverted U-shaped curve. Stimulation in the form of white noise is generally held to raise the level of arousal. The degree of stimulation received will then depend on the initial arousal level of the subject. According to Eysenck's theories, which have been confirmed by repeated experimentation, introverts are more aroused than extroverts. In quiet conditions, the former will perform better than the latter but the latter will show a greater improvement when noise is added while the introverts will show no decrement nor improvements under conditions of noise.

Another confounding factor originating from the test situation is the time of the day in which reaction times are administered since their relationship with the level of arousal makes this parameter extremely sensitive to the individual vigilance state. We cannot underestimate the arousal rhythms of the subjects, their being, as Kletman noted, a "morning" or "evening" type with consequent fluctuations in mental performance.

The table below is a tentative list of the data we think should be known in all administration paradigms and reported in all articles.

SIMPLE VISUAL AND AUDITORY REACTION TIMES: Administration Paradigm

- Structure of the test situation	individual group	
- Number of stimuli	visual	practice series; number of signals: test series; number of signals:
	auditory	practice series; number of signals: test series; number of signals:
- Signal duration	visual auditory	
- Intertrial interval	fixed variable	
- Visual signal luminance		
- Auditory signal frequency		
- Signal size		

- Width of the perceptive field
- Warning signal present / absent
- Room illumination
- Distance of the subject from the RT equipment.

These data, together with all other information concerning the population usually studied in epidemiological investigations, help to make reaction times a highly reliable tool. Through this review, we have seen that the number of factors producing an inhibiting or facilitation effect is high. Their number and their capability of inducing both positive and negative consequences explain the imperfect reproducibility of the results of this test in individual subjects, but also its sensitivity to changes in the internal or external environment of the subject. Of course, the problem of interfering factors is also present in every other test situation with other methods, some of which are less sensitive to environmental factors or technical variations, but which give the subject the possibility of resorting to compensatory mechanisms in case of increased difficulty in coping with the test requirements. In reaction times this possibility is minimal and is thus an advantage when reading the results because it makes the administration paradigms easier to control and makes rigorous standardization of application and interpretation feasible.

ACKNOWLEDGEMENTS

Thanks are extended to Mrs Giovanna Radaelli for expert secretarial assistance and Mr. Gianni Ciboldi for technical assistance in making the tables suitable for printing.

REFERENCES

Bevan, W., D.L. Hardesty, L.L. Avant (1965). Perceptual and Motor Skill, 20, 969-972.
Cattell, J.M. (1886). Brain, 8, 512-515.
Eysenck, H.J. (1971). Readings in Extraversion-Introversion: Bearing on Basic Psychological Processes, Staples Press Ltd.
Kletman, N. (1963). Sleep and Wakefulness, Chicago, Ill. University of Chicago Press.
Luce, R.D., D.M. Green (1972). Psychological Review, 79, 14-57.
Pekka,Niemi, Risto Näätänen (1981). Psychological Bulletin, 89, 1, 133-162.
Piéron, H. (1920). Année Psychologique, 22, 58-142.
Teichner, W.H., M.J. Krebs (1972). Psychological Review, 79, 4, 344-358.

Hand Tremor Measurements: Methodology and Applications

R. F. Fawer*, Y. de Ribaupierre**, M. P. Guillemin*, M. Berode* and M. Lob*

*Institute of Occupational Health and Industrial Hygiene, University of Lausanne, César-Roux 18, CH-1005 Lausanne
**Institute of Physiology, University of Lausanne, Rue du Bugnon, CH-1011 Lausanne, Switzerland

ABSTRACT

Occupational hand tremor has been described among workers exposed to metallic mercury, lead, manganese, DDT, lindane, lithium, methyl bromide. The various ways of measurement are reviewed with reference to metallic mercury. In order to quantify the tremor with precision, various behavioral tests have been developed, such as integrator counter giving numerical scores proportional to the duration, amplitude and frequency of the tremor, or hole tremormeter tests. It is now however agreed that tremors can be conveniently recorded by some form of mechanical device measuring either displacement, velocity or acceleration; the tremor signals can then be subjected to Fourier transformation in order to obtain a spectral analysis. The present method consists of records of hand tremor by means of an accelerometer attached to the dorsal face of the hand. Analysis of the records include the determination of the highest peak frequency (frequency corresponding to the highest acceleration) and, in order to take the whole spectrum into account as far as possible, three moments have been defined: the first, second and third moments consist, for each frequency band, of the products of the acceleration by respectively its corresponding frequency at its first, second and third power. This method has been used in a study of 51 workers exposed to metallic mercury. Measurements show good relationships with indices of exposure (defined by the duration and intensity of exposure). A pathophysiological explanation is attempted, taking into account the well-known accumulation tendency of metallic mercury in the cerebellum and in the basal ganglia.

KEYWORDS

Hand tremor; acceleration recording; spectral analysis; frequency distribution; metallic mercury.

INTRODUCTION

Occupational hand tremor has been described among workers exposed to metallic mercury, lead, manganese, DDT, lindane, lithium, methylbromide (Doull et al, 1980). But metallic mercury has certainly been more extensively studied than any other substance.

The purpose of quantifying hand tremor measurements was to develop tests which were more reliable, more sensitive and better suited to quantitative statistical analysis than standard neurological tests. This paper will review the different ways of measurement of hand tremor; then a method of recording and analysis will be illustrated with its application to a population of workers exposed to metallic mercury.

1. REVIEW OF THE DIFFERENT WAYS OF MEASUREMENT WITH SPECIAL REFERENCE TO METALLIC MERCURY

The first record of tremor was published more than eighty years ago (Taylor, 1901). In a study of workers exposed to mercury in felt hat factories, some records were obtained by fixing to a finger a needle which was applied on a tambour covered with black smoke.

Later, Kazantzis used the following method (Kazantzis, 1968): the subject was seated with his elbow resting on a table and was asked to point his finger at a target for five minutes. Tremor was measured with an angular accelerometer strapped to the dorsum of the index finger of the dominant hand. The electrical signals from the accelerometer were then picked up through a receiver and an amplifier passed frequencies around five cycles/s. Then an integrator-counter gave a numerical score proportional to the duration, amplitude and frequency of the tremor signals. Results were expressed as the total tremor count over the five minutes observation period. However no spectral analysis of the tremor was performed.

In a very recently published paper (Roels et al, 1982), the authors used an entirely different method with an apparatus called "hole tremormeter". To perform this test, the subject was seated with the arm and the hand unsupported, and was asked to insert a stylus vertically in a series of holes with decreasing diameters. A counter recorded the number of contacts with the edge of the hole for 15 seconds. Again, no spectral analysis was performed.

It is now agreed among neurophysiologists that tremor is conveniently recorded by some form of mechanical device measuring either displacement, velocity or acceleration; the tremor signals can then be subjected to Fourier transformation in order to obtain a spectral analysis (Marsden, 1978).

Wood et al, (1973) studied hand tremor due to industrial exposure to metallic mercury vapor in two patients. The subjects were instructed to maintain a force on a finger-trough attached to a strain gauge between a lower and an upper limit as signaled by two lights. The trials lasted 20 or 40 seconds, with the arm either supported or not. A description of the frequency characteristics was followed with power spectral analysis. The results are given in arbitrary units of power spectral density in terms of frequency.

More recently, Schuckmann (1979), developed the following procedure in his study of workers exposed to inorganic mercury in chloralkali plants: the subject was seated with his forearm fully supported and the hand freely movable from the carpal joint. The middle finger touched the sensor point of a piezo-electric receiver. The voltage changes induced by the tremor were recorded and a power spectrum obtained with the aid of Fourier analysis. The test lasted twice 20 seconds and was repeated several times.

Chaffin and Langolf's group have published many studies. In their first study (Chaffin *et al*, 1974), the following method was used: the subject was seated and his dominant arm was supported by an elbow rest which allowed a 90° elbow angle with the forearm unsupported. A padded wrist cuff was strapped around the wrist to which both a force-sensing transducer and a 6.8 kg load were attached. The subject was instructed to pick up the load by flexing his arm to a 90° elbow angle and hold it. Then, by performing a power spectral analysis on the forearm tremor for the first 15 seconds of the test, it was possible to quantify each worker's tremor frequency distribution and amplitude.

In their next studies (Langolf *et al*, 1977; Langolf *et al*, 1978), tremor was also recorded under very light muscular loading with the forearm supporting a light-weight pointer (285 g). The subject was asked to hold it as steadily as possible in an illuminated target zone. Recorded tremor signals were then analysed to produce power spectra as described above.

Briefly summarizing: one author, one method. Therefore any comparison of their respective results would be rather difficult. Only their methods of analysis of the records of the power spectra will be discussed later under the third section.

2. PRESENT METHOD

The subject was seated with his right arm supported by an elbow which allowed a 90° elbow angle with the forearm and hand unsupported. The hand was in the supine position. An accelerometer (Brüel and Kjaer 4366, weight 28.7 g) was fixed to the dorsum of the third metacarpal bone with a cuff. The electrical signals from this accelerometer are proportional to the acceleration; they were then amplified (Brüel and Kjaer 2204) and then transformed by Fourier techniques to obtain a spectrum (EMR/Schlumberger 1510-02 Real time spectrum analyser), which was subsequently recorded on paper (Sefram TRP 668). Calibration was performed before each test (Brüel and Kjaer 4291). The frequency range was 0.2 Hz to 51.2 Hz and each frequency band was 0.2 Hz wide. The acceleration range was 0.0 to 100.0 mm/s^2. A first measurement was carried out at rest for a period of 80 seconds. A second measurement was then performed under load (1250 g) attached to the hand for a period of 80 seconds, starting 5 seconds after attachment of the load to the wrist.

3. ANALYSIS OF THE RECORDS

The highest peak frequency (HPF) was the frequency corresponding to the highest acceleration. When more than one peak was shown, the highest one was chosen provided it differed from any other peak by more than 1.5 Hz and by more than 5.0 mm/s^2. When these criteria were not met, no HPF was determined (Fig. 1).

In order to take the whole spectrum into account as far as possible, the following moments were defined:

a) the products of the acceleration by the corresponding frequency for each band from 2.0 Hz to 30.0 Hz were obtained. Their sum divided by the number of bands was defined as the first moment (M1). ($M1 = \Sigma a_i f_i / n$, where a = acceleration, f = frequency, n = number of bands, i = 1,2,...n).

b) the products of the acceleration by the square of its corresponding frequency for each band from 2.0 Hz to 30.0 Hz were obtained. Their sum divided by the number of bands was defined as the second moment (M2). ($M2 = \Sigma\ a_i f_i^2/n$).

c) the products of the acceleration by the third power of corresponding frequency for each band from 2.0 Hz to 30.0 Hz were obtained. Their sum divided by the number of bands was defined as the third moment (M3). ($M3 = \Sigma\ a_i f_i^3/n$).

Frequency bands for which the frequency was less than 2.0 Hz were not taken into account because of possible interferences from cardiac beats. All those for which the frequency was more than 30.0 Hz were also not taken into account because the corresponding accelerations were so low that they can be considered as negligible (Marsden, 1978).

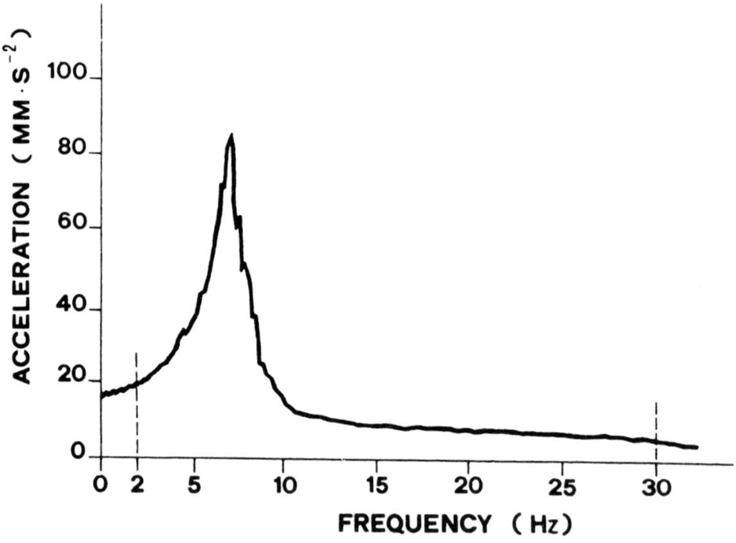

Fig. 1. Typical record of tremor at rest. The lower and upper limits of 2.0 and 30.0 Hz used for the calculation of the moments are shown. The highest acceleration is 85 mm/s^2 and its corresponding highest peak frequency (HPF) is 7.0 Hz.

These moments were calculated for two reasons:

a) HPF cannot always be easily determined. For example, no peaks were shown in 3 out of 61 spectra recorded in normal adults in a study on physiological tremor (Marsden et al, 1969); and in our own study 7 out of 102 HPFs could not be determined with the criteria described above (Fawer et al, 1982). On the other hand, the moments can always be determined.

b) HPF is by no means representative of the whole spectrum; it is only a part of it (i.e., frequency corresponding to the highest acceleration), and therefore almost the whole information given by the tremor record is ignored. Moreover, the computation of the three moments allows the following inference: a shift towards the right (high frequencies) of the spectrum, even with a HPF in the "normal range" will bring about increased corresponding values of M1 to M3, with M3 being relatively the most increased (M1, M2 and M3 are related to respectively the first, the second and the third power of the frequency).

However, in all studies reviewed above (Wood et al, 1973; Schuckmann, 1979; Chaffin et al, 1974; Langolf et al, 1977; Langolf et al, 1978) no analysis of the whole spectrum was performed; only the peak frequency was taken into account and therefore a certain amount of information provided by the power spectrum analysis is lost.

4. APPLICATION TO WORKERS EXPOSED TO METALLIC MERCURY

26 workers exposed daily to metallic mercury and 25 workers from the same factories were studied within the framework of an investigation of the effects of metallic mercury on health. The results of the evaluation of occupational exposure have been published in detail elsewhere (Berode et al, 1980). In brief, the duration of exposure (mean $\pm$ SE) was 15.3 $\pm$ 2.6 years, the TLV-TWA (personal exposure) 0.026 $\pm$ 0.004 mg/m^3, and the concentration of mercury in urine 11.3 $\pm$ 1.2 µmol/mol creatinine.

The clinical examination revealed that tremor was more frequent among the exposed workers (9/26) than among the controls (1/25) (Fisher's exact test: $p = 0.013$). However, clinical examination was performed with knowledge of the kind of exposure and was therefore probably biased.

Writing has often been described as a good discrimination test. In our study, the subjects were asked to write a standard sentence on a card after a light effort. After coding, the cards were given a score from 1 (tremor absent) to 3 (tremor present) by an experienced occupational physician. Two interpretations performed at a 3-week interval showed an overall agreement rate of 63%. It can thus be concluded, at least in subjects with low exposure, that writing is probably not an appropriate test.

In the case of metallic mercury, the tremor is of the intentional type, completely subsiding during sleep, hardly noticeable at rest and triggered off by voluntary use of the affected muscle (Doull et al, 1980). Therefore, two measurements of hand tremor were performed, the first at rest, the second under load, with the method described earlier in this paper. Preliminary results have already been presented (Fawer et al, 1979). Full details of the study will be published elsewhere (Fawer et al, 1982). Only some results are presented here:

The HPFs at rest (mean $\pm$ SE) were higher in exposed workers (7.60 $\pm$ .22 Hz) than in controls (6.40 $\pm$ 19 Hz); multiple regression analysis showed that age and duration of exposure were the best set of predictor variables. The HPFs changes (mean $\pm$ SE) between rest and load were again higher in exposed workers (3.62 $\pm$ .29) than in controls (2.69 $\pm$ .19 Hz); again multiple regression analysis showed that age and duration of exposure were the best set of predictor variables.

The three moments were very closely related to each other and the lowest observed correlation coefficient was $r = .968$ ($n = 51$). This indicates that no noticeable shift on the right of the acceleration towards high frequencies was observed. Therefore, only the results of the second moment (M2) are presented.

The mean M2s (mean ± SE) at rest were not higher in exposed workers (13.3 ± 0.9 x 10^2 mm/s^4) than in controls (10.9 ± 1.1 x 10^2 mm/s^4). On the other hand, the M2s changes (mean ± SE) between rest and load were higher in exposed workers (4.1 ± 1.2 x 10^2 mm/s^4) than in controls (0.9 ± 1.1 x 10^2 mm/s^4); multiple regression analysis showed that duration of exposure and the $\log_e$ of the concentration of mercury in urine were the best set of predictor variables.

These results might have the following physiological explanation: it is well known that metallic mercury accumulates in the cerebellum and in the basal ganglia (Gerstner et al, 1977); it has also been shown that a perturbation of these brain structures can lead to an intention tremor (Marsden, 1978).
With light load, only some muscular fibers will be recruited and therefore the "intoxicated" central nervous system on the mercury exposed workers might not adequately control the amount of tremor generated by the added mass; a shift in moments should thus be observed in exposed workers, but not in controls. This study provides some evidence of shift in M2s between rest and load in exposed workers, but not in controls.

The results of this investigation are based on a small number of workers and have to be confirmed; they are nevertheless consistent with the hypothesis that metallic mercury even at concentrations very probably below the current TLV of 0.05 mg/m^3 can lead to neurological disorders.

ACKNOLEDGMENTS

We wish to thank all workers for their kind participation, and the Safety and Medical Officers for their help in the organisation of the study. We are particularly indebted to Prof. M.J.R. Healy, Professor of Medical Statistics at the London School of Hygiene and Tropical Medicine for his advice in the statistical analysis of the data.

REFERENCES

Berode, M., M.P. Guillemin, B. Martin et al (1980). Mechanisms of toxicity and hazard evaluation. Elsevier/North-Holland Biomedical Press, 371-374.

Chaffin, D.B. and J.M. Miller (1974). In: Behavioral Toxicol.Early detection of occupational hazards. NIOSH, 214-239.

Doull, J., C.D. Klaassen and M.O. Amdur (1980). Toxicology. The basic science of poisons, Macmillan, New York.

Fawer, R.F., Y. de Ribaupierre and M. Lob (1979). Arch. Mal. Prof., 40, 1016-1019.

Fawer, R.F., Y. de Ribaupierre, M.P. Guillemin et al. To be published.

Gerstner, B.G. and J.E Huff (1977). J. Toxicol. Environ. Health, 2, 491-526.

Kazantzis, G. (1968). Working paper for the conference on MAC values, Institute of Hygiene, Karolinska, Stockholm.

Langolf, G.D., D.B. Chaffin and H.P. Whittle (1977). Clinical Chemistry and Chemical Toxicology of Metals, Elsevier/North-Holland Biomedical Press, 213-219.

Langolf, G.D., D.B. Chaffin, R. Henderson and H.P. Whittle (1978). Am. Ind. Hyg. Ass. J., 39, 976-984.

Marsden, C.D., J.C. Meadows, G.W. Lange and R.S. Watson (1969). Electroencephal. Clin. Neurophysiol., 27, 168-178.

Marsden, C.D. (1978). Prog. Clin. Neurophysiol., 5, 1-16.

Roels, H., R. Lauwerys, J.P. Buchet et al (1982). Int. Arch. Occup. Environ. Health, 50, 77-93.

Schuckmann, F. (1979). *Int. Arch. Occup. Environ. Health, 44*, 193-200.
Taylor, J.G. (1901). *Guy's Hosp. Rep., 55*, 171-190.
Wood, R.W., A.B. Weiss and B. Weiss (1973). *Arch. Environ. Health, 26*, 249-252.

Pupillography in Neurotoxicology

F. Trimarchi, P. E. Bianchi and C. Gelmi

Department of Ophthalmology, University of Pavia, Italy

ABSTRACT

The Authors suggest pupillography as a simple and reliable method for the early diagnosis of neurotoxic involvement of the visual system.

KEYWORDS

Pupillography; toxic optic neuropathy.

The study of the photomotor reflex allows evaluation of optic pathway function from the retinal receptors to the pre-geniculate tract,thus pupillography can be employed to evaluate central nervous system damage due to exposure to toxic agents (Levatin, 1969; Lowenstein *et al*, 1964; Lowenstein, 1954; Morone *et al*, 1969). Many toxic substances with central nervous system effects cause early lesions especially to the central fibres of the optic nerve, so early diagnosis of intoxication can be made by assessment of optic nerve function. The study of iris dynamics, as an indicator of optic pathway function, can be used in addition to other subjective examinations such as visual acuity, colour tests, visual fields and subjective critical fusion frequency, and objective tests such as electroretinography (ERG) and visual evoked responses (VER). The photomotor reflex allows functional evaluation of the optic pathway, because it is a reflex response elicited from the retinal receptors. Many anatomic-functional studies have indeed demonstrated the optic and pupillary pathway receptors to be one and the same, and that this identity extends to the ganglion cells and fibres of the first two-thirds of the optic tract (Fig. 1). At this point, optic pathway fibres synapse in the geniculate ganglion and leave the pupillary fibres which synapse in the pre-tectal nucleus. From the pre-tectal nucleus, direct and crossed fibres go to the Edinger-Westphal nucleus and synapse there; efferent parasympathetic fibres go from the nucleus to the iris sphincter. It will be remembered that in the iris there is also a dilatator muscle with sympathetic innervation, which maintains pupillary size in continuous functional equilibrium (Fig. 2). Eventual involvement of the efferent paths in alterations of photomotor reflex characteristics is of secondary importance in the early diagnosis of intoxication because the toxic effect occurs initially on the optic fibres. Evaluation of pupillary response to photomotor stimulation is carried out with a special apparatus, consisting of a closed-circuit television system sensitive to infrared radiations. It allows visualisation of the iris and is connected to an electronic computer for evaluation of changes in the pupil area as a function of time. The computer is connected to a signal averager. The signal averager is very important in photomotor reflex studies. The synchronization of the memory of the averager to stimulus eliminates all the pupillary movements due to cortical and subcortical afferent stimuli reaching the Edinger-Westphal nucleus (Fig. 3). It also cancels humoral information that, by overlapping the light stimulus evoked movements and resulting as low frequency background, could complicate the evaluation of photomotor reflex. Since background signals are not synchronized to the light stimulus, the signal averager does not recognize their evoked responses and evaluates the response of synchronized stimulus only. The stimulus to evoke photomotor reflex can be global or localized. As toxic lesions affect the central bundle of the optic nerve, a global stimulus was used in this study, since it is simple to perform and the response is particularly sensitive to the functional condition of the posterior retinal pole and of the first tract of the optic nerve. The stimulating apparatus consists of a square yellow surface composed of 25 small yellow LED lamps controlled by an intermittent function generator. The increase rate of the lamps is constant: 20 m.sec. Frequencies could be varied from 0.5 Hz to 5 Hz with increases of 0.5 Hz. LED lamps were chosen because the increase rate of these diods is very fast, so the impulse pattern is nearly a square wave. In addition, the stimulus luminance is constant and undergoes no deterioration. LED emitted light is rigorously monochromatic and the type we used is equivalent to a wavelength of 560 nm. This wavelength stimulus attains greatest efficacy when the retina is in photopic adaptation conditions. The pattern obtained by the described method is a wave showing changes in pupil diameter as a function of time. After a latency period the pattern shows a contraction phase followed by a redilatation phase. The latency period and the amplitude of contraction are usually assessed. When optic fibres are damaged, the latency period increases and contraction amplitude is reduced. This pupillography method seems to eliminate instability problems which rendered pupillography of little use for diagnostic purposes. The possibility of assessing responses induced by photomotor stimulus only and of using a stimulator guaranteeing stability, could allow this method to be used in diagnostic routine. The methodology has many advantages: it allows a large number of patients to be examined in a day, as it takes a very short time; it causes no discomfort or harm and only a minimum of collaboration from the patient is required.

REFERENCES

Levatin, P. (1969). Arch. Ophtal., 62, 768-779.

Lowenstein, O. and H. Kawabata (1964). Am. J. Ophtal., 57, 569-595.

Lowenstein, O. (1954). Arch. Ophtal., 52, 385-403.

Morone, G. and F. Trimarchi (1969). Ann. Ott. e Clin. Ocul., 95, 761-771.

Trimarchi, F., P.E. Bianchi and C. Gelmi (1976). Simp. Intern. sull'UVEA (Pavia), 127-171.

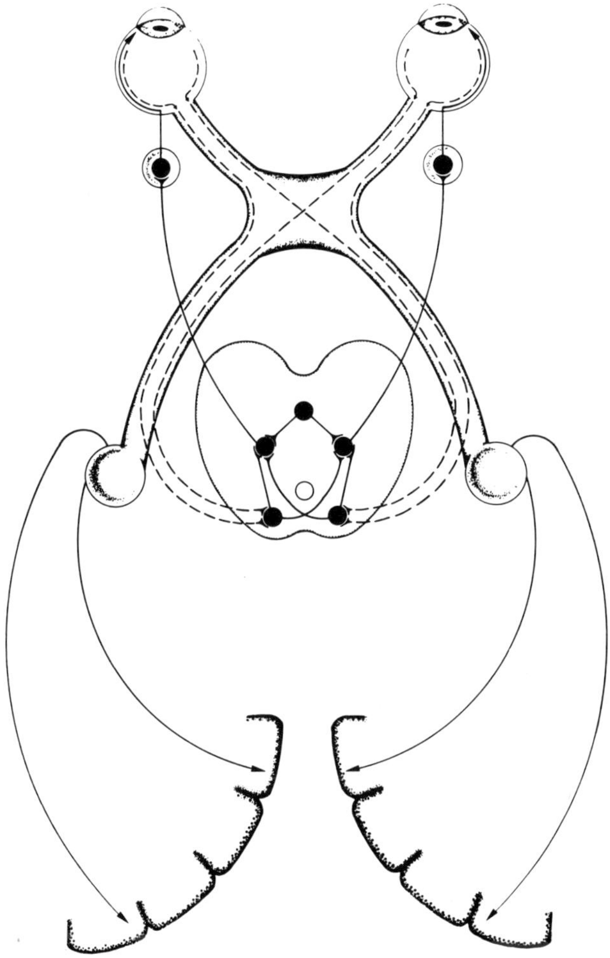

Fig. 1. Afferent and parasympathetic efferent pupillary pathways.

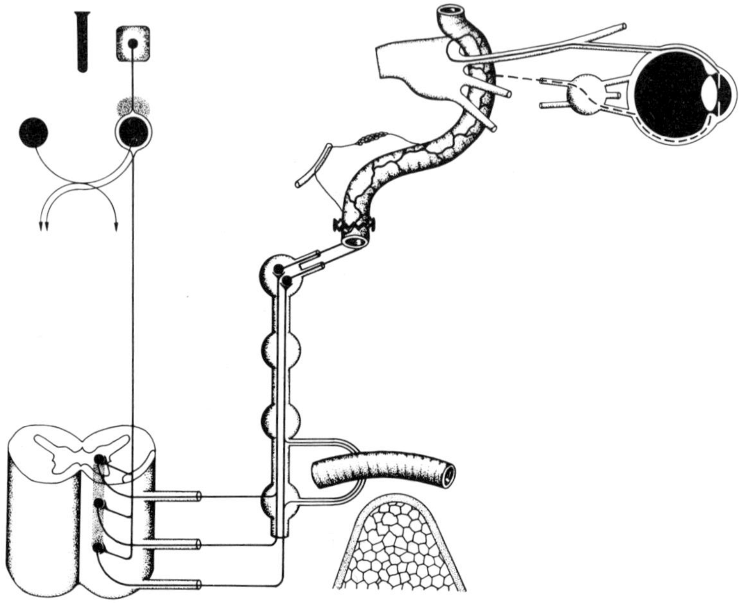

Fig. 2. Sympathetic efferent pupillary pathways.

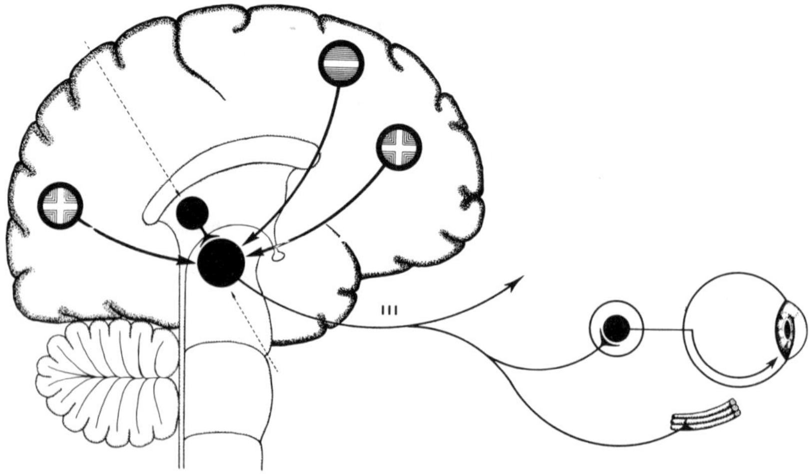

Fig. 3. Cortical and subcortical afferent pathways to the Edinger-Westphal nucleus.

Computerized Recording of Multichannel Biological Functions

R. Buzzi and K. Bättig

Swiss Federal Institute of Technology, Institute of Behavioral Science, Behavioral Biology Laboratory, Turnerstrasse 1, ETH-Zentrum, CH-8092 Zürich, Switzerland

ABSTRACT

A computerized system for simultaneous recording and analysis of several bioelectrical signals is presented and the statistical analyses of two research projects are outlined as examples of the adopted technique.

KEYWORDS

Computerized data recording; psychophysiology; cluster analysis; discriminant analysis; caffeine; noise.

INTRODUCTION

A recent survey (Jones & Buckley, 1981) revealed that 69% of 301 psychophysiological laboratories in the USA were equipped with computers. Nearly all laboratories without computers viewed this lack as seriously handicapping their ability to carry out complex experiments. Although a variety of "ready to use" polygraphic or physiographic equipment is commercially available, it usually lacks adequate and flexible possibilities of gaining access to the data in their digitalized form. The fully computerized system developed at our laboratory was specifically aimed at the analysis of phasic changes in different parameters of the vegetative tonus. The main goal of the system was, therefore, to sample and average data parallel for different vegetative functions and for adjustable time periods.

Two research projects will be presented as examples of the adopted technique. The first study was a field experiment (Bättig, et al, 1980; Bättig & Buzzi, 1981), in which airport area residents were psychophysiologically monitored in their private homes. The second study was an experiment carried out under laboratory conditions (Feierabend & Bättig, in preparation), in which the physiological effects of caffeine during conceptual learning were investigated.

DATA ACQUISITION

Since no on-line analysis of the bioelectrical signals was required, a stand-alone acquisition system was designed, which allows complete independence in the realization of the experiments and additionally permits the carrying out of field experiments.

The series of incoming bioelectrical signals includes so far: electrocardiogram (ECG), respiration (RESP), skin conductance (SC), electromyogram (EMG), plethysmogram (PLET), pulse transmission time (PTT), and electroencephalogram (EEG).

The raw values for all bioelectrical signals, except EMG and PTT, were collected on magnetic tape.

EMG would require a particularly high sampling rate, because the spectral content of EMG extends up to several hundred hertz, with most of the signal power concentrated from 30 to 100 Hz (Geister & McCall, 1975). Since a high sampling rate runs opposite to a design guideline of low system costs, a preprocessing technique was applied to EMG, as described in Fig. 1.

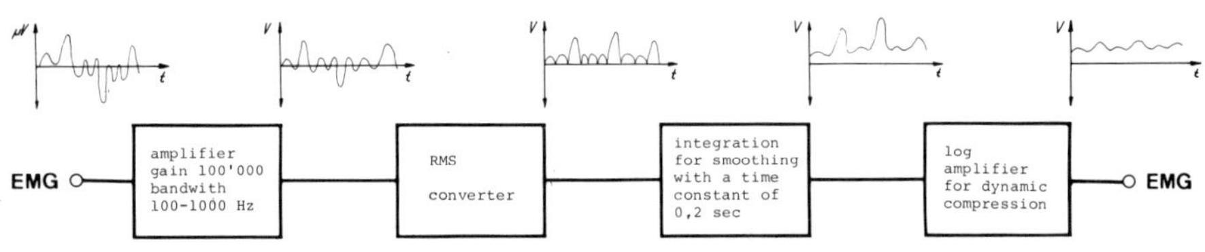

FIG.1 METHOD OF PREPROCESSING THE EMG SIGNAL

PTT was measured (Fig. 2) as the time interval between the occurrence of the R-wave in the ECG and the arrival of the peak of the associated pulse wave at peripheral sites such as the finger (Geddes, et al, 1981), as obtained by an infrared miniature transducer (Spaiser, 1977). Ongoing experiments have indicated, however, that a major deviation of the plethysmographic response from the normal shape, such as can occur at the beginning of noise exposure or during intense respiration, may occasionally invalidate the measurement of PTT.

The pulse code modulation (PCM) technique was selected as the recording system for all these bioelectrical signals over that of frequency modulation (FM) for several reasons. As a particular advantage, the dynamic range of the PCM technique amounts to 60 dB, which permits signal resolution up to 0.1%, whereas the dynamic range of the FM technique is 45 dB, allowing signal resolution of only about 1 to 2%. Further, the direct transfer of the signals into a computerized data processing system is facilitated by the fact that the PCM procedure includes A/D conversion. Finally, over the last ten years the PCM system costs have become gradually lower than the FM system costs, and the PCM system also easily allows the realization of field experiments.

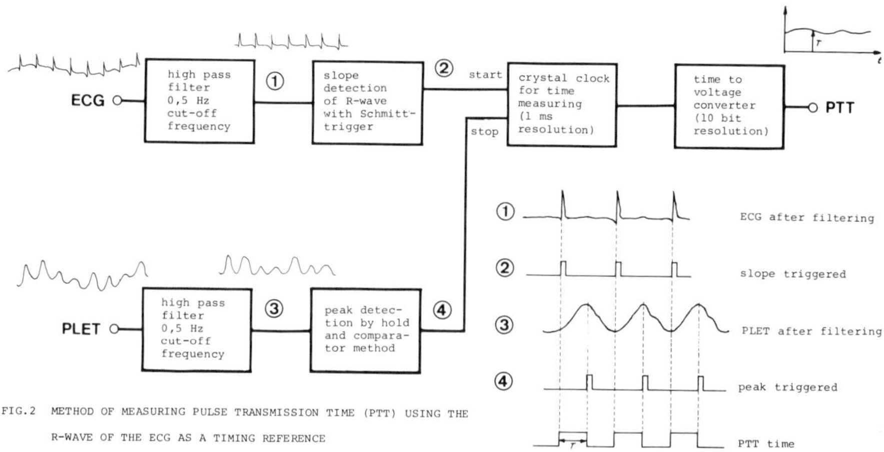

FIG.2 METHOD OF MEASURING PULSE TRANSMISSION TIME (PTT) USING THE R-WAVE OF THE ECG AS A TIMING REFERENCE

DATA PROCESSING

The raw values of all collected bioelectrical signals, except EEG and ECG, are read for data processing directly into a minicomputer (PDP 11/34) with a specially developed interface. For ECG the R-peaks are detected and subsequently read into the PDP 11/34 by a special hardware module, which includes amplification and filtering of the raw signals. The raw values of the EEG are restricted to the frequency range between 1.5 and 40 Hz. Then, the rms values of the EEG, obtained by hardware converters for the six classical frequency bands (delta, theta, alpha 1, alpha 2, beta 1, beta 2) and for the entire 1.5 - 40 Hz band, are read into the PDP 11/34.

The computer interface allows fractionating of the data protocol using time codes of selectable duration and has the option of selecting the required channels, scanning frequency, and data input rate. A data processing program establishes the time block structure by decoding the information of the time codes for the experimental phases and includes a peak-detection algorithm for RESP, PLET, and SCR, which neglects peaks due to small nonphysiological fluctuations. The appropriate algorithms were validated by comparing the results with the visual analysis of some selected polygraphs. The number of falsely recognized peaks and the number of nondetected physiological peaks were smaller than 5% of the number of all existing physiological peaks.

A data reduction program determines the following parameters for each successive time block: - ECG: mean value and standard deviation of pulse intervals. - RESP, SC: number of peaks, mean value, and standard deviation of the peak amplitude and the peak intervals. - EMG: mean values and standard deviation of rms values. - PLET: mean values and standard deviation of successive pulse amplitudes. - EEG: mean value and standard deviation of the rms values separately for each of the six classical frequency bands and for the entire 1.5 - 40 Hz band.

Whenever no valid parameters exist, a special code, to be identified by subsequent statistical analysis, is assigned to indicate a missing value. Special keywords describing the experimental structure are assigned to each single time block in order to allow automated interindividual analysis. Then the data protocols are read into a mainframe computer (CYBER 170) for further statistical analysis.

INTRAINDIVIDUAL AND CLUSTER ANALYSIS

The statistical analysis has to be restricted to intraindividual procedures whenever methodological disparities between individual testing sessions do not allow direct intergroup comparisons, as was the case in the field study on the vegetative effects of aircraft noise. As a special multivariate technique for intraindividual data analysis, the cluster analysis procedure was successfully applied in this study (Bättig, *et al*, 1980; Bättig & Buzzi, 1981), where ECG, SCR, RESP, and EMG were continuously recorded together with the actual sound in the private apartments of 49 experimental subjects. Each session lasted roughly 1-1/2 hours and was divided into different activity phases, such as assessing noisiness on a scalometer, performing a letter cancelling test, relaxing, and instructive conversation with the experimenter. This test program served the double purpose of providing control conditions in comparison to the noise effects and of studying the possible interactions between ongoing activity and noise. Splitting the recordings of each session into time blocks provided about 200 - 400 data packages per session for the subsequent clustering procedure.

Clustering seeks to group measurements that are closely related in terms of some distance measure. The procedure thus extracts typologies for the data sets, which in turn represent a reduction of data complexity and may lead to conceptual simplifications. Since the aim of the classification was to find clusters with minimum variance, the K-means algorithm was applied as an iterative relocation procedure, and program RELOCATE of the CLUSTAN package (Wishart, 1975) was selected for classifying the measurements by the combined values of the set of parameters considered. The K-means method was applied for different initial cluster distributions, for different similarity coefficients, and for different numbers of clusters, which varied from two to ten. The optimal cluster distribution was determined for every fixed number of clusters, and the number of clusters finally considered was determined as the point of inflection of the minimal error functions corresponding to the derived optimal cluster distributions. The results showed that the optimal cluster distribution was independent of the initial cluster distributions and varied little when different similarity coefficients were considered; this indicates that the optimal cluster distribution was not only locally, but also globally optimal.

Submitting the data of the noise experiments to the cluster procedure resulted in 4 - 5 clusters per session. The comparison of the cluster distributions between distinct experimental test phases was done in a following step using the conservative procedure of applying CHI^2 tests. As a first result, the expected influences of ongoing activity turned out to be highly significant. Filling out forms, assessing noisiness, letter cancelling, relaxation, and conversation produced significant shifts of the cluster distribution for at least one physiological variable in all 49 subjects. Heart rate and pulse variance were affected significantly by these ongoing activities in about 85% of all cases, whereas skin resistance and respiration were affected by these activities somewhat less frequently, with the exception of conversation, which also altered these two physiological functions in most of the cases. Aircraft and accidental noise also produced significant shifts of the cluster distributions, but in fewer cases. Significance was found in 20 of the 49 cases for accidental noise and in 27 subjects for aircraft overflight.

A further step of the analysis concerned the dose response relationship. Toward this goal, a correlational analysis was applied individually to the physiological parameters and noise intensity. Heart rate, skin conductance changes, and respiratory amplitudes and variances increased in most cases significantly with increasing noise intensity levels.

In summary, the adopted procedure revealed that physiological effects of noise may not be entirely subject to habituation, as would be expected from laboratory studies. Rather, it seems that such effects may persist. But whether this is due to the consistently varying conditions which contrast the homogeneity of laboratory conditions or to some other reasons remains an open question.

INTERINDIVIDUAL AND DISCRIMINANT DATA ANALYSIS

As a special multivariate technique for the detection of interindividual differences, the discriminant procedure was applied in the experiment on effects of caffeine, carried out under laboratory conditions (Feierabend & Bättig, in preparation). In this experiment, letter cancelling performance, concept formation with a mental maze, and concomitant readings of ECG, EMG, SCR, RESP, and PLET were compared for 16 subjects who received decaffeinated coffee, 16 subjects who received decaffeinated coffee with 300 mg pure caffeine, 8 subjects who received the same amounts of warm water, and 8 subjects who received no treatment at all. All these experimental treatments were given after the subjects had mastered the first mental maze task and before they were required to perform the letter cancellation task and a second mental maze problem.

The discriminant analysis is a multivariate technique which may be applied to a set of measurements belonging to different groups of subjects. A set of weights, called discriminant functions, is derived, which builds a new composite variable, such that the group means on the composite variable will be maximally, although not necessarily significantly different. If the hypothesis to be tested is that there may be a subset of independent measurements along which the groups differ, then the discriminant analysis should be selected; if overall significance of the group means should be tested, then multivariate analysis of variance would be the preferred technique. A further attractive aspect of discriminant analysis is the possibility of comparing the measurements in terms of their relative ability to distinguish the groups.

A methodological problem is encountered by the fact that increasing the number of measurements frequently leads to a greater amount of "noise" in the discriminant functions, which decreases the discriminating efficiency. Since, additionally, the number of subjects in psychophysiological experiments is usually small compared with the number of measurements, a stepwise discriminant procedure was selected (program P7M of the BMDP package).

The drawback of a stepwise method is the introduction of the number of selected variables as a new random variate, which is not accounted for by the test statistic. However, a study by Weiner and Dunn (1966) shows that the stepwise selection procedure was considerably better than other methods considered and roughly comparable to the t-test selection procedure for smaller sets.

Another methodological problem is the estimation of classification errors, since they depend on unknown population parameters. As a solution, the "jackknife" error rate and the "hold-out" error rate are considered as alternatives.

The "jackknife" error rate is the proportion of misclassification associated with a classification rule, where each subject is left out once and considered as a new unknown, not classified subject. This gives a biased estimation of the actual error rate, since the entire data set is used in previous steps of the calculation.

The "hold-out" method reserves a portion of the sample for cross-validation of the performance of the assignment rule. Although the "hold-out" method gives a much better approximation of the actual error rate, it was not used in the analysis of the caffeine project due to the high number of subjects required.

In the experiment on the effects of caffeine, the bioelectrical parameters were first transformed into percentage scores of the baseline for the statistical analysis across the subjects in order to reduce the effects of individual anatomic or physiological factors. The application of discriminant analysis to the physiological data revealed no intergroup differences for the test phases preceding the experimental treatments, but significant group classifications were obtained for all successive phases of the experiment. The only variable separating the groups turned out to be the pulse amplitudes (PLET), and the highest levels of correct classification were obtained consistently for the caffeine treatment group. Although the vasoconstrictions observed in the caffeine group were modest, they also turned out to be significant, using univariate statistical criteria (t-tests) due to the high degree of interindividual consistency of this phenomenon. Besides this effect, an increase in the EMG potentials was also seen, but this reached statistical significance only marginally due to the large interindividual variance.

On the behavioral level, the administration of caffeine resulted in a decrease of performance variance with the letter cancellation task, but it had no differential effects on learning with the second mental maze task.

CONCLUSIONS

The analysis of polygraphic psychophysiological recordings confronts the experimenter with the difficult task of detecting and localizing external effects in a highly complex data set. As a first step, this task requires the elimination of recording artifacts, which remain unavoidable with this type of experimentation. The following steps require appropriate and flexible data reduction and statistical treatment of the data. The fully automated system presented here permitted for the first time in two experiments the detection and characterization in continuously recorded vegetative polygrams of effects of environmental noise measured under field conditions and the subtle effects of a moderate dose of caffeine.

ACKNOWLEDGEMENT

The authors wish to express their thanks to Mr. J. Wespi and Mr. J. Zwimpfer for their excellent technical assistance.

REFERENCES

Bättig, K., and R. Buzzi (1981). Zeitschrift für experimentelle und angewandte Psychologie, 28, 1-14.
Bättig, K., H. Zeier, R. Müller and R. Buzzi (1980). Archives of Environmental Health, 35, 228-235.
Feierabend, J. M., and K. Bättig (in preparation). Sozial- und Präventivmedizin.
Geddes, L. A., M. H. Voelz, C. R. Babbs, J. D. Bohrland and W. A. Tacher (1981). Psychophysiology, 18, 71-74.
Geister, E. D., and W. D. McCall Jr. (1975). Proceedings of the IEEE, 63, 1404-1414.
Jones, G. E., and M. F. Buckley (1981). Psychophysiology, 18, 331-340.
Spaiser, L. H. (1977). Psychophysiology, 14, 75-77.
Weiner, E. A., and O. J. Dunn (1966). Biometrics, 22, 268-275.
Wishart, D. (1975). CLUSTAN 1C User's Manual, University College, London.

Conclusions

Perspectives of Application of Neurobehavioral Methods in Developing Countries

P. K. Suma'mur

National Centre for Industrial Hygiene, Occupational Health and Safety, Jakarta, Indonesia

ABSTRACT

In developing countries, neurotoxic chemicals are used and lead to potential as well as real problems of occupational health. Therefore, those countries pose needs for the adoption of neurobehavioral methods.

Developing countries are not a single entity and present a broad spectrum of conditions. Application of neurobehavioral methods should be tailored to each country concerned.

Perspectives of application of neurobehavioral methods in dynamic developing countries are good. They may proceed to decide on principles and methods which are suitable for themselves.

Emphasis is laid on epidemiological assessment by employing simple techniques and relatively inexpensive equipment. This is of particular significance for those countries which are just initiating occupational health programmes.

Technical cooperation is required to develop programmes of neurobehavioral methods in occupational health in developing countries.

KEYWORDS

Neurobehavioral toxicology; neurobehavioral methods; occupational health; developing countries.

INTRODUCTION

This topic "Perspectives of Application of Neurobehavioral Methods in Developing Countries" concerns the possible use and future development of Neurobehavioral Methods in Occupational Health. The content of this presentation is a response to the state of the art and emerging trends in Neurobehavioral Methods in Occupational Health.

To fully understand the situation in developing countries and study the perspectives of application of neurobehavioral toxicology there are several aspects to be discussed:

1. Occupational health in developing countries.
2. Problems and needs in neurobehavioral toxicology.
3. Feasibility of application and programme development.

OCCUPATIONAL HEALTH IN DEVELOPING COUNTRIES

Developing countries are not a homogeneous entity. They are quite heterogeneous and present a broad spectrum of conditions reflecting the socio-economical progress of the countries concerned. They may differ very much in attitude towards occupational health in general. Occupational health in these countries ranges from just starting to rather advanced stages. In many cases, occupational health faces the problems of being too low on the scale of national priorities, early development of legislation, shortage of personnel, relatively small programmes, etc. Due to limited existing resources, they have to decide on priorities.

In many developing countries, working populations are underserved with respect to occupational health. Particularly, those engaged in small agricultural and industrial undertakings do not even enjoy preemployment and periodical health examinations.

The lack of standards in occupational health, hygiene and safety practices in many developing countries results in the use of hazardous materials rather than the less toxic ones, exposure to high concentrations and doses in handling those chemicals and improper protection of the workers from occupational hazards. Such situations may lead to ill-health conditions and occupational diseases, but due to ignorance and no adequate law enforcement on notification and compensation of these illnesses, the problems remain unnoticed but will be revealed more and more clearly if systematic assessment by field studies is made. The diseases occuring in the working population may not be of typical occupational nature but in many cases are a deterioration of ill-health conditions.

In developing countries where occupational health receives a good priority, or at least the Government has considerable interest and develops programmes, this field is advancing rapidly. The economic development of the country and the inclusion of occupational health in the reconstruction plan help promote occupational health in the country concerned. Occupational health which aims at better health, welfare and productivity of the working population is of paramount importance from the socio-economical point of view. Unfortunately, the number of countries which have included occupational health in national development is still small and efforts should always be made to persuade them to adopt a more favourable policy.

As in other health fields, occupational health is a sort of technology. Developing countries differ among themselves in their capabilities in terms of actual procedures as well as potentialities in adopting and adapting technology through transfer from developed to developing countries. Existing technologies in occupational health, safety and ergonomics constitute the wide range from the most traditional to the most sophisticated and advanced ones. In situations where national policy on occupational health is well formulated, developing countries have a broad spectrum of choice for the most appropriate technologies in occupational health, including the most economic ones. The selected technologies should fit with the available manpower that carries out the programme.

Developing countries have to learn from both positive and negative experience of the developed nations. They should not repeat the mistakes made in the advanced countries but attempt to take shorter ways to reach the modern stage of occupational health development. This is, of course, not an easy task. For developing countries have to catch up with the progress of industrialized countries which initiated occupational health 150 years earlier. The question is, how to shorten the time to really come to the present level of occupational health practice, for example in just one generation.

PROBLEMS AND NEEDS IN NEUROBEHAVIORAL TOXICOLOGY

Many neurotoxic chemicals are used in developing countries. They consist of solvents, pesticides, heavy metals, and other chemicals possessing neurotoxicity.

Neurotoxic solvents utilized in developing countries are trichlorethylene, toluene, xylene, carbon disulfide, etc. Trichloroethylene is widely used as a degreaser for metal parts, a thinner for paints, lacquers, varnishes and tar, a solvent for rubber, a solvent for extraction of oils and fats from vegetable and animal products. Toluene is a solvent for oils, resins, natural rubber, coal-tar, asphalt, and solvent and diluent for cellulose paints and varnishes, etc. Xylene is widely used as a thinner for paints, lacquers, etc., as a cleaning agent and degreaser, as a solvent in the printing, rubber and leather industries, etc. Carbon disulfide is used as a solvent for fats, lipids, resins, rubber, etc. and in the fumigation of grain. In the production of viscose rayon fibres which also exists in some developing countries, carbon disulfide plays a significant role. Developing countries may also use other solvents in addition to these chemicals.

Due to lack of standards in occupational health, hygiene and safety practices, there is a trend in developing countries to use rather hazardous materials. Workers may be exposed to relatively high doses of neurotoxic solvents without adequate protection. Because of ignorance and possibly no way to cope with the problems, very little or no attention is payed to early neurobehavioral changes. The effects may appear as serious neurobehavioral symptoms, if the problems are totally unknown and nobody is interested in the matter.

Organophosphate pesticides are now extensively used in developing countries. They are essential in the overall attempts to increase grain and other food crops as well as other agricultural products. Developing countries, of course, have to accept both advantages and disadvantages of the application of these chemicals. Most organophosphates are very toxic and some of them have neurobehavioral and neurophysiological effects. In developing countries, where control on the use of these chemicals is not properly exerted, these pesticides have created enormous problems in terms of intoxications, other ill-health conditions and also deaths. Severe neuropathological symptoms and signs may be found in intoxicated cases due to massive exposures, accidental ingestion, etc.

Heavy metals are mined, extracted, produced, and processed and may also be part of industrial use in developing countries. Some heavy metals such as mercury, lead and manganese have neurobehavioral effects. Mercury occurs in industrial processes like extraction and production of pure metallic mercury, extraction of silver and gold by employing amalgams, production of barometers and thermometers and other mercury-containing instruments, manufacture of mercury lamps, etc. Lead exposure may take place in lead battery manufacture, printing shops, painting works, rubber vulcanization, glass and ceramics production, welding, manufacture of electric cables, children's toys and metal alloys. Manganese exposure occurs especially in mining and handling the ores and in its use in different metal works.

There are hazards and problems of neurobehavioral and neurophysiological effects of these metals in developing countries. The effects go together with the signs and symptoms of their intoxications. As is true for occupational diseases in general, they are usually underreported, while research is limited in number and scope.

Problems of alcoholics and narcotics differ from one country to another in developing countries. The problem of drinking, which is very closely related to neurobehavioral toxicology, may or may not be considerable depending on the availability of alcoholic beverages, on culture, and many other factors. The problem of narcotics is similar. In short, there are problems of drug abuse in many developing countries and these countries endeavour to minimize them.

In a broader context, neurobehavioral toxicology is an instrument for early detection and preventive measures in occupational health. The needs of developing countries as regards neurobehavioral toxicology should come from the occupational health needs of developing countries. This field of health is needed by developing countries as part of occupational health practice.

There is a need in developing countries for methods of early detection of the various effects of chemicals in the working environment. These methods will permit early recognition of the problems and enable action to be taken in good time. Technical knowhow in assessing occupational health problems in developing countries usually constitutes the weakest part of occupational health practice. There should be transfer of technology in early detection methods from developed to developing countries.

There are three main needs of developing countries in the field of neurobehavioral toxicology (Suma'mur, 1982) as follows:

1. Need for Training and Education.

The most important factors indicating the need for training and education in neurobehavioral toxicology are the lack of awareness of the neurotoxic chemicals and the shortage of trained personnel in developing countries. Like occupational health training and education in general, the three levels of education and training, i.e., (a) awareness; (b) training for specific tasks; (c) specialization, are required by developing countries. It is well recognized that developing countries differ widely in their ability to fulfil these needs.

To create awareness of neurobehavioral toxicology is the basic level of education with the objective of providing an understanding of the principle of this particular field in order to build up correct attitudes at work with exposures to neurotoxic materials and to permit proper participation in any action taken. The target groups include legislators and policy-makers, managers and workers. The persons for specific task training may include administrators, managers, workers, architects and designers, medical and paramedical personnel, chemists and engineers, and occupational health and safety technicians. Specialization should create specialists and professionals in the field of neurobehavioral toxicology.

Objectives, technology, methodology, and education and training programmes in neurobehavioral toxicology should be adjusted to the existing problems and the level of practice of occupational health in general and neurobehavioral toxicology in particular.

2. Need for Inclusion of Neurobehavioral Toxicology in Occupational Health in Developing Countries.

Occupational health growth differs very much in developing countries, and ranges from just a very early to a rather advanced stage. Despite this fact, neurobehavioral toxicology needs to be included as a part of occupational health practice. This principle can only be followed if the practical aspects of neurobehavioral toxicology are carefully studied and well selected to make their implementation possible. Initial phases of their application may be characterized by development and use of psychological and neurological test batteries adopted to the national and local needs.

Since health delivery infrastructures in developing countries differ widely from one to another, the system of service in neurobehavioral toxicology should be well suited to the existing conditions. The assessment should meet the need for health maintenance of the working population in developing countries.

Developing countries need instruments in neurobehavioral toxicology which can be made available as grants or technical assistance from developed countries or by their own resources.

3. Need for Applied Research.

To clarify the actual magnitude of the problems of neurotoxic materials, applied research needs to be carried out in developing countries. The scope of the research should focus on applied rather than basic questions. They may be assisted by developed countries with the arrangement of international organizations. The findings of such research are the grounds for further development of the service as well as for use in education and training.

FEASIBILITY OF APPLICATION AND PROGRAMME DEVELOPMENT

All developing countries would agree that neurobehavioral toxicology has developed very fast in some advanced countries. As a new field in toxicology, its progress has been only very recent in the last decade. It is focused on the effects of chemicals on nervous system functions by using psychological and neurological test batteries.

In industrialized countries, there have been several reasons for its development (Wands, 1971). It is an important instrument to assess the effects of neurotoxic chemicals on the working population. This new field of toxicology has also been used to detect early effects of chemicals to humans. It is also hoped that it will, and is in fact utilized in practice to improve health and safety protection of workers by having better safety and health standards based on neurobehavioral signs and symptoms of exposure to chemicals.

Developing countries may learn that an achievement in neurobehavioral toxicology such as in industrialized countries has resulted from the high level of occupational health practice and research together with advanced practices and research in neurology and psychology.

The technical know-how in industrial hygiene laboratory analyses of minute concentrations of chemicals in the work room air, the accuracy in biological test batteries, the production of highly accurate instruments and other factors have been the key to the present practice of neurobehavioral toxicology.

In the light of the said background, an attempt is made to analyze the relevance of this special occupational health technology and to examine its feasibility of application in developing countries.

Neurobehavioral toxicology may or may not be known in developing countries. Some were familiar with it only in the middle of the last decade because some publications had rechead them during that time. Many of them, however, until recently had not even heard of the terminology.
These countries may be in Asia, Africa or America. At present, there are more people from developing countries who know and attempt to implement this branch after seminars, workshops and training courses on neurobehavioral toxicology have been organized by developed countries in cooperation with international organizations.

Let us first speak about countries where occupational health develops with good national support. In such countries the perspectives are positive. There is no doubt that for those countries which emphasize the need for occupational health development and proceed with the creation of programmes and are not too handicapped by resources, application of neurobehavioral methods in occupational health is only a matter of time. If those countries are capable of adopting industrial hygiene and control technology methods which are no less advanced than neurobehavioral methods, they will be in a good position to utilize these new techniques. The state of the art in neurobehavioral methods covering neurological examination and assessment of the syndromes caused by exposure to neurotoxic agents, electromyography, electroneurography, current behavioral techniques, and behavioral performance assessment as a biological control of occupational exposure to neurotoxic substances can be transferred to professionals in dynamic developing countries without creating difficulties in terms of inclusion in the existing programmes. There should be transfer of knowledge and skill in neurobehavioral methods from the developed to developing nations. But finally, it should be the developing countries which decide on certain aspects to be applied in their countries concerned. They should use neurobehavioral techniques stepwise to suit the existing conditions.

Both in neurological and behavioral methods, there have been very rapid recent developments in addition to the previously used methods. In neurological techniques, there has been progress in screening for peripheral neuropathies, electro-oculography, EEG spectral analysis, regional cerebral blood flow measurement, and computed axial tomography. There have also been developments in behavioral methods which include psychological test batteries, reaction time, hand tremor measurement, pupillography and computerized recording of multichannel biological functions. All these methods involve both simple and sophisticated procedures, reasonable or very expensive instruments. It is preferable for developing countries at present to start first with simple assessment and rather cheap apparatus. Application of single techniques and a comprehensive approach can be used as long as the activities are feasible in terms of available personnel, time and funds.

Epidemiological assessment by using simple techniques and equipment should be particularly stressed for developing countries.

To illustrate a case in adopting neurobehavioral toxicology, Indonesia is an example. Some information on the subject reached the country in 1973 and it was considered at that time too much of a luxury to take any action, except to ask those concerned to be aware that such a new multidisciplinary assessment existed. Occupational health people concentrated their efforts on how to develop industrial hygiene, occupational health, ergonomics and occupational health nutrition in the country. It was only in 1980 that a decision was taken to include the development of neurobehavioral toxicology in the occupational health programme. A committee has been set up at the National Centre for Industrial Hygiene, Occupational Health and Safety to plan and implement programmes in this special field. This committee has discussed and agreed on steps to be taken. In Bandung, West Java, the Technical Meeting on Occupational Health and Safety agreed on the development of neurobehavioral methods by the National Centre and resolved to ask Regional Offices to start supplying information to the communities and related professionals on the matter. Psychological test batteries have been decided to assess the neurobehavioral effects of manganese, organo phosphate pesticides and carbon monoxyde as the first substances for trial. Further activities will expand to other neurotoxic chemicals. A project budget is available to purchase a Disa Electromyograph and a reaction time measuring instrument. The plan will grow in the future by having an EEG. After certain achievements have been made, a workshop inviting occupational health professionals, neurologists, psychologists, psychiatrists, etc., will be organized. Neurobehavioral toxicology in now also included in education and training curricula on occupational health for different categories of trainees. At the National Centre, a simple instrument for measuring tremors is being made and tested.

Developing countries, in which occupational health is dynamically progressing will be proceeding in this way. Some of them may need more technical cooperation and look forward to having more support from other countries.

Simple methods and relatively inexpensive equipment may also be used by developing countries where occupational health is just in embryo form or nonexistent. Such countries should, of course, develop occupational health but also might start with the activities of neurobehavioral methods. To do so, neurologists, psychologists and psychiatrists in the country should be informed and trained in neurobehavioral methods. General practitioners should also participate in preventing ill-health conditions in the working population exposed to neurotoxic chemicals.

The development of neurobehavioral toxicology has been the result of recent experience in several developed countries. This experience gained by developed countries should be transferred to developing nations. In fact, both industrialized and developing countries have common problems of neurobehavioral toxicology with representative examples of worker exposure to carbon disulfide in viscose rayon production, inorganic mercury in the manufacture of instruments containing mercury, inorganic lead in storage battery factories, organic solvents in paints, manganese in mining of ores, and pesticides in agriculture.

The governments in both developed and developing countries should particularly assess their problems and needs and accordingly develop neurobehavioral toxicology in their programmes. Occupational health practice should be given good priority and neurobehavioral toxicology is one aspect of it.

Programmes in developing countries should include education and training, services and research work.

Exchange of experiences at international, regional, as well as national levels, in occupational health in general and neurobehavioral toxicology in particular is believed to be essential to promote further growth of this field in developing nations.

REFERENCES

Suma'mur, P.K. (1982). Problems, Needs and Programme Development of Neurobehavioral Toxicology in Developing Countries. National Centre of Ind. Hyg., Occ. Hlth and Safety, Jakarta.

Wands, R. (1974). Future of Behavioral Toxicology, in: Xintaras, C., Johnson B.L. and de Groot, I. Behavioral Toxicology. Early Detection of Occupational Hazards, NIOSH.

Perspectives of Application of Neurobehavioural Methods in the Industrialized and Developing Countries

C. Xintaras

World Health Organization, Division of Noncommunicable Diseases, Office of Occupational Health, Geneva, Switzerland

The World Health Organization is pleased to co-sponsor the "International Symposium on Neurobehavioural Methods in Occupational Health" (Como and Milan, Italy, 1982).

The Thirty-third World Health Assembly in May 1980 requested the Director-General "to support the developing countries in ensuring safe working conditions and effective protective measures for workers' health in agriculture, mining and industrial enterprises which already exist or which will be set up in the process of industrialization by using the experience available in this field by both industrialized and developing countries" (Resolution WHA 33.31).

Dr M. El-Batawi, Chief Medical Officer, Office of Occupational Health sends his personal wishes for a very successful conference. He asked me to express the immediate need to apply on a world wide basis the information exchanged at this conference to prevent and control neurotoxic effects of workplace chemicals. WHO has been active in the field of occupational health for many years. However, it looks to you and the Collaborating Centres in Occupational Health for guidance and direction in methods development for the early detection and prevention of occupational disease.

During this first session I was impressed by the elegance of the instrumentation that is currently in use for bio-medical research. In several presentations instrumentation was described with a view to potential application in the early detection of occupational disease. In general, however, I was left with the impression that the methods presented and discussed have the greatest utility in the clinic for diagnostic purposes. It was not clear, and perhaps not even practical, how these methods can be used under field conditions - at the "workplace". Increasing the resolution of computer tomography through the use of a contrast medium was truly impressive. One should, however, ask immediately what the added risk is to the patient or worker. Is the risk worth taking? Will the worker be aware and understand the risk as explained to him by the physician or other medical support staff?

Dr Cassitto raised a very important point in her presentation and asked "What are we looking for and why?" We should ask ourselves this question more frequently. The need for applied research in occupational health is great. There is a need also for adapting our clinical methods for field use so that a large number of workers can be screened and where quick intervention can be initiated.

Institutes of Occupational Health that provide in-house diagnostic capability and where workers are examined for compensation purposes have accumulated considerable experience in diagnosis and treatment of occupational disease. This information must be made available to those responsible for the workers' health. A critical component in identifying the etiological factor associated with an occupational disease is the occupational history of the worker. Very often this information is not available. Worker exposure to multiple chemicals and the potential interaction of these chemicals adds another dimension in trying to identify causal factor(s) in occupational disease.

Chemical factors at the workplace have been discussed during this session. Very little has been said however, about physical factors. Nothing was mentioned about work stress. The importance of work stress, early signs and symptoms of stress and mechanisms to cope with stress must be considered. The worker should recognize the symptoms of work stress. It is also our responsibility to bring the workers into active participation in our discussions. WHO has initiated a new programme in psycho-social factors and the workplace with primary emphasis on developing countries. This programme, however, impacts on all working populations. There are implications of work stress on performance efficiency, the health of workers, and injury prevention. In field studies where comparisons are made between exposed workers and a reference group, the level of work stress may not be identical. The results of such studies should be interpreted cautiously.

In recommending tests for evaluating health hazards associated with exposure to chemical, physical and psycho-social factors, recommendations should be based and guided by a desire to provide tests that can be characterized in four ways: 1) the tests should be simple to administer, 2) unambiguous in interpretation (understanding the significance of a change as methods become more sensitive), 3) the measure must be an indicator of nervous system dysfunction, and 4) comprehensive.

Excellent selection criteria were presented during this session. Furthermore, one doesn't simply create a method and assume instant reliability. The key to achieving reliability is standardization. I am addressing this issue to the neurologists, psychologists and electrophysiologists who are with us today. Without standardization of methods we will be faced with data that lack a basis for comparability, the need for costly replicate studies and agencies that will be reluctant to take regulatory action. Although we have not focused on standardization during our discussions the forum for such discussions perhaps could be the "round table". The output of such a "round table" will be standardized test batteries of increasing level of complexity for field use (screening). Test batteries developed in this manner obviously should consider the selection criteria noted earlier; simple, unambiguous, valid, comprehensive. The use of simple test batteries allows us to examine a larger number of workers. Administration of the test battery (questionnaire, etc.) can be entrusted to trained Primary Health Care workers. Special training modules must be developed however, to implement such an outreach programme. Referral to higher level health support components is implicit in this approach. Workers in developing countries and migrant workers in industrialized countries who are not covered under any health delivery system could receive under the Primary Health Care approach a higher level of health care than is presently available. Total health care of workers is obviously the ultimate goal. It is essential though that all workers have a level of awareness of occupational hazards.

The WHO programme in neurobehavioural toxicology has focused on the need to develop standardized test batteries and in the exchange of information on early detection and treatment. This information is sent to WHO Collaborating Centres in Occupational Health and to Ministries of Health. Health care is usually the responsibility of the Ministry of Health. In some countries the Ministry of Labour assumes this responsibility. There is a need therefore for closer cooperation not only between and among the International Agencies but also within member countries. This exchange of information is critical for future development of occupational health and in particular the field of neurobehavioural toxicology.

The information of this International Symposium and the "International Workshop on Neurotoxicity of Workplace Chemicals: Detection and Prevention in Working Populations", scheduled to be held in the USA in 1983 and co-sponsored by WHO must reach the field where work problems exist and where intervention can be effected quickly. This information must reach all workers, especially those in the agricultural sector who traditionally have been on the periphery of health care systems.

In summary, to detect and prevent occupational neurologic disease and behavioural disorders what is needed is a high degree of <u>clinical suspicion</u>. Physicians must be sensitized also to recognize specific occupational, environmental and psycho-social factors in the etiology of disease. A first step in this direction is the need for in-depth occupational histories. The worker's health file must also contain relevant baseline data, especially pre-exposure information. Health dossiers will raise issues of medical confidentiality and worker acceptability. The worker must play an active role in the development of this approach.

Early detection and prevention is essential to ensure worker protection from exposure to workplace chemicals. Several factors are important and should be emphasized such as: 1) <u>workers' awareness</u> (many workers don't recognize the signs and symptoms of overexposure to neurotoxic chemicals and workers should be informed about the chemicals they are manufacturing or using), 2) <u>physician sensitivity</u> is high during a neurotoxic epidemic but in many areas is very low, 3) <u>biological monitoring</u> is essential for screening purposes and to provide assurance to workers and management that control techniques and good work practices are effective, 4) <u>medical surveillance</u> to detect a trend and to provide for early intervention and to avoid irreversible damage to workers requires good baseline data and exposure information.

The usefulness of animal experimental data must not be underestimated. This type of information will provide the clue(s) in what to look for in the working population. There are many neurotoxic chemicals in the workplace. Our need is for simple, specific and reliable tests to apply in specific work situations. This approach will also reduce the cost in operating medical monitoring programmes. Medical monitoring will, however, probably require specific legislation.

Acceptability of the field of neurobehavioural toxicology will require meaningful interpretation of our results. The need exists to demonstrate a reduction in occupational disease and accidents that results from the application of neurobehavioural toxicology principles and practices at the workplace.

PART TWO APPLICATIONS OF METHODOLOGIES

Application of Single Techniques to the Study of Exposed Populations

Chairmen: V. Foà (*Italy*)
R. Lauwerys (*Belgium*)

Evaluation, Limits and Perspectives of Neurobehavioral Toxicology in Occupational Health

V. Foà

Institute of Occupational Health, University of Milan,
via S. Barnaba 8, 20122 Milan, Italy

One of the main purposes of Occupational Health and specifically of Industrial Toxicology is to investigate the interactions between noxious substances present at the workplace and the human organism, and then to identify the potential health hazards for man due to occupational and/or environmental exposure to toxic agents.
The data supplied by the investigators of these disciplines have practical, highly relevant applications: i.e., the fixing of environmental exposure limits based on health criteria as well as the development of methods suitable for medical surveillance of exposed populations.

Monitoring of individuals occupationally exposed to neurotoxic agents can be either environmental or biochemical.

Biological monitoring employs two basic techniques: internal dose tests that measure the chemical agents or their metabolites in the organism (internal dose indicators), and the effect tests, that measure the alterations in biological functions following exposure to toxic substances present in the environment (indicators of effect).

The internal dose indicators supply information on current or past exposure and its degree, while the indicators of effect clarify the interactions between the external agent and the recipient organism. They can thus evaluate the total effect of a given substance. The extent of the effects is also determined by individual differences, past functional alterations of body organs or apparatuses, and synergic or antagonistic effects due to simultaneous exposure to other chemical agents.

Therefore, indicators of effect rather than indicators of internal dose more adequately supply the physician with the necessary information. However, these tests must comply with some basic requirements in order to be fully utilized in preventive medicine and in prevention of occupational hazards.

Indicators of effect must detect early, still reversible effects (namely, when they cannot be defined as "adverse effects") and there must be a relationship between external and/or internal dose and the intensity of the measured effect.

Moreover, these tests must be sensitive, that is, they must be capable of detecting the effects also of low-level exposure; and specific, that is, they must not give too many false positive results. Moreover, they must be easy to use, non-invasive and not too expensive in order to be routinely used in occupational health on large groups of subjects. Unfortunately, only very few tests satisfy these requirements at present, and with the sole exception of inorganic lead, no indicators of effect complying exactly with the above requirements are available for hardly any of the industrial neurotoxic agents. Moreover, recent progress in toxicology, improved legislation on prevention in the various countries better technology and production, along with greater awareness among individuals of the possible consequences of their work on their health, means that we will increasingly have "cleaner" workplaces. It follows that the already reduced levels of exposure to harmful substances that are generally found in the more industrialized countries, are bound to be further reduced, thus making it increasingly difficult to evaluate these situations. At the same time, multiple combined exposure sources are becoming more and more frequent and their effects on the organism are still difficult to evaluate.
All the above reasons led to the great effort now being made in the field of industrial toxicology towards the research and development of new indicators of effect complying with the requirements of sensitivity and specificity considered adequate for present and future situations.

One of the more promising new trends in this direction has certainly been the adoption in occupational health of neurophysiological and behavioral methodologies first used in experimental investigations and then transferred to human studies in the last ten years. It is now time, and this is one of the main reasons which led us to organize this congress, to carry out an analysis and evaluation and, if possible, draw the logical conclusions of these ten years' experiences.

First of all, we should answer the question of which of the many neurobehavioral parameters can better comply with the sensitivity, reliability and specificity requirements; second, at which levels will neurobehavioral alterations represent an adverse effect, and third, whether or not it is possible to correlate these alterations, measured by means of these instruments, to indicators of dose or effect.

Therefore, the core of the problem is to know whether the alterations in some neurobehavioral parameters can be regarded as an adverse effect or whether they should rather be considered as indicators of a new level of adaptation of the central and/or peripheral nervous system to changed conditions of life such as normally occur as a consequence of a new type of work.

These questions are not negligible ones and the presence of the WHO at this Congress, not only as a supporting institution but through the presence of Dr. Xintaras, who is a member of the Occupational Health Office and who organized the First Behavioral Toxicology Workshop for Early Detection of Occupational Hazards in 1973, proves the relevance of the problem. In fact, we should only reflect that if we maintain that any alteration in parameters as sensitive as neurobehavioral parameters must be regarded as an adverse effect, we will be obliged to continually reduce the exposure limits at the workplace as neurophysiological and behavioral techniques become more and more refined.
But if this effort does not lead to better health conditions of the occupationally exposed populations owing to a misinterpretation of the significance of these results, we will then not only have failed to obtain an

effect but will have wasted energy and resources which might have been better employed to reach the same or even better goals.

Of course this problem is even more keenly felt in developing countries, where the generally low resources must be managed very carefully.

We therefore thought it would be appropriate to divide this Symposium into two Sections. In the first, more technical, section, neurophysiologists, psychologists and other specialists have discussed the state of the art and the emerging methodological trends with a view to establishing, after exhaustive and critical discussion, which parameters used in their different disciplines can be regarded as reliable, and to what extent.
Today, we will start discussing and hopefully validate their conclusions on the basis of our practical experience at the workplace.

Before opening the second Part of the Symposium, I will sum up the main content of the first part in Como for the benefit of those who were not present.

In the first introductory session, current neurophysiological and behavioral techniques were reviewed. It was interesting to note the caution and critical awareness accompanying the analysis of the state of the art. After stressing the importance of the clinical part of the neuropsychological examination, the reliability and limits of the techniques currently used were discussed. It was once again emphasized that the use of these methods and the interpretation of the data must be confined to responsible and skilful specialists in this discipline. An important aspects of this session was, in my view, the discussion on the need to use test batteries for clinical purposes, and especially to confirm or otherwise a diagnosis based on clinical data, which differ in characteristics and range from batteries used in field studies aiming at detecting an indicator of effect in occupationally exposed subjects. This is a wellknown and relevant issue since the occupational health physicians willingly entrust the suspect clinical case to neurologists in order to obtain a diagnosis of location, extent and severity, while they regard the interpretation of the results and the evaluation of their relevance in prevention as their own task.

I feel that light has been shed on this problem, that is, the necessity of selecting tests according to their modality and scope in order to obtain the best answer to the question.

The most impressive papers, almost resembling science fiction, I would say, were given in the sessions devoted to the recent developments in neurological and behavioral techniques.
The methods and instruments presented in these sections had, in some cases, such a high power of analysis of the anatomical and functional properties of the CNS structures as to appear unbelievable and even unimaginable
Most of these methods have and probably always will have only clinical applications, but attempts have been made, sometimes with success, to use them in neurotoxicology. This is proof of the vitality of research in our discipline.

Conclusions were drawn by Dr. Xintaras, representing the WHO, and by Dr. Suma'mur who spoke on behalf on the developing countries. I shall first report our Indonesian colleague's remarks.

I think it is important that, far from regarding behavioral toxicology as a luxury for the wealthy countries, Dr. Suma'mur thinks it necessary to further elucidate the methods and techniques being discussed here, so that they can be introduced in the less industrialized countries, notwithstanding the fact that some of these countries are still not aware even of the bare terminology we use. His belief is based on the conviction that neurobehavioral toxicology is an integral part of occupational health as a complex discipline and therefore must become current practice and be developed as a whole in the developing countries. These countries must understand that the practice of neurobehavioral toxicology in the industrialized countries is the result of long experience and high research standards in occupational health. Their objective must be to reach these standards and therefore they must include neurobehavioral toxicology right from the beginning. To this end, support in terms of instruments, technicians and teaching centers, is required from those countries which have been practising occupational health, with more or less success, for the last 150 years.

Dr. Xintaras was impressed by the new methodologies developed for central and peripheral nervous system investigation and evaluation and he fully acknowledged the clinical relevance demonstrated by the speakers, but he was doubtful about their utilization in occupational medicine.

In fact, it is difficult to attribute a nervous disease to a single agent when this agent is present in low concentrations together with other chemicals and/or physical agents, noise, stress factors, etc. Such situations are very frequent at the workplace and can result in multiple lesions which taken separately cannot be identified clinically and/or instrumentally, but become measurable due to an addition or summation effect.

Dr. Xintaras objects to the practice of control groups only apparently similar to the exposed groups but for the chemical agent investigated. I agree with him that the preliminary hypothesis whereby different workplaces share the same stress factors is wrong.

Finally, he once more stressed the importance of the requirement that tests be simple to administer, unambiguous to interpret (the significance of the measurements performed must be clear) and capable of revealing real nervous system disfunctions,if they are to be used as reliable tools.

The main problem remains test standardization, especially in psychodiagnostic testing: in fact each research group uses its own test battery, thus making results poorly comparable.

The WHO feels that the recommended tests should be administered by adequately trained non-graduate technicians. This applies not only to the developing countries, but also to the USA and Europe, where we all know how long it takes to obtain a consulting specialist, whether ophtalmologist or neurologist.

In conclusion, the first Part of our Symposium has produced an interesting comparison between long since established methodologies and techniques and the new emerging ones in the investigation of the central and peripheral nervous systems.

Today and tomorrow we shall review the data obtained through the practical application of these techniques in occupational health investigations. Our hope is that at the end of this Symposium the criteria for their correct utilization will be more clearly defined and become standard for all research groups working in this field.

The Use of EMG Techniques in Solvent Exposure

A. M. Seppäläinen

Institute of Occupational Health, Haartmaninkatu 1,
SF-00290 Helsinki 29, Finland

ABSTRACT

Electromyographic findings and nerve conduction studies among solvent-exposed workers are reviewed.

KEYWORDS

Electromyography; nerve conduction velocities; occupational exposure to solvents; n-hexane; carbon disulfide.

INTRODUCTION

Various solvents and their mixtures are widely used in industry in many applications. Glues, paints and degreasing agents are the most usual sources for solvent vapors in the workroom air, and inhalation is the route of exposure, although some exposure may occur also through intact skin. Hydrocarbons, especially those with six carbons atoms - hexacarbons - have been shown to be neurotoxic (Spencer and Schaumburg, 1977) and other solvents have been suggested to be dangerous for the nervous system. Carbon disulfide causes neuropathological changes similar to those produced by hexacarbons (Seppäläinen and Haltia, 1981).

Workers exposed to solvents often complain of paresthesiae, numbness and pain in their hands, arms, feet and legs and these symptoms may indicate peripheral neuropathy. As the nervous system has a limited capacity of regeneration, an early diagnosis is important in order to prevent permanent damage. The use of improved electronic devices has enlarged our possibility of detecting already small lesions.

Acute solvent poisonings rarely occur and in a single attack they seldom affect the peripheral nervous system. Chronic effects of occupational exposure to solvents are of greater practical and research interest. Electromyography (EMG) with various nerve stimulation studies is a good way of studying patients with suspected toxic neuropathy and these methods have also been succesfully applied in larger epidemiological studies among exposed populations.

METHODS

Electromyography means the recording of the electrical activity of muscles and nerves. An impulse is propagated in a nerve in the form of electrical potential changes, the actual potential change happens momentarily in the axon membrane and changes the ionic permeability of the membrane. These membrane potential changes occur at the same moment in a large quantity of axons and the compound action potential can be recorded with plate electrodes fixed to the skin above the nerve trunk or with small needle electrodes inserted close to the nerve. During voluntary contraction potential changes occur also in the muscle and these can be recorded with needle electrodes inserted in the muscle.

In electromyography the electrical activity of the muscle is studied during maximal voluntary contraction to evaluate the recruitment pattern, during a weak voluntary contraction to study the duration and shape of motor unit potentials and during muscle relaxation to detect possible spontaneous activity, like fibrillations, fasciculations or positive sharp waves. This type of electrical activity has to be greatly amplified and displayed in a suitable form. The potential changes are very short lasting and occur with a high frequency, consequently the amplification system has to be of great capacity and the visual display with an oscilloscope is completed with loudspeakers, since the human ear easily detects changes in the frequencies of muscle action potentials. Several good quality electromyographs are available commercially also equipped with an electrical stimulating device for the study of nerve conduction velocities and often with some optic, electrostatic or UV graphic units for permanent recording.

In the nerve conduction studies, motor or sensory nerves are eletrically stimulated and the electrical response of an appropriate muscle is recorded (motor conduction velocity, MCV) or of a nerve trunk (sensory conduction velocity, SCV). The latencies from the stimulus to the response are directly measured from the oscilloscope display, which in modern apparatuses contains a memory. When we divide the distance passed in the nerve by the latency in question, we can calculate the conduction velocity, which is expressed in meters/second (m/s).

RESULTS

n-Hexane was incriminated as the causative agent for several polyneuropaties in Japanese sandal makers (Yamamura, 1969; Sobue et al, 1978) and in American cabinet finishers (Herskowitz et al, 1971). The most striking feature of the EMG in n-hexane poisoning was the appearance of denervation activity, fibrillations and positive sharp waves (Yamamura, 1969); the number of motor unit potentials in maximal voluntary contraction was often reduced too, and abnormally slow MCVs of the median, ulnar, tibial and peroneal nerves were noted in more than half of the patients.

These electrophysiological findings are typical of axonal neuropathy. Fibrillations are found in axonal neuropathy, not in purely demyelinating neuropathy. Loss of motor unit potentials, which is reflected as a reduced interference pattern during maximal voluntary muscle contraction, is also typical of neuropathy (Hopf and Struppler, 1974). Nerve conduction velocities may slightly slow down in axonal neuropathy; a greater reduction in conduction velocities indicates demyelinating processes, which may be added as secondary phenomena to long-standing axonal neuropathy. Neuropathological studies revealed axonal changes in patients with n-hexane neuropathy

(Herskowitz et al, 1971) and later experimental studies have shown that a typical finding in hexacarbon neuropathy is accumulation of 10 nm neurofilaments within axons (Spencer and Schaumburg, 1977). Widespread microscopic changes of this type in multifocal fashion could be detected already in the early stages of poisoning without major disturbances being observed in the affected animal (Cavanagh and Bennetts, 1981). Usually the neuropathological changes start in the distal areas and spread to proximal regions, thus the first EMG abnormalities are to be expected in the distal hand and foot muscles. Neuropathy in n-hexane poisoning is usually sensory or mixed motor and sensory neuropathy (Yamamura, 1969), and the sensory disorder can be shown by measuring the sensory conduction velocity in the distal branches,for example from finger to wrist. The amplitude of the sensory action potential may already be decreased before the SCV becomes abnormally slow (Kaeser, 1975). This occurs because several axons are not able to conduct the signals, consequently the reduced number of acting axons can generate only a low amplitude action potential.

Methyl-butyl ketone has a common metabolite with n-hexane and it caused a similar type of neuropathy in an epidemic in an American industrial plant producing plastic-coated and colorprinted fabrics (Allen et al, 1975; Billmaier et al, 1975). EMG was proven a sensitive and useful method of identifying and diagnosing patients with mild toxic neuropathy. Allen et al (1975) analyzed in detail 194 workers from the coated fabrics plant; these workers were suspected of having a peripheral neuropathy or other neurological disorders based on a screening procedure. Eighty-six workers were identified as having a toxic polyneuropathy after careful differential diagnosis. Among these, 11 had characteristic and disabling manifestations of peripheral neuropathy, while 38 were considered mild with physical findings indicative of neuropathy and with characteristic electrodiagnostic abnormalities. In 37 no objective neurological findings were noted, but even they had characteristic electrodiagnostic abnormalities. EMG studies were thus more sensitive indicators of mild toxic neuropathy than clinical examination alone. EMG abnormalities were approximately symmetrical and were either greater in degree in distal than in proximal muscles or entirely restricted to a distal distribution. The most usual abnormalities in methyl-butyl ketone neuropathy were also fibrillations and positive sharp waves. Initially motor unit abnormalities, mainly a reduction in their number, were less frequent, but in serial examinations denervation activity tended to subside, while polyphasic and abnormally large motor unit potentials became more abundant. The latter changes are indicative of regeneration phenomena; innervation of muscle fibers deprived of their original nerve supply is taken care of by sprouting terminal fibers from nearby axons. The MCV of the ulnar, peroneal and/or tibial nerves and the SCV of the sural nerve was often reduced and the distal latencies increased in affected patients; in certain cases the slowing-down was of the order of 20% or more, which could indicate concomitant demyelinating processes. In a considerable number of workers the neurologic and electrodiagnostic findings deteriorated further during the first 7 months after cessation of exposure.

Similar EMG findings and slowing-down of the MCVs have been described in Italian shoe and leather workers exposed to mixtures of solvents in glues or cleaning mixtures (Cianchetti et al, 1976; Buiatti et al, 1978). The motor distal latency of the peroneal nerve was often clearly prolonged, and in more severe cases it was not possible to elicit a motor response from the extensor digitorum brevis muscle by stimulating the peroneal nerve at the ankle, although a stimulus at the knee level still resulted in a response (Cianchetti et al, 1976). In many of these cases the reduction in the MCV of the peroneal nerve also deteriorated further during the first months after cessation of occupational exposure, even to the point where no response was evokable with stimulation at the knee level. In 12 to 18 months after cessation of exposure, a clear improvement was noted in the MCVs usually, although symptoms like weakness, paresthesia, and cramp-like muscular pain might remain unchanged for months or even years (Cianchetti et al, 1976). The duration of the evoked motor action potential from the peroneal nerve stimulation was abnormally long in many patients and it further lengthened during the follow-up period (Cianchetti et al, 1976). The EMG also continued to show loss of motor unit potentials and fibrillations in the workers, when the MCV had returned to a normal level (Perticoni and Cianchetti, 1976). Other authors have also reported that recovery was not complete; for example Giuliano et al (1975) reported in 15 patients with shoe-workers neuropathy, that although EMG abnormalities disappeared in eight, they stayed the same in three subjects and showed some improvement in four subjects.

A recent follow-up study by our institute has confirming that EMG abnormalities, especially reduction in the number of motor units and denervation activity, are long-lasting after solvent poisoning (Seppäläinen and Antti-Poika, 1982). 87 patients with solvent poisoning were followed up with serial EMG examinations for 6 years on average (range 3-9 years). Among them, 43% showed some improvement in their EMG findings, but 33% showed deterioration, usually mild, during the follow-up. At the latest EMG examination, 64 out of 87 patients still showed some EMG findings indicative of neuropathy, and the most usual finding was loss of motor units, alone or together with denervation activity; denervation activity was still noted in 46% of the patients (Table 1).

Table 1 - EMG findings among patients with solvent poisoning

Examination	N	Loss of motor units	Fibrillations	Fibrillations + loss of MU's
Initial	69	18	7	19
Follow-up	87	27	14	26

These patients had usually been occupationally exposed to various mixtures of solvents (66), of whom 13 had been exposed to both solvent mixtures and tri- or perchloroethylene, and 21 had been exposed solely to tri- or perchloroethylene. Those exposed to both mixtures and tri- or perchloroethylene tended to have polyneuropathy more often than the others.

Even workers without overt poisoning may show abnormalities in the EMG examination. Among 117 solvent exposed workers who had come for periodic health examinations at the Institute of Occupational Health in Helsinki and who had some mild symptoms like paresthetic feelings or muscular pain, 61 or 52% had at least one abnormally slow nerve conduction velocity, often in combination with some neurogenic EMG abnormalities like reduced number of motor units and/or denervation activity (Seppäläinen, 1973). Among those persons whose exposure level was the highest - although usually not exceeding the accepted safety limit - abnormal findings were more frequent than among those with slight or moderate exposure.

Epidemiological studies have been reported on worker groups with long-term occupational exposure to various solvents. Workers with exposure to 20-80 ppm of carbon disulfide for 1-30 years had slower MCVs and longer motor distal latencies than age-matched referents (Gilioli *et al*, 1978; Knave *et al*, 1974; Seppäläinen and Tolonen, 1974). Abnormal fatigability in certain patients with chronic carbon disulfide poisoning was probably due to changes in the myoneural junction, since a repetitive stimulation test showed a clear decrement in the motor response during two minutes of stimulation at the rate of 20 Hz (Seppäläinen *et al*, 1972). Carbon disulfide thus causes multifocal lesions within the neuromuscolar system, which is typical of many neurotoxicants. The application of various EMG techniques is an aid in the detection of such lesions.

Mean MCVs and SCVs of several nerves were significantly slower among 80 Swedish industrial and car painters exposed to solvent mixtures in comparison to a reference group of 80 workers matched for age, education and job qualification (Elofsson *et al*, 1980). Finnish car painters more often had abnormally slow MCVs and SCVs in the arm nerves than railroad engineers (Seppäläinen *et al*, 1978). These reports show an exposure-related impairment of peripheral nerve function among healthy active workers.

Neuropathological studies have been performed for only a limited number of solvents. Present knowledge indicates that at least n-hexane alone or potentiated by methyl-ethyl ketone (Spencer and Schaumburg, 1977; Altenkirch *et al*, 1977), methyl-n-butyl ketone (Spencer and Schaumburg, 1977) and carbon disulfide (Seppäläinen and Haltia, 1981) cause axonal changes, namely accumulation of 10 nm neurofilaments within the axons, and that the same solvents also cause clinical signs and symptoms of neuropathy among exposed workers. EMG studies have revealed neuropathic findings compatible with axonal neuropathy in clinically neuropathic workers as well as in their fellow-workers with mild symptoms and no current abnormal neurological signs. EMG was more sensitive than clinical examination, as early changes can already be demonstrated objectively. EMG techniques can and should be applied also to follow up the patients, where they can reveal regeneration after injury and reflect the possible continuation of the neuropathic process.

Etiology of neuropathy is not revealed by EMG, and the actual diagnosis of toxic neuropathy can be made only after careful consideration of the individual's exposure and relevant differential diagnostic measures. EMG studies combined with epidemiological studies also offer etiological information, if the reference population is relevant. The role of various solvents in the solvent mixtures is still unclear. I feel that many other solvents than just hexacarbons have neurotoxic properties. Further studies are needed to prove that, and experimental studies may add to our knowledge.

REFERENCES

Allen, N., J.R. Mendell, D.J. Billmaier, R.E. Fontaine and J. O'Neill (1975). *Arch. Neurol. (Chic.)*, *32*, 209-218.

Altenkirch, H., J. Mager, G. Stoltenburg and J. Helmbrecht (1977). *J. Neurol.*, *214*, 137-152.

Billmaier, D., H.T. Yee, N. Allen, B. Craft, N. Williams, S. Epstein and R. Fontaine (1974). *J. Occup. Med. 16*, 665-671.

Buiatti, E., S. Cecchini, O. Ronchi, P. Dolara and G. Bulgarelli (1978). *Brit. J. Ind. Med.*, *35*, 168-173.

Cavanagh, J.B. and R.J. Bennetts (1981). *Brain, 104*, 297-318.

Cianchetti, C., G. Abbritti, A. Perticoni, A. Siracusa and F. Curradi (1976). J. Neurol. Neurosurg. Psychiatr., 39, 1151-1161.
Elofsson, S.A., F. Gamberale, T. Hindmarsch, A. Iregren, A. Isaksson, I. Johnsson, B. Knave, E. Lydahl, P. Mindus, H.E. Persson, B. Philipson, M. Steby, G. Struwe, E. Söderman, A. Wennberg and L. Widén (1980). Scand. J. Work Environ. Health, 6, 239-273.
Gilioli, R., C. Bulgheroni, P.A. Bertazzi, A.M. Cirla, M. Tomasini, M.G. Cassitto and M.T. Jacovone (1978). Medicina del Lavoro, 69, 130-143.
Giuliano, G., O. Ronchi, L. Rossi and L. Focardi (1975). Lav. Um., 27, 33-41.
Herskowitz, A., N. Ishii and H. Schaumburg (1971). N. Engl. J. Med., 285, 82-85.
Hopf, H.C. and A. Struppler (1974). Elektromyographie, Georg Thieme Verlag, Stuttgart.
Kaeser, H.E. (1975). Nervous and muscular evoked potentials. Handbook of electroencephalography and clinical neurophysiology, 16A, Elsevier Amsterdam.
Knave, B., B. Kolmodin-Headman, H.E. Persson and J.M. Goldberg (1974). Work-Environ. Health, 11, 49-58.
Perticoni, G. and C. Cianchetti (1976). Electromyograph. Clin. Neurophysiol. 16, 493-506.
Seppäläinen, A.M.(1973). Electroenceph. Clin.Neurophysiol., 34, 702-703.
Seppäläinen, A.M. and M. Antti-Poika (1982). Electroenceph. Clin. Neurophysiol., 53, 72 p.
Seppäläinen, A.M. and M. Haltia (1981). G. Ital. Med. Lav., 3, 87-89.
Seppäläinen, A.M., K. Husman and C. Martenson (1978). Scand. J. Work Environ. Health, 4, 304-314.
Seppäläinen, A.M. and M. Tolonen (1974). Work-Environ-Health, 11, 145-153.
Seppäläinen, A.M., M. Tolonen, P. Karli, H. Hänninen and S. Hernberg (1972). Work-Environ-Health, 9, 71-75.
Sobue, I., M. Iida, Y. Yamamura and T. Takayanagui (1978). Int. J. Neurol. 11, 317-330.
Spencer, P.S. and H.H. Schaumburg (1977). J. Neuropath. Exp. Neurol., 36, 300-320.
Yamamura, Y. (1969). Folia Psychiatr. Neurol. Jpn, 23, 45-57.

Benton V.R.T. and Raven P.M. 38: Qualitative Analysis of the Answers

D. Camerino and E. Arzuffi

Institute of Occupational Health, Department of Preventive Medicine, University of Milan, Milan, Italy

ABSTRACT

Performance testing data in Behavioral Toxicology are usually analyzed by means of quantitative criteria.
This paper reports a qualitative analysis carried out on logic reasoning and perceptual memory testing administered to subjects occupationally exposed to solvents and lead and to a control group. RAVEN P.M. 38 and BENTON V.R.T. were examined according to the different categories in terms of intellectual reasoning and mnestic schemes referred to in the literature. Namely, Raven errors were classified according to incomplete solutions, arbitrary lines of reasoning, overdetermining choices and repetitions, while Benton V.R.T. errors of mnestic storage, mnestic uncertainty, space disorientation, vertical instability and right or left prevalence were examined.
Thirty subjects were examined for each group, from 30 to 50 years of age, all of only primary educational level.
Data were subjected to a frequency distribution analysis which indicated the existence of significantly relevant differences among the groups examined.
No causal interpretation of these differences is suggested owing to the quantitative and qualitative nonhomogeneity of the groups considered. However the possible relevance of qualitative data analysis is stressed for a better understanding of the mechanisms of action of the neurochemicals investigated.

KEYWORDS

Qualitative analysis of performance tests; neurotoxic effects of lead; neurotoxic effects of solvents.

INTRODUCTION

In recent years, the validity of psychodiagnostic testing as an instrument for studying alterations in mental efficiency in subjects exposed to neurotoxic agents has been demonstrated by means of quantitative analyses.

Behavioral testing has proved capable of discriminating exposed subjects from controls and in some cases of establishing a relationship between tests results and some biochemical parameters of metabolic alterations consequent to exposure.

The above can satisfy the necessary requirements to recognize as adequate a certain number of psychodiagnostic indices to be used in epidemiological investigations.
However a qualitative analysis of test results, whenever possible, might consistently increase their diagnostic discriminating power and supply more precise information on the nature and level of some alterations observed in the parameters studied and thus make individual diagnosis in clinical settings easier.

MATERIALS AND METHODS

From these premises we carried out a qualitative analysis of the data obtained with Raven P.M. 38 and Benton V.R.T. from subjects hospitalized in our Institute for suspected intoxication from lead and organic solvent mixture.

The controls were selected among other groups of inpatients with diseases not referable to neurotoxic agent exposure, nor to head injuries or serious psychopathological disorders. Each of these groups consisted of 60 workers, all of primary school level, divided into two groups of 30 subjects aged between 30 and 39 and between 40 and 49 years respectively.

The exposed subjects were non-homogeneous as regards health conditions and level of exposure.

Table 1 shows the anamnestic data of the groups considered relevant for our investigation.

Both Raven P.M. 38 and Benton V.R.T. belong to the testing battery routinely employed in our laboratory and are certainly the two tools which can better allow a qualitative analysis.

Raven P.M. 38 is a well-known test used in the evaluation of the subject's ability to understand the relationship existing between geometrical shapes namely: similarity, symmetry, proportion, progression, addition and algebric substraction.

Each item shows a meaningless incomplete drawing whose missing part must be found among the 6-8 shapes given below. It consists of five series whose first item presents no difficulty while all the others have different degrees of solving complexity.

The test is highly sensitive to a depression of the intellectual efficiency and to difficulties in organizing perceptual material (Urmer *et al*, 1960; Campbell *et al*, 1981).

The Benton Visual Retention Test is a clinical tool developed to evaluate short-term memory; it consists of ten tables of graduated difficulties which illustrate one or three geometrical figures. The subject is requested to reproduce each of the tables on separate sheets, after examining it for ten seconds. It solicits not only visual memory but also perceptual, intellectual and motor abilities and it is obviously influenced by attention and concentration efficiency.

TABLE 1 a. YEARS OF EXPOSURE FOR SOLVENT AND Pb GROUPS.

GROUPS		X	S.D.	RANGE
Solvent	30-39 years	9.79	7.25	7 m. - 25 years
Solvent	40-49 years	14.05	10.08	8 m. - 36 years
Lead	30-39 years	10.89	6.65	3 m. - 26 years
Lead	40-49 years	15.20	9.33	3 - 29 years

TABLE 1 b. RESULTS OF EXPOSURE TESTS IN LEAD-EXPOSED GROUP.

		30 - 39 YEARS			40 - 49 YEARS		
		X	S.D.	RANGE	X	S.D.	RANGE
Pb B	μg/100 ml	70.70	62.81	11-334	56.50	26.36	18-134
Pb U	μg/l	87.35	67.46	10-226	108.35	106.66	10-468
COPRO	μg/l	163.06	83.04	1-703	71.00	114.12	5-470
ALA-D	mU/ml RBC	17.06	15.42	3.1-55.2	14.39	11.48	1.9-31.5
ALA-U	mg/l	11.79	11.61	1.7-34.7	10.05	7.90	1.7-27.7

TABLE 1 c. ENZYMATIC TEST RESULTS IN SOLVENT-EXPOSED GROUP.

		30 - 39 YEARS			40 - 49 YEARS		
		$\overline{X}$	S.D.	RANGE	$\overline{X}$	S.D.	RANGE
GOT	U/l	24.75	23.23	2-85	17.09	10.96	6-40
GPT	U/l	23.05	19.81	4-68	19.13	16.27	3-60
GT	U/l	63.41	61.73	7-190	45.50	42.96	6-170

Both Raven P.M. 38 and Benton V.R.T. activate both cerebral hemispheres (Campbell *et al*, 1981).

Table 2 illustrates variables and factorial categories which have been used for the qualitative analysis.

TABLE 2. VARIABLES AND FACTORIAL CATEGORIES CONSIDERED IN THE QUALITATIVE ANALYSIS OF RAVEN P.M. 38 AND BENTON V.R.T. RESULTS.

Variables	Factorial categories	
RAVEN P.M. 38		
Incomplete solutions		
Arbitrary lines of reasoning		
Overdetermined choices		
Repetitions		
BENTON V.R.T. (Form C)		
Substitutions Omissions Internal Omissions Perseverances	▶ MNESTIC STORAGE	▶ Left prevalence Right prevalence
Distortions Internal details location	▶ MNESTIC UNCERTAINTY	
Displacements Misplacements Size errors Rotations	▶ SPACE DISORIENTATION	
Table VII - X Rotations	▶ VERTICAL INSTABILITY	

In Raven P.M. 38 analysis, we calculated the number of errors in each item series and, according to the instructions supplied by Forbes (1964), we evaluated the type of errors according to the following scheme:

a) Incomplete solutions due to disregard of the variables necessary for a correct solution.
b) Solutions resulting from principles extraneous to the matrix variables.
c) Solutions which reflect the difficulties in discriminating relevant from irrelevant characteristics as well as a confluence of different elements in one image.
d) Solutions which are a repetition of one shape already present in the matrix.

For the Benton Visual Retention Test we followed the instructions given by Cristante (1976) in a factorial analysis of this test.

According to his scheme, the errors are distributed into their different factorial categories, i.e.:

a) Omissions, substitutions and perseverations belong to a mnestic retention difficulty factor.
b) Distortions and internal detail location errors belong to a mnestic uncertainty factor.
c) Displacements, misplacements, rotations and wrong size belong to a poor space orientation factor.
d) Rotations of the figures lying at one angle belong to a vertical instability factor.

Moreover since the error lateralization is relevant in discriminating between subjects with brain damage and controls and since our analysis takes different groups into account, we also considered these categories.

For Raven P.M. 38 as well as for Benton V.R.T., we calculated the frequency of each type of error in the three groups: solvent, lead and control, and for each group, the two age subsets were analyzed separately.

The Chi Square test was used to evaluate significant differences in error frequencies and their respective weight in the three groups and in their two age subgroups.

RESULTS

Table 3 shows the differences observed, the frequency distribution and the levels of significance. Above, variables and factorial categories are given, below on the left, groups and age subgroups. Small letters refer to subjects aged 30-39 years, capital letters to subjects aged 40-49 years.

TABLE. 3 OVERALL ERRORS DISTRIBUTION IN THE GROUPS CONSIDERED.

The groups indicated in the cases have significantly higher values in the errors listed on top of the column

GROUP CONSIDERED	RAVEN P.M.38: OVERDETERMINED CHOICES	REPETITIONS	BENTON V.R.T.: SUBSTITUTIONS	OMISSIONS	PERSEVERATIONS	DISTORTIONS	INTERNAL DETAILS LOCATION	MISPLACEMENTS	SIZE ERRORS	Factorial categories: MNESTIC STORAGE	MNESTIC UNCERTAINTY	SPACE DISORIENTATION	VERTICAL INSTABILITY	LEFT PREVALENCE	RIGHT PREVALENCE
s S - c C		s	s	S	cC		C		S					c	
l L - c C			L		cC				l			l			
s S - l L	S	L	L	S		L	L			S	L				s
s - S			s	S						S			s		
l - L	l	L		L				l			L				L
c - C	c	C													

SMALL LETTERS REFER TO 30-39 YEARS GROUPS

CAPITAL LETTERS REFER TO 40-49 YEARS GROUPS

S – SOLVENT GROUP

L = LEAD GROUP

C = CONTROL GROUP

The intra-group differences show that the answers given by the subjects in their 30's are different from those given by the subjects in their 40's.

As shown in the table, errors made in the Benton V.R.T. by the solvent group consist mainly of substitutions and vertical instability for the age 30-39 subgroup, while mnestic storage difficulties are prevalent in the age 40-49 subgroup (p .005).

In the control group no significant differences were found in the Benton V.R.T., while a prevalence of errors due to poor discrimination between relevant and non relevant elements can be observed with the Raven P.M. 38 in the 30's subgroups and more repetitions in the 40's subgroup.

The same is true in the Raven for the lead-exposed group. In the 30-39 years group more size errors were observed with the Benton V.R.T. while the 40-49 years group more frequently made errors of omissions and in mnestic uncertainty, with a prevalence of errors on the right of the table.

If we now compare the performance of the solvent group to the control group, we can observe that both age subgroups of the controls make more perseveration errors as well as more errors on the left of the table in the 30-39 years subgroup and internal detail location errors in the 40-49 years subgroup.

The solvent-exposed group showed more repetitions in the Raven P.M. 38 and substitutions in the Benton V.R.T. in the 30-39 years subgroup and omissions and size errors in the 40-49 years subgroup.

The lead-exposed subjects compared to the control group made more size errors and space disorientation errors in the 30-39 years subgroup and more substitution errors in the 40-49 years subgroup. In both age subgroups the controls were significantly different in making more perseveration errors.

The comparison between the solvent-exposed subjects and lead-exposed subjects showed more lateral distribution with a prevalence in the right size in 30-39 years solvent subgroup. The 40-49 years solvent subgroup on the contrary made more omissions and internal detail omission errors. This means, as far as the factorial categories are concerned, that the solvent-exposed group have more retention difficulties, while the lead exposed subjects with their prevalence in substitutions, distortions and internal detail wrong locations, have more difficulties in the mnestic uncertainty factorial category. The Raven Test does not supply data discriminating between lead-exposed subjects and controls at the variables related to the error's qualitative characteristics.

Error frequency evaluation for each of the four item series appears much more discriminating. Significant values varied from below .001 to below .0005. As shown in Table 4, differences between controls and exposed groups are consistently higher with a prevalence of errors in columns A (last items) and D, while the two exposed groups show a high significant prevalence in columns B and C.

Raven B and C column items solicit a right perception of the whole symmetry and closure, a correct evaluation of dimensions and space orientation among the elements. These items are in fact more loaded in space factor. On the contrary, A (last items) and D series stimulate a syncretic perception and an abstract reasoning.

TABLE 4. SERIES ERRORS DISTRIBUTION IN THE GROUPS CONSIDERED (TEST USED X^2). THE GROUPS INDICATED HAVE SIGNIFICANTLY HIGHER VALUES IN THE ERRORS MADE AT DIFFERENT SERIES OF RAVEN P.M. 38.

SERIES	s S - c C	l L - c C	s S - l L
A	c C	c C	
B	s S	l L	
C	s S	l L	
D	c C	c C	

S = Solvent group; L = Lead group; C = Control group

Small letters refer to 30-39 years group
Capital letters refer to 40-49 years group

The greatest difficulties in this case arise from the necessity of considering multiple elements and combining more reasoning at the same time. A higher number of errors in the B and C series for the subjects exposed to the toxic agent partially confirm the results of Dills, who, in an investigation carried out on patients with brain injury, noticed a higher number of errors in those items which stimulate symmetrical and closure solutions. These variables in fact are considered a sound index of cerebral damage. Referring again to Forbes schema, the solvent-exposed group differs from the control and lead-exposed groups due to greater similarity of results in the two age subgroups. In comparison with the other two groups they show more important and unimportant elements, discrimination errors and greater confusion in organizing these elements.

On the contrary, the control group shows no significant differences in the Benton V.R.T. A prevalence of perseveration errors was characteristic in this group. A resistance in the storage of new elements after the reproduction of the preceding onset can be assumed. This fact might be referred to the prevalence of a dysforic syndrome in the control group. It must also be stressed that the data given by Petris seem to demonstrate that the brain-injured subjects make more omission and size errors as well as vertical instability and lateral distribution errors, while more perseveration errors are characteristic of the control group.

A significant difference between the exposed groups seems to exist for the 40-49 years subgroups. In fact, the solvent-exposed subgroup makes more omissions and mnestic storage errors while the lead-exposed subjects seem to have a greater mnestic uncertainty.

DISCUSSION

Since our population was not homogeneous as regards jobs and exposure levels, it is impossible to verify whether there is a progression between the variables chosen and the factorial categories considered. For instance, it is impossible to verify whether a poor space orientation, vertical instability and lateral distribution would lead to mnestic uncertainty and storage difficulties reflecting a deficit progression or rather whether these dimensions are independent, structural units, equally or selectively present in the deterioration processes. The absence of other investigations of this kind on other populations exposed to neurotoxic agents also precludes any generalization of our results and therefore excludes possible casual explanations.

The same is true for individual error discriminant analysis as well as any possibility to establish a relationship between Raven P.M. 38 and Benton V.R.T. and spatial factor difficulties.

However, in consideration of the extent and importance of the differences observed in our groups, we feel that the qualitative analysis of these tests might be a useful tool for a better understanding of the deficit levels quantitatively observed and of the possible relationships existing among these variables.

REFERENCES

Campbell, A.L., J.E. Bogen and A. Smith (1981). Brain, 104, 493-511.
Cristante,F. and L. Petris (1977). Bollettino di Psicologia Applicata, 141, 185-204.
Dils,C.W. (1960). J. Clin. Psychol., 16, 414-416.
Forbes, A.R. (1964). British J. Educ. Psychol., 34, 223-236.
Petris,L. (1980). Bollettino di Psicologia Applicata, 153, 79-87.
Urmer, A.M., A.B. Morris and L.V. Wendland (1960). J. Clin. Psychol., 16, 182-185.
Valsecchini, S. and F. Del Ton (1973). Contributo critico alla taratura su 1123 soggetti e considerazioni sulla validità ed attendibilità della prova: Raven P.M. 38. Organizzazioni Speciali, Firenze.
Weiss, A.A. (1974). Perceptual and Motor Skills, 38, 623-635.

The Use of Reaction Time in Solvent Exposure

N. Cherry, H. Venables and H. A. Waldron

Institute of Occupational Health, University of London, London, UK

In this paper we report on a number of field studies we have carried out in Britain to investigate the effects of organic solvents on men exposed to these substances in the course of their daily work. In describing these studies we concentrate on results from a single behavioural test, that of the time taken to respond to a light stimulus presented approximately eight times a minute during a ten minute test period: the inter-trial interval is randomly varied with a range of 2.5 to 12.5 seconds from the previous response. The test is described in detail elsewhere (Wilkinson and Houghton, 1982).

There were good reasons for choosing this test as the mainstay of our programme of studies in industry. Gamberale (1979) had concluded, in summarising the results of a series of studies of different solvents in which subjects had been experimentally exposed, that 'tests of reaction time, especially those of simple reaction time, appeared to give the most consistent results throughout the different experimental conditions studied'; in these experiments he found, using a test similar, but not identical, to the one described here, that reaction time slowed down as the solvent concentration increased.

In setting up our own work we were fortunate in being able to adopt the simple reaction time test developed by Wilkinson and Houghton which had, most importantly for our purpose, been developed for testing at the work place. Each man in our studies is tested on a separate machine (an adapted cassette recorder) and his responses recorded on tape and decoded at a later point in the laboratory. Several men can be tested together in one session. The test has proved to be robust and simple to administer and to be sensitive to factors such as sleep deprivation and anxiolytic drugs which might be expected to have similar effects on behaviour as the anaesthetic properties of volatile organic solvents.

In this paper we refer to a number of the studies we have carried out to investigate the effects of solvents on behaviour. In each of the studies our interest has been to establish whether men exposed to solvents had a poorer performance on the test than men in a non-exposed referent group; where this proved to be the case we attempted to establish whether the difference could legitimately be attributed to solvent exposure (rather than some other difference between the groups) and if so, whether the effect was transient or indicative of some long term damage.

The first study we carried out was of workers exposed to styrene in a boat building factory (Cherry *et al*, 1980). Each Friday for six weeks in the spring of 1979 we visited the factory and briefed a group of up to six men and their age matched controls. Each man was given a questionnaire to complete during the weekend, asking for details of his work history, his smoking and drinking habits (to check the adequacy of the matching procedure) and of past and recent symptoms. At the Friday briefing each man was given a short test on the reaction time machine to overcome initial effects due to unfamiliarity with the test and to ease administration on the day of the test. On the Monday morning following the briefing each man carried out the test at the beginning and end of the shift, completing also a sheet reflecting the way he felt on dimensions of sleepiness, physical and mental tiredness and general good health. During the day, and until he went to bed that night, each man collected all the urine he passed. Each specimen was taken into a separate container which was labelled with the time of collection. A sample of blood was taken at the end of the shift to measure the styrene concentration. This study is described in detail elsewhere (Cherry *et al*, 1980).

It was found that men working with styrene started the shift with a significantly slower reaction time than the men in the referent group but, at the end of the day, reaction time was not slower than that of the non-exposed group. Within the exposed group, however, those with the highest blood styrene concentrations at the end of the day had not shown any improvement, and their feelings of alertness and exhaustion had deteriorated more than those of the referent group or of the exposed men with low blood styrene concentrations. The deterioration in mood, found in all our studies of exposure during a workshift even to low levels of organic solvent was, in this first study of styrene workers, highly correlated with a slow reaction time; deterioration in feelings of mental alertness having, for example, a correlation of 0.47 ($p < 0.01$) with slowing of reaction time in the group.

The slowing of morning reaction time after a sixty hour break from exposure was of particular interest in that it appeared to indicate that the solvent had more than a transient effect on behaviour. To investigate this possibility further we first identified a second group of men using the same solvent. The study design was identical with the first study except that in the Friday briefing session the men were given a container in which they were asked to collect the first urine specimen passed on Monday morning after the sixty hour break from work. On analysis, the early morning urine from these men was found to contain appreciable amounts of mandelic acid (a metabolic end product of styrene), the level of mandelic acid being positively correlated with start of shift reaction time ($r = 0.58$, $p < 0.01$). We then confirmed this result by examining the fourteen exposed men in the first sample who had passed urine soon after arrival at work. A substantial correlation ($r = 0.75$, $p < 0.001$) was again found, men who, after sixty hours away from work were still passing large amounts of mandelic acid, having a significantly slower reaction time. These results have been published in detail elsewhere (Cherry *et al*, 1981) together with such information as could be obtained from these studies on the reversibility of the slowing in reaction time. It was found, in a company with exposure levels of the order of 20 ppm, that early morning reaction time did not speed up after a two week break from exposure; in a company with levels of exposure close to the threshold limit value of 100 ppm, however, three men initially excreting large amounts of mandelic acid speeded up considerably after several months at lower levels whereas the men with low initial levels of mandelic acid in an early morning urine specimen showed no improvement. The suggestion from these findings (that exposure to styrene may, in some men, cause slowing of reaction time that persists for many days but is reversible with reduction of exposure) is in good agreement with the results reported by Kjellberg *et al* (1979); three men tested during exposure were slower in a simple reaction time test than non-exposed referents, but, after thirty-five days without exposure, any difference between the exposed and non-exposed had disappeared.

The results presented so far refer to the short term effects of a single solvent, styrene. In moving on to consider factors that may interfere with the interpretation of such behavioural effects of solvents, we turned first to data from a study we have carried out to investigate the longer term effects of exposure to paint solvents. In this study, having matched for age and educational level, we found a marked difference in reaction time between exposed and control workers, the exposed men being much slower (369 msec) than the referent group (321 msec). The exposed men, however, also did much less well on a reading test designed to reflect 'pre-morbid' intellectual ability; having excluded pairs badly matched on this dimension the reaction time difference was reduced to insignificance (369 msec for the exposed group compared with 362 msec for the referents). The magnitude of this confounding effect (reading score) was so unexpected that we turned our attention to the large amount of information we had collected on non-exposed groups. How far, we wondered, could we identify important confounders that should be accounted for in the design of future studies?

We had available test information from 148 men in manual work not involving solvent exposure. These men had been tested at the factories under the same conditions as the solvent workers and it was reasonable to suppose that factors associated with reaction time score in these men would apply equally to the solvent exposed group.

Three findings from the non-exposed population are particularly relevant to the choice of referent group in future studies.

(i) Mean scores on the reaction time test were significantly related to scores on a reading test reflecting educational achievement. This result, that reached statistical significance in two of the three main study groups ($r = -0.34$, $p < 0.001$) appears not to be simply a chance effect; non-exposed men in the lowest 20% of the reading test had a mean reaction time (of 345 msec) that was 27% slower than non-exposed men in the top 20% in the reading test (with a mean reaction time of 272 msec).

(ii) Mean scores on the test of reaction time are *not* related to age; this is perhaps unexpected but it is repeated in several of our study groups of non-exposed workers. The age range of the groups was from 17 to 65, all the men carrying out manual jobs requiring a certain amount of skill and physical fitness. Even with this finding, however, we would be sceptical of results from studies in which age was inadequately controlled.

(iii) The reaction time test reflects feelings of well-being even at the start of shift of non-exposed men. Those coming to work feeling tired do badly in the test and change in performance on the test during the day reflects change in physical tiredness. A referent group in a physically more (or less) demanding job than the exposed group would not, on this result, be expected to show a comparable change in reaction time across a shift.

Other results from the non-exposed workers are relevant to the design of the study rather than the choice of referent group.

When we adopted the Wilkinson and Houghton test of reaction time for use in the field there was no published information on the practice time needed to overcome familiarisation effects. From the nature of the test it appeared likely that practice would bring little improvement, and the study design, with a brief test session at the end of the week before the study, was judged likely to reduce considerably any variation in morning test score due to initial unfamiliarity. In the course of a shift work study, however, it proved possible to test eighteen men on four occasions, the first and last testing session being at the same time of day for each man. The mean score at the start of shift was 245 msec on the first occasion and 244 msec on the fourth. At the end of the shift the scores were 246 msec on the first occasion and 244 msec on the fourth. Neither of these differences approaches significance in a matched pairs t-test, and any improvement due to increased familiarity with the test would, in these studies, be at most 2%. While it remains possible that any improvement due to familiarity may be masked by an equivalent decrease in performance with increasing boredom with the test, the results from this study of shift workers in encouraging; the test appears to be readily learnt even by men in less skilled manual work and changes in score during the working day in a study design requiring repeated testing are unlikely to be attributable to changes in perception or familiarity with the test.

This study also provides information on the effect of shift on performance. Men on all three shifts started work with very similar mean scores (243 msec at 0700, 241 msec at 1500 and 242 msec at 2100) but there were marked differences at the end of the shift, with performance deteriorating significantly ($p < 0.05$) on the night shift. For men leaving work having completed the shift the mean reaction time was 254 msec at 0700, 244 msec at 1500 and 246 msec at 2100. Although such differences would be expected by psychologists studying the effects of shift work, findings of large shift effects in non-exposed populations on tests sensitive to solvent effects must serve as a caution, pointing to the need for the utmost care in designing studies that can be unambiguously interpreted as providing information on effects truly attributable to solvent exposure.

CONCLUSIONS

In the first part of this paper we have shown that exposure to one solvent (styrene) may have a marked effect on simple reaction time. This effect is relatively long lasting but appears to be reversible when exposure is discontinued or substantially reduced. In the studies on which these conclusions are based, each man has been acting as his own control over a work shift or over a longer period at differing levels of exposure. In this way, any ambiguities in interpretation due to the effects of confounding factors have been minimised.

In the second part of the paper, by contrast, we show that the effect of some confounding factors may be considerable. In one study, for example, an apparent difference in reaction time (thought to represent the long term effect of exposure to paint solvents) disappeared when the exposed and controls were adequately matched on a test of pre-morbid intellectual ability.

The effects of solvents on behaviour at the exposure levels found in British industry are small and difficult to detect. If systematic bias or uncontrolled variation is introduced by the study design, real and significant effects of solvents may go undetected or be attributed to chance. Experimental psychologists have a justified reputation for the design of their laboratory studies; if we are to produce clear and replicable results using well designed tests, we must do all we can to ensure that our work in the factory stands up to similar rigorous scrutiny.

REFERENCES

Cherry, N., H.A. Waldron, G.G. Wells, R.T. Wilkinson, H.K. Wilson and S. Jones (1980). *Brit. J. Ind. Med.*, *37*, 234 - 240.

Cherry, N., B. Rodgers, H. Venables, H.A. Waldron and G.G. Wells (1981). *Brit. J. Ind. Med.*, *38*, 346 - 350.

Gamberale, F. (1979). Behavioural effects of exposure to solvent vapours. Experimental and field studies. *Adverse Effects of Environmental Chemicals and Psychotropic Drugs*, *2*, Elsevier, Amsterdam.

Kjellberg, A., E. Wigaeus, J. Engström, I. Astrand and E. Ljungquist (1979). *Langtidseffekter an styrenexposition vid en plastbatsindustri*, Arbete och Halsa No 18, Stokholm.

Wilkinson, R.T. and D. Houghton (1982). *Human Factors*, in the press.

Brain Evoked Potentials in Solvent Exposure

I. Franchini*, F. Ferri**, G. Lommi**, S. Lotta**, S. Lucertini* and A. Mutti*

*Institute of "Semeiotica Medica", University of Parma, Parma, Italy
**Local Health Unit No. 3, Fiorenzula d'Arda (Piacenza), Italy

ABSTRACT

Fifteen females from a shoe factory were examined clinically and their cerebral evoked responses to 256 electrical stimulations of the median nerve at the wrist were averaged. Neurophysiological investigations included maximal motor (MCV) and distal sensory (dSCV) nerve conduction velocity measurement on ulnar, median and peroneal nerves. A referent group was composed of 15 healthy age-matched females without exposure to neurotoxic chemicals. MCVs and dSCVs of the exposed workers were significantly reduced vs. referents, while P15 and N20 components of the somatosensory evoked potentials (SEP) showed an increased latency. A negative linear relationship was found between dSCV and P15 latency. However, two subjects showed an abnormally low dSCV but normal SEP latency, and two others displayed an abnormally increased SEP latency concomitant with dSCV in the normal range. Therefore, SEP investigation may supply additional information on nervous system function, even in subjects with peripheral neuropathy. The later SEP components were much flatter in the exposed than in the referent group, suggesting some toxic effects on the central nervous system, too.

KEYWORDS

Neurophysiology; neurotoxicity; organic solvents; polyneuropathy; somatosensory evoked potentials; hexanes.

ABBREVIATIONS

EMG = Electromyography; ENeG = Electroneurography; MAP = Motor Action Potential; MCV = Maximal Motor Conduction Velocity; SNAP = Sensory Nerve Action Potential; SEP = Somatosensory Evoked Potentials; TLV = Threshold Limit Value; TWA = Time-Weighted Average

INTRODUCTION

Long-term exposure to organic solvents may cause impairment to both central and peripheral nervous system functions (Hernberg, 1980). Among workers exposed to hexacarbon mixtures, e.g., technical grade hexane, several outbreaks have been reported of acute polyneuropathy characterized by symmetrical, predominantly distal motor deficit (Spencer et al., 1980). In most instances sensory involvement has been described and groups of patients with purely

sensory polyneuropathy have been identified (Sobue et al., 1978).
In addition to nerve damage, lesions of the central ascending tract have been found in experimental toxic neuropathy (Spencer, 1978). Hexane-induced changes have also been described in EEG (Giuliano et al., 1974) and in visual evoked potentials (Seppäläinen et al., 1979). Damaged central pathways can be evaluated easily by somatosensory evoked potential (SEP) latencies (Dorfman, 1977; Hume and Cant, 1978; Obeso, 1980). SEP investigation in patients with hexane-induced polyneuropathy gave conflicting results. Zappoli et al. (1978 and 1979) were not able to find changes in SEP of polyneuropathic as compared to referent subjects. On the contrary, most of polyneuropathic workers examined by Serra et al. (1979) showed an increased SEP latency.
Recently, we reported an exposure-related reduction of motor nerve conduction velocity in workers exposed to hexacarbon mixtures (Mutti et al., 1982). We have now investigated both motor and sensory nerve conduction velocities and SEP in workers occupationally exposed to technical grade hexane.

SUBJECTS AND EXPOSURE

The subjects were 15 females from a shoe factory. Their age ranged from 19 to 53 y (mean 26.6, SD 11.4) and their exposure time from 2 to 8 y (mean 4.5 SD 2.3). The main source of exposure was solvent evaporation from glued surfaces and open cans containing solvents. A detailed description of the methods of exposure measurement has been given elsewhere (Mutti et al, 1981) while the results of the hygienic survey are in press. (Mutti et al, 1982). In summary, over 50% of the examined 8h TWA samples exceeded the current ACGIH recommended TLV for the mixture, which was mainly composed of n-hexane and cyclohexane, methyl-ethyl ketone and ethyl acetate.
Fifteen healthy age-matched females without any exposure to toxic substances were selected from 4 shoe factories and served as control group.

METHODS

ENeG measurements included MCV of the median, ulnar and peroneal nerves, and dSCV of the median and ulnar nerves. All conduction velocities were measured using skin electrodes and a Disa 1500 EMG digital system, according to routine methods (Smorto and Basmajan, 1972). No systematic difference could be detected in skin temperature of exposed subjects as compared to the referents. Many laboratory tests enabled us to exclude the presence of chronic metabolic diseases, such as chronic renal or liver failure and diabetes.
SEP were obtained by stimulation of the median nerve at the wrist by means of silver cup electrodes. Supramaximal square stimulation pulses of 0.1 ms were generated at rates of 0.5/s. Evoked potentials were recorded by means of two monopolar electrodes inserted in the scalp 7 cm lateral to the midpoint of the nasion-inion line, and the reference electrode was placed midfrontally. Input from recording electrodes was led to differential amplifiers and the responses to 256 stimulations were averaged with analysis time of 100 ms. Peak latencies and peak-to-peak amplitudes were calculated for statistical comparison between exposed and referent workers. The latencies of the early SEP components (P15 and N20) were adjusted to 65 cm constant arm length.
Parametric tests were used for statistical handling of data. Student's t test and Fisher's exact treatment of 2 x 2 tables were used for the comparison between the groups, and Pearson's correlation coefficient for the analysis of the linear relationships.

RESULTS

Table 1. summarizes the ENeG measurements on median, ulnar and peroneal nerves. All nerve conduction velocities were significantly slower in the

Table 1: Nerve conduction velocities in 15 females exposed to technical grade hexane and in 15 age-matched referents.

ENeG parameters	Referents mean	Referents SD	Exposed mean	Exposed SD	[a]Outliers No.
Median nerve					
MCV (m/s)	59.5	5.4	[b]53.9	5.6	3
MAP duration (ms)	11.4	1.7	14.7	3.3	1
MAP amplitude (mV)	13.6	5.0	12.7	6.9	-
Distal latency (ms)	3.1	0.3	3.4	0.4	-
dSCV (m/s)	51.5	8.5	[c]43.5	8.1	[f]4
SNAP peak latency (ms)	2.7	0.4	[c]3.2	0.6	[f]4
SNAP amplitude (μV)	27.6	10.2	26.5	10.1	1
Ulnar nerve					
MCV (m/s)	64.0	4.6	[b]56.0	5.2	[f]4
MAP duration (ms)	12.5	2.3	13.6	2.4	-
MAP amplitude (mV)	12.6	4.7	12.1	5.7	-
Distal latency (ms)	2.8	0.4	2.6	0.3	-
dSCV (m/s)	55.2	7.2	[b]46.9	8.9	[f]4
SNAP peak latency (ms)	2.5	0.3	[b]3.0	0.5	[e]7
SNAP amplitude (μV)	12.5	5.3	13.0	6.0	-
Peroneal nerve					
MCV (m/s)	51.1	4.0	[d]48.1	3.9	[f]4
MAP duration (ms)	11.6	1.8	11.6	2.3	-
MAP amplitude (mV)	8.7	4.4	8.4	3.7	-
Distal latency (ms)	4.0	0.7	4.3	0.6	-

[a]Number of values outside of the mean ± 2SD of reference values.
Student's *t* test: b = $p<0.01$; c = $p<0.02$; d = $p<0.05$ vs. referents.
Fisher's exact test for 2x2 tables: e = $p<0.01$; f = $p<0.05$ vs. referents.

exposed workers as compared to the referents. Motor (MAP) and sensory (SNAP) action potential peak-to-peak amplitudes were reduced in the exposed workers, but the differences were not statistically significant.

Ten workers had one or more nerve conduction velocities slower than the mean -2SD or sensory peak latencies longer than the mean +2SD of the reference values. The prevalence of abnormal findings was statistically significant for the ulnar sensory peak latency, MCV and dSCV, and for the median sensory peak latency. It should be noted that abnormal values were more frequent for the sensory peak latencies than for the respective dSCV calculated at the beginning of the response.

We also checked whether nerve conduction velocities were reduced according to the overall exposure. Probably because the number of examined subjects was quite small, no relationship was found between nerve conduction velocities and exposure time. However, all nerve conduction velocities were age-dependent in the referents, but not in the exposed group (Fig. 1).

In Figure 2 we plotted the mean latencies and amplitudes of the SEPs obtained in referents (continuous line) and in two subgroups of exposed workers (broken lines). In most of the subjects, SEPs were composed of ten peaks, which were identified according to their mean latency in the referents. In five of the exposed workers, P23 and N26 were lacking (Fig. 2, dashed line). The same peaks were only identified as deflection points in four exposed and two referent subjects. Thus, ten exposed and two referent workers were deleted from the comparison between groups concerning these

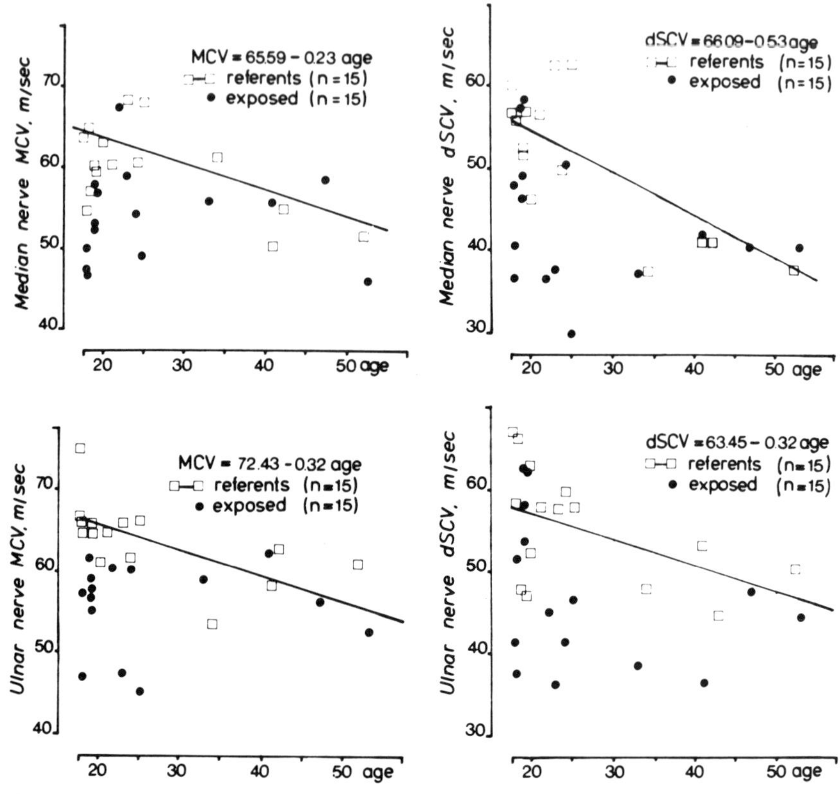

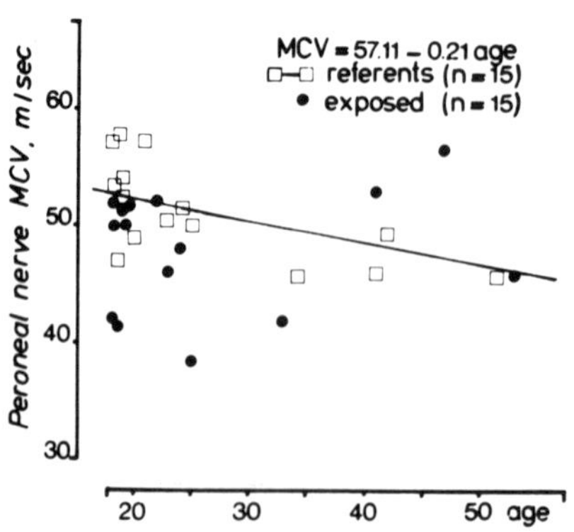

		Exposed		Referents	
		r	p	r	p
Median nerve					
	MCV	0.02	n.s.	0.54	0.05
	dSCV	0.19	n.s.	0.70	0.01
Ulnar nerve					
	MCV	0.08	n.s.	0.75	0.01
	dSCV	0.40	n.s.	0.52	0.05
Peroneal nerve					
	MCV	0.15	n.s.	0.65	0.01

Fig. 1: Negative linear relationships between age and nerve conduction velocities in the referent group (squares). No relationship was found between these variables in the exposed workers (dots). The lack of linear relationships in the exposed group seems to be due to a marked decrease in conduction velocities in the younger exposed subjects.

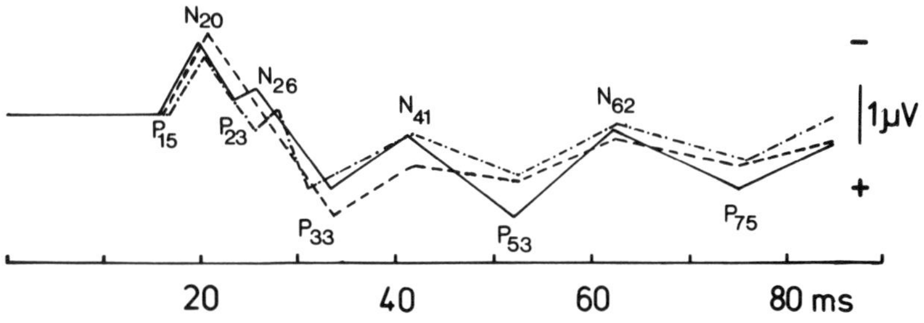

Fig. 2: Schematic presentation of mean latencies and amplitudes of the SEP components elicited by median nerve stimulation in 15 referents (continuous line) and in n-hexane-exposed workers (line with dashes and dots). Dashed line represents SEPs in which P23 and N26 were lacking.

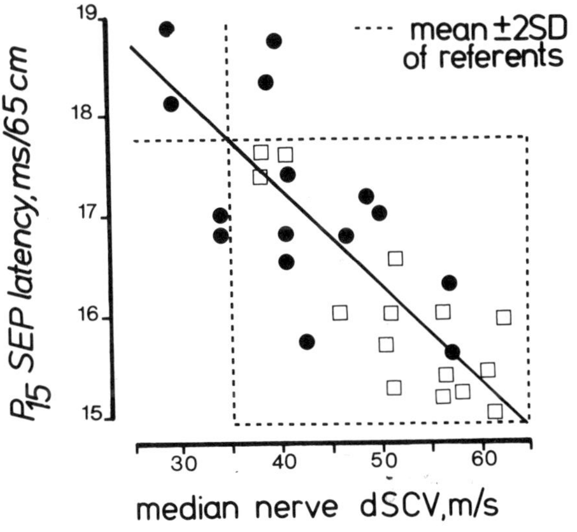

Fig. 3: Negative linear relationship between dSCV and P15 SEP latency, adjusted to 65 cm constant arm length, obtained by stimulation of the same nerve (median nerve) in referent (squares) and exposed (dots) subjects. The regression parameters for both exposed and referents were: No. = 30, $r = 0.81$, $p < 0.01$; P15 latency = 21.99 (±0.67) - 0.09 (± 0.01) dSCV.

SEP components.
In comparison with the controls, the exposed workers showed a slight increase in the latency of early SEP components, while the later phase tended to decrease in amplitude in the heavily exposed workers.
We also checked whether P15 latency was related to the dSCV of the same nerve (Fig. 3). A linear negative relationship was found, after adjusting the SEP latency to 65 cm constant arm length. Two workers showed an abnormally increased SEP latency concomitant with a dSCV slower than the mean - 2SD of the reference values. However, two subjects showed an abnormally increased SEP latency while the dSCV was in the normal range. On the contrary, two other subjects displayed a normal SEP latency while the dSCV was slower than the mean -2SD of the reference values.

DISCUSSION

The neurotoxicity of 2,5-hexanedione, a common putative metabolite of n-hexane and methyl-n-butyl ketone, is enhanced by simultaneous administration of methyl-ethyl ketone (Abdel-Rahman et al., 1976). Only the combination of n-hexane and methyl-ethyl ketone might cause polyneuropathy in glue-sniffers (Altenkirch et al., 1977). Cyclohexane-induced changes in the myelinic sheet were found after subcutaneous treatment of chickens (Franchini et al., 1978). Methyl-ethyl ketone and cyclohexane were present in the mixture used by our study subjects. n-Hexane neurotoxicity might be enhanced by the simultaneous exposure to these other two organic solvents.
Slowed nerve conduction velocity in n-hexane-exposed workers has already been described (Abbritti et al., 1976; Cianchetti et al., 1976; Buiatti et al., 1978; Sanagi et al., 1980; Mutti et al., 1982); these findings were confirmed in the present study. Among ENeG parameters, the sensory peak latency of the distal tract of median and ulnar nerves seemed to be the most sensitive sign of the neurotoxic effects of organic solvents. It should be noted that an increased latency of sensory peak was found even in subjects with normal dSCV of the same nerve, indicating an increased temporal dispersion in the elicited response. All subjects with impaired sensory nerve function also had one or more slowed MCV.
Somatosensory evoked potentials by peripheral nerve stimulation are a useful technique to evaluate the conduction through the somesthetic pathways (Desmedt, 1973). However, the neuronal source of these potentials has not yet been fully elucidated
Serra et al. (1979) and Zappoli et al. (1978 and 1979) gave conflicting answers to the question of whether SEP latencies are delayed in workers with peripheral polyneuropathy. In our study, early SEP components in workers exposed to hexacarbon mixtures were significantly increased, but only after adjusting to 65 cm constant arm length. Moreover, the later phase (N41-P75) of the SEP was much flatter in the exposed compared to the referent subjects.

The increased latency of the first SEP peak (P15) could reflect a predominantly peripheral neuropathy. In fact, it was related to the dSCV of the same nerve. Despite this relationship, two exposed subjects with normal median nerve dSCV, but with other ENeG signs of impaired nerve function, showed an increased latency of the first SEP peak. Thus, SEP evaluation may furnish additional information on conduction through impaired somesthetic pathways, and integrate ENeG examination.

A central conduction blockage might explain our findings of flatter SEP peaks in n-hexane-exposed workers. On this basis, a toxic effect of n-hexane on the central nervous system could be suggested. It should be noted that our study started at least three months after the last exposure. Consequently, the observed effects on the SEP were not due to the acute prenarcotic effect of organic solvents.

In addition to the well established hexacarbon-induced polyneuropathy, we would draw attention to the possible toxic effects on the central nervous system and to the need for reducing the current ACGIH recommended TLVs (ACGIH, 1981).

ACKNOWLEDGEMENT

We are indebted to Dr. C. Arcari for exposure monitoring and to the Dept. of Social Security, Emilia-Romagna Region, for their financial support.

REFERENCES

Abbritti, G., A. Siracusa, C. Cianchetti, C.A. Coli, F. Curradi, G.F. Perticoni, F. De Rosa (1976). Br. J. Ind. Med. 33, 92-99.

Abdel-Rahman, H.S., L. Hetland, D. Couri (1976). Am. Ind. Hyg. Ass. J. 37, 95-102.

Altenkirch, H., J. Mager, G. Stoltenburg, J. Helmbrecht (1977). J. Neurol. 214, 137-152.

American Conference of Governmental Industrial Hygienists (1981). Threshold limit values for chemical substances and physical agents in the workroom environment. Cincinnati, Ohio.

Buiatti, E., S. Cecchini, O. Ronchi, P. Dolara, G. Bulgarelli (1978). Br. J. Ind. Med. 35, 168-173.

Cianchetti, C., G. Abbritti, G. Perticoni, A. Siracusa, A. Curradi, (1976). J. Neurol. Neurosurg. Psychiatry 39, 1151-1161.

Desmedt, J.E., and P. Noel (1973). New developments in electromyography and clinical electrophysiology, Vol. 2, 352-372, Karger, Basel.

Dorfman, L.J.(1977). Electroenceph. Clin. Neurophysiol.42, 26-34.

Franchini, I., A. Cavatorta, M. D'Errico, M. De Santis, R. Gatti, S. Magnani, G. Juvarra, G. Palla (1978). Experientia 34, 250-253.

Giuliano, G., A. Iannacone, R. Zappoli (1974). Lav. Umano 26, 33-42.

Hernberg, S. (1980). Neurotoxic effects of long-term exposure to organic solvents. Proceedings 2nd Intern. Congr. on Toxicology. Brussels, Belgium.

Hume, A.L., and B.R. Cant (1978). Electroenceph. Clin. Neurophysiol.45, 361-375.

Mutti, A., C. Pedroni, M. Falzoi, A. Cavatorta, I. Franchini (1981). J. Appl. Toxicol. 1, 220-224.

Mutti, A., A. Cavatorta, S. Lucertini, G. Arfini, M. Falzoi, I. Franchini (1982). Scand. J. Work Environ. Health 8, Suppl. 1, 136-141.

Obeso, J.A.(1980). Electromyogr. Clin. Neurophysiol. 20, 75-83.

Sanagi, S., Y. Seki, K. Sugimoto, M. Hirata (1980). Int. Arch. Occup. Envir. Health 47, 69-79.

Schaumburg, H. H., and P. S. Spencer (1978). Science 199, 199-200.

Seppäläinen, A.M. (1978). Diagnostic utility of ERPs and other neuroelectric measures in environmental and occupational medicine. Multidisciplinary Perspectives in Event-related Brain Potential Research, EPA, U.S. Govt. Printing Office, Washington, DC.

Serra, C., A. Rossi, D. Facciolla, A. Ruocco, A. Biondi (1979). Riv. Ital. Elettroencefalogr. Neurofisiol. 2, 135-139.

Smorto, M.P., and J.V. Basmajan (1972). Clinical Electroneurography. Williams and Wilkins, Baltimore, Maryland.

Sobue, I., I. Mitsuo, Y. Yamamura, T. Takayanagui (1978). Intern. J. Neurol. 11, 317-330.

Spencer, P.S., D. Couri, H.H. Schaumburg (1980). n-Hexane and Methyl-ethyl ketone. Experimental and Clinical Neurotoxicology. Williams and Wilkins, Baltimore/London.

Zappoli, R., G. Giuliano, L. Rossi, M. Papini, O. Ronchi, A. Ragazzoni, A. Amantini (1978). Multidisciplinary Perspectives in Event-related Brain Potential Research. EPA, U.S. Govt. Printing Office, Washington, DC.

Zappoli, R., G. Giuliano, L. Rossi, M. Papini, O. Ronchi, A. Ragazzoni, A. Amantini (1979). Riv. Pat. Nerv. Ment. 100; 189-200.

Vestibular Function in Solvent Exposure: Clinical Criteria

S. Binaschi* and L. Cantù**

*Laboratory of Occupational Health, Varese; Institute of Occupational Preventive Medicine, University of Pavia, Pavia, Italy
**Department of Otorhinolaryngology, Regional Hospital, Varese, Italy

ABSTRACT

Vestibular function was investigated in patients with occupational exposure to organic solvents, using clinical criteria. The results in 35 patients with chronic exposure to single organic solvents or to a solvent mixture show that the vestibular system was affected in nearly all cases, with an elective localization of the lesion in the nucleo-reticular substance of the brain stem (chronic syndrome), often associated with slight diffuse slow waves in the EEG and a neurasthenic syndrome.

KEYWORDS

Organic solvents; vestibular functions; EEG, neurasthenic symptoms; reticular substance.

INTRODUCTION

Several industrial solvents cause vertigo and dizziness, nausea and feeling of drunkenness. In the rabbit, xylene, styrene, methylchloroform and trichloroethylene cause positional nystagmus (Savolainen et al, 1980), and in exposed workers alterations in equilibrium were revealed. The maintenance of equilibrium in man is an integral function of the vestibular apparatus, vision and proprioceptors in the peripheral parts of the body. Arslan and Sala (1956) described a vestibular syndrome, the "toxic nucleo-reticular vestibular syndrome" in which impairment of the vestibular apparatus at the level of the central vestibular nuclei and reticular pathways in the brain stem results in an acute and chronic form. The latter consists of: a) slight, usually short-lasting vertigo influenced by head movements; b) horizontal, or horizontal-rotatory spontaneous nystagmus (Class 1 in Alexander's classification), the beat direction of which is opposite to the lesion in the vestibular nuclei; c) symmetric optokinetic nystagmus (OKN); d) slight tonic segmentary deviations; e) clear signs of impaired vestibular reflectivity, bilaterally decreased (hypo-excitability), with rhythm disturbances consisting of a slow phase increase always present (Girardi *et al*, 1967; Dufour *et al*, 1979). The involvement of the nervous pathways in the brain stem is also revealed by decay of the stapedial acoustic reflex (Bosatra, 1978). Effects on central vestibular pathways similar to the toxic nucleo-reticular syndrome have been described for several solvents, like carbon disulfide (Cis *et al*, 1964; Merluzzi *et al*, 1981; Corsi *et al*, 1981), xylene (Savolainen *et al*, 1981), trichloroethylene (Szulc-Kuberska *et al*, 1976; Molinari *et al*, 1974), aliphatic hydrocarbon solvents (D'Andrea *et al*, 1979). Yet often, especially for the purposes of epidemiological surveys, researchers have used easy and rapid techniques to study vestibular function (Veits stimulation, even only with warm or cold water). We therefore decided to evaluate vestibular function in 35 patients with long-term exposure to solvents using exhaustive clinical criteria.

PATIENTS AND METHODS

Patients

Occupational disease caused by organic solvents was suspected in 35 patients examined at the Occupational Medicine Service or admitted to other Divisions of the Hospital of Varese. The quality and the intensity of exposure was determined from the occupational history given by the patient, the description of the job and of the working conditions supplied by the employer, the observations made visiting the places of work, and available results of solvent measurements in workroom air (Table 1.).

Table 1. Type of exposure

SOLVENT	No. of patients
Trichloroethylene	9
Toluene and Xylene	11
Mixture of solvents (Alcohols, Esters, Ketones)	15
Total	35

For all the patients considered the concentration of solvents in the working environment was under the TLV-TWA for the individual solvents and for the solvent mixture; but it is more than likely that on some occasions the solvent concentration exceeded the ACGIH limits for a short time. The mean duration of exposure was 7 years; with a minimum of 3 years for 5 patients. All the patients had subjective symptoms (Table 2.); other non-vestibular diseases were excluded.

Table 2. Early subjective symptoms.

Symptom	Total
Fatigue	30
Psychic disturbances	27
Dizziness	23
Vegetative symptoms	11
Vertigo	7

Methods

1. The vestibular apparatus was examined on the basis of spontaneous and provoked vestibular symptoms.

1.1. Nystagmus (Ny).

1.1.1. Spontaneous Ny, classified according to Alexander.

1.1.2. Revealed or Positional Ny. Nylen's classification: type I changing direction positional Ny; type II fixed direction positional Ny; type III irregular positional Ny. Rose's position: dorsal decubitus with head leaning backward.

1.1.3 Opto-kinetic Ny: saccadic and tracking movements.

1.1.4. Provoked Ny.
Fitzgerald-Hallpike method: the patient sits with his head bent forward at an angle of 30°; 250 cc of water, at 30 and 44°C, flow in 40 seconds from a thermostatic calorimeter to his meatus acustic externus. Electronystagmographic registration is performed.
Rotatory stimulation according to the Buys-Fisher-Arslan method: subliminar acceleration lower than 1 degree per square second; rotation up to 90° degrees per second is maintained for 1 minute; sudden stop followed by electronystagmographic registration.
Pendular stimulation according to Hennebert: excursion of 180 degrees round the vertical axis of a chair, alternatively clockwise and counter-clockwise; progressive amplitude dampings, during 15 periods of 20 second each. Electronystagmographic registration for each period. The test is made twice: the second time 2 minutes later than the first.
Electronystagmographic registration was performed with ENG Racia, 5 channels, 3 Ag's electrodes.

1.2. Tonic segmentary deviations (Romberg, indication and deviation tests).

1.3. Evaluation of the pure tone threshold measurement.

2. Electroencephalogram (EEG) was recorded in all patients with the usual, classic method.

3. Psychiatric examination was based on interviews, psychiatric history and prevalence of 37 different symptoms, using the scale proposed by Knave (1979).

RESULTS

1. The vestibular system was affected in nearly all cases, with an elective localisation of the lesion in the nucleo-reticular substance and pathways of the brain stem, appearing as a chronic syndrome. The characteristics of this syndrome that we call, like Arslan, "toxic vestibular nucleo-reticular chronic syndrome", were: a) subjective vertigo present only in 20% of the subjects and increased after head movements. b) Spontaneous Ny in 80% of the subjects: 1st class horizontal or horizontal-rotatory, bilateral with small arrhythmic jerks. c) Revealed Ny: positional Ny of 1st and 2nd type according to Nylen. d) Opto-kinetic Ny: present and symmetric in 80% of cases. e) Provoked Ny: evident signs of impaired vestibular reflectivity, markedly bilaterally reduced in duration and in number of jerks. Bilateral symmetrical hyporeflexia (75%), bilateral symmetrical hyperreflexia (15%); arrhythmias with constant amplitude, variation of frequency and velocity with prevalence of the slow phase;

disrhythmias also with amplitude variation; groups of jerks of the same amplitude with pauses due to the slow phase prevalence. f) Tonic segmentary deviations: harmonic in 80% of the cases, tendency to disharmony in 20% with deviation of the arms in opposite direction to the slow phase of Ny. g) The pure tone threshold measurement was always within normal limits. In 10% of the patients the vestibular investigation revealed a labyrinthic monolateral syndrome.

2. EEG changes included essentially slight diffuse slow waves (50%) and slight focal slow waves (10%).

3. The psychiatric examination revealed a mild psycho-organic syndrome of neurasthenic nature in 71% of the subjects. In agreement with Knave (1979) we diagnosed a neurasthenic syndrome if fatigue and emotional instability (including anxiety and depression) were recorded as prominent mental symptoms, together with decreased capacity (memory and concentration difficulties).

DISCUSSION

The association of subjective and objective results of medical studies, the interpretation of the importance of statistical differences rather than the biological significance of clinical observations, are still the most important problems in the analysis and interpretation of data obtained in the study of occupational neurotoxicity, especially in the case of organic solvents (Spencer et al, 1980). We think all the results should be viewed cautiously, because individual differences in susceptibility to neurotoxic agents also complicate the evaluation of a cause-effect relationship and the basic biological phenomena underlying the manifestation of chronic organic solvent intoxication are such that epidemiological studies are difficult (Juntunen et al, 1980a). We believe that all the research methods should be first clinically and exhaustively validated on the basis of well defined physiopathological hypotheses; only later is it possible to choose the most suitable tests (according to easiness, rapid performance and sensitivity) for epidemiological investigations in large groups of exposed workers. As regards the vestibular examinations, we observed in our patients that chronic exposure to low doses of industrial solvents causes symmetrical lesions in the vestibular nuclei and in the reticular substance, while the vestibular receptors are unimpaired. The lesions are similar to those of the toxic vestibular nucleo-reticular syndrome described by Arslan and supported by the classical theoretical work of Barany and Lorente de Nô on the function of the reticular substance and synaptic chains; these are both places where the nystagmus rapid phase is induced and the slow phase is periodically inhibited.
The marked bilateral decrease in vestibular reflectivity, both in duration and in the number of jerks, may be due to metabolic conditions of the nucleo-reticular substance impaired by neurotoxics (Girardi et al, 1967). This hypothesis is also supported by the high incidence of brain stem chronic nucleo-reticular syndrome in diffuse vascular disease of the CNS (Felletti et al, 1980). Also Odkvist (1978) suggested that the action site of solvents, due to their chemical rather than physical properties, may be in the central vestibular pathways, because the solvent readily reaches the well perfused central parts of the vestibular system.

The EEG abnormalities, the characteristics and incidence of which are similar to those found by other investigators in workers with solvent exposure (Juntunen et al., 1980a; Seppäläinen, 1981), were probably related to decreased vigilance.

The mild psycho-organic neurasthenic type syndrome that we found in our patients was also observed in long-term exposure to organic solvents, with basically toxic effects on the CNS, and signs of diffuse brain damage appearing also as a psycho-organic syndrome (Husman et al, 1980; Juntunen et al, 1980b; Arlien-Soborg et al, 1979).

The findings in our patients all point to a single pathogenetic action, that can also be demonstrated by different methods. The neurologic and psychic signs of diffuse brain damage and the involvement of the nucleo-reticular substance are in our opinion evidence of effects on the CNS as a result of long-term low exposure to different organic solvents. The involvement of the reticular substance is very important as regards the integrity of the vestibular apparatus response and the changes in vigilance and subjective fatigue. In a neuro-physiological interpretation these two phenomena point to an antagonism between the activating reticular pathways and the cortical activity inhibiting system (Grandjean, 1979).

REFERENCES

Arlien-Soborg, P., P. Bruhn, C. Gyldenstedt and B. Melgaard (1979). Acta Neurol. Scand., 60, 149-156.

Arslan, M. and O. Sala (1956). Proc. 44th Nat. Congr. Ital. Soc. Lar. Otol. Bologna, Italy.

Bosatra, A. (1978). Nuovo Arch. Ital. Otol., 6, 181-186.

Cis, C. and G. Perani (1964). Med. Lav., 55, 198-211.

Corsi, G., and G. Picotti (1981). G. Ital. Med. Lav., 3, 93-94.

D'Andrea, F., M. Cavazzini, L. Perbellini, P. Apostoli and P. Zampieri (1979). Med. Lavoro, 1, 16-20.

Dufour, A., F. Cocchini, and A. Formenti (1979). Nuovo Arch. Ital. Otol., 7, 217-228.

Felletti, V., M. Lazzaroni and C. Russo (1968). Ann. Lar. Otol., 67, 262-274.

Girardi, G., C. Cis and A. Piatti (1967). Arch. Ital. Otol., 78, 756-770.

Grandjean, E. (1979). Brit. J. Industr. Med., 36, 175-182.

Husman, K. and P. Karli (1980). Scand. J. Work Environ. Health, 6, 33-39.

Juntunen, J., V. Hupli, S. Hernberg and M. Luisto (1980a). Int. Arch. Occup. Environ. Health, 46, 219-231.

Juntunen, J., S. Hernberg, P. Eistola and V. Hupli (1980b). Eur. Neurol., 19, 366-375.

Knave, B., P. Mindus, and G. Struwe (1979). Acta Psychiat. Scand., 60, 39-49.

Merluzzi, F., A.M. Cirla, T. Terrana and N. Di Credico (1981). G. Ital. Med. Lav., 3, 95-97.

Molinari, G.A., B. Saia and G. Marcer (1974). Nuovo Arch. Ital. Otol., 2, 315-326.

Odkvist, L.M., B. Larsby, R. Tham, S.R.C. Liedgren, and G. Aschan (1978). In: Hood J.D. Ed. Vestibular Mechanisms in Health and Disease. Academic Press, New York, pp. 188-194.

Savolainen, K., V. Riihimaki, E. Vaheri, and M. Linnoila (1980). Scand. J. Work Environ. Health, 6, 94-103.

Savolainen, K. and V. Riihimaki (1981). Acta Pharmacol. Toxicol., 48, 279-283.

Seppäläinen, A.M. (1981). Scand. J. Work Environ. Health, 7, suppl. 4, 29-33.

Spencer, P.S. and H.H. Schaumburg Eds. (1980). Experimental and Clinical Neurotoxicology. William & Wilkins, Baltimore, pp. 663-674.

Szulc-Kuberska, J., J. Tronczynska, and B. Latkowski (1976). Min. Oto., 26, 108-112.

Regional Cerebral Blood Flow in Subjects Occupationally Exposed to Organic Solvents

S. Hagstadius and J. Risberg

Laboratory of Neuropsychology, Department of Psychiatry, University Hospital, Lund, Sweden

ABSTRACT

Regional cerebral blood flow (rCBF, ^{133}Xe inhalation method) was measured in 50 male paint-factory workers with a mean of 18 years of exposure to a mixture of organic solvents. A group of 50 workers in a sugar refinery, matched for age and education, served as controls. The measurements were made during resting and during activation by mental tasks (word-pair learning). The results showed slightly lower mean flow level (4 %; $p<0.05$) in the exposed group. The largest differences were seen in frontotemporal areas. The difference between the exposed group and the controls increased at higher dose levels. The largest rCBF increases during mental activation were seen in the exposed group, especially in the highest exposed subjects. This finding might indicate mechanisms compensating for a somewhat defective brain function. Although the differences between the groups were generally small with considerable overlap, the results give some evidence of disturbances of brain blood flow and brain function probably related to the influence of organic solvents.

KEYWORDS

Regional cerebral blood flow; organic solvents; brain function; neurastenic syndrome.

INTRODUCTION

The effects upon the central nervous system of long-term exposure to organic solvents have attracted considerable interest during the past decade. The chronic mental symptoms displayed after several years of exposure include memory disturbances, tiredness, irritability and emotional instability (Axelson, et al, 1976 and others). This cluster of symptoms is often referred to as the neurastenic or psychoorganic syndrome. Psychometric studies have shown that exposed workers have poorer results than unexposed controls in tests measuring memory, visual perception and psychomotor functions (Götell, et al, 1972, Gamberale, et al, 1975, Hänninen, et al, 1976, Hane, et al, 1977). The neurastenic syndrome can, however, arise after long-term exposure to other toxic agents, e.g., alcohol as well as in several types of organic brain disease, especially presenile dementia (Alzheimer's disease). This makes the clinical evaluation difficult in many cases. It is therefore important that additional objective examinations of the nervous system be used as a complement to clinical and psychometric data.

Among the methods tried are pneumoencephalaography and computerized tomography (CT). The results obtained so far have shown only tentative indications of pathological changes in brain morphology in exposed workers (Gregersen, et al, 1978, Juntunen, et al, 1979, Elofsson, et al, 1979). Studies of EEG have shown results varying from an increase in slow wave activity to higher occurrence of beta activity (Seppäläinen and Härkonen, 1976, Rosén, et al, 1978). Thus, so far, these methods have not demonstrated any consistent or conclusive evidence of CNS pathology.

The aim of the present study was to investigate whether measurements of the regional cerebral blood flow (rCBF) would be an aid in the description of changes of brain function related to chronic solvent exposure. The use of the rCBF-method as an indicator of cerebral function is based on the finding of a close relationship between rCBF and cerebral metabolism of oxygen and glucose (Raichle, et al, 1976). This relationship holds in the normal brain as well as in most cases of brain disease. The method has been used successfully in the study of the cerebral organisation underlying mental functions (see Risberg, 1980, for a review and also Risberg and Hagstadius, this volume). It has also been possible to relate different patterns of flow reductions in patients with symptoms of dementia to the clinical picture and to neuropathological findings (Gustafson and Risberg, 1974, 1979). The method has been shown to be of considerable clinical value for evaluation of patients with symptoms of dementia. It was thus quite logical to continue our studies of dementia with the present investigation focusing on disturbances of brain function due to long term solvent exposure. When we started the study very little was known about rCBF changes in exposed subjects. The only evidence was the results of a pilot study of 32 exposed workers (average 25 years of exposure to a mixture of solvents) indicating a slight reduction (6 %) in the average flow level compared to normal subjects of similar age (Maximillian, et al, 1980).

MATERIAL

The exposed group consisted of 50 male workers from a paint factory ranging in age from 26 to 62 years (mean age 42), exposed to a variety of organic solvents (ketones, aromatic hydrocarbons, ethers, alcohols, etc) for a minimum duration of ten years (mean exposure: 18 y, SD: 10 y).

The control group consisted of 50 workers from a sugar refinery, who were not exposed to organic solvents in the working environment. The controls were matched to the exposed subjects by age and educational background. The alcohol and smoking habits did not differ between the groups. Subjects who were on drug therapy, or were suffering from epilepsy, diabetes, psychiatric disorders, pulmonary disorders or alcohol/drug abuse were not included in the study. None of the subjects had evidence of brain damage due to trauma or vascular disorder. They were all paid volunteers.

METHODS

All participants underwent thorough medical examination, psychiatric and electrophysiological examinations as well as psychometric testing and rCBF. Only the results from the rCBF measurements will be reported here.

Determinations of concentration levels of the most frequently used organic solvents in different parts of the factory had more or less been systematically carried out during earlier periods. On the basis of these data and other information the total dose of each individual was estimated. The exposed subjects were then divided into three dose groups (low, medium, high) of about equal size.

The regional cerebral blood flow was determined by the non-invasive 133-Xenon inhalation technique, as outlined by Mallet and Veall (1965) and further developed by Obrist, et al, (1967, 1975) and Risberg, et al, (1975). In our system each hemisphere is covered by 16 scintillation detectors, which record the radiation from the tracer used, 133-Xe. The patient inhales the tracer mixed with air for one minute through a face mask and a rebreathing spirometer system, followed by ten minutes of normal air breathing. The rate of isotope washout recorded forms the basis for calculation of flow values. In this paper the ISI-index of grey matter flow will be used (Risberg, et al, 1975). For a more detailed description of the technique and measurements procedure, the reader is referred to Risberg (1980) and Risberg and Hagstadius (this volume).

Arterial pCO_2 was determined by recording of end-tidal CO_2 concentrations by means of a capnograph. The arterial pCO_2 level has a great influence on the flow level recorded and has to be taken into account in all comparisons of flow data.

Each subject was first measured during "resting", a condition where the subject was lying supine on a comfortable bed, with eyes closed and with minimized auditory stimulation. The subject was asked to relax without falling asleep. A second measurement was also carried out during which mental activation was induced by a verbal memory test. The tasks consisted of learning visually presented word-pairs (6 slides with 10 word-pairs on each). Recall was checked immediately after the flow measurement.

RESULTS

Table 1 shows the mean flow levels during rest. The mean flow was 46.5 in the exposed group and 48.5 in the controls, thus the difference was 4%. In spite of the limited size of this difference, it was statistically significant ($p<0.05$ by t-test). Table 1 also shows the results from a subdivision of the group into three dose groups. There is a tendency to larger differences between exposed and controls at higher doses, the high dose group showing a 7 % lower value. The arterial $apCO_2$ values show very little variation within or between groups and this parameter has thus not influenced the results obtained.

Figure 1 illustrates that there is a considerable overlap of the individual rCBF values between the two groups at all ages. The general decline of flow with age is also evident.

Figure 2 shows the regional flow differences between the groups. The exposed group had generally lower flows in all regions. The most significant differences were seen in frontotemporal areas of both hemispheres as well as in one parietal area of the left hemisphere.

Changes in mean flow level from rest to activ tion by the verbal memory test are shown in Table 2. There is a tendency for mean flow to increase

Group	n	ISI (SD)	Diff	pCO_2 (SD)	sign
Total exposed	50	46.5 (5.5)	2.0	40.0 (3.1)	p=<.05
control	50	48.5 (5.6)		39.9 (3.4)	
Low dose exposed	17	49.4 (4.6)	0.8	41.2 (2.0)	ns
control	17	50.2 (5.7)		40.7 (2.5)	
Med dose exposed	18	47.4 (5.9)	2.1	39.8 (3.2)	ns
control	18	49.5 (7.2)		40.2 (3.2)	
High dose exposed	15	41.8 (4.3)	3.0	39.0 (3.6)	ns
control	15	44.8 (8.2)		38.5 (4.0)	

Table 1. Resting mean flow (ISI) and arterial pCO_2 in the exposed group compared to the control group. The exposed group and matched controls are subdivided into three groups according to level of exposure. Test of significance by t-test.

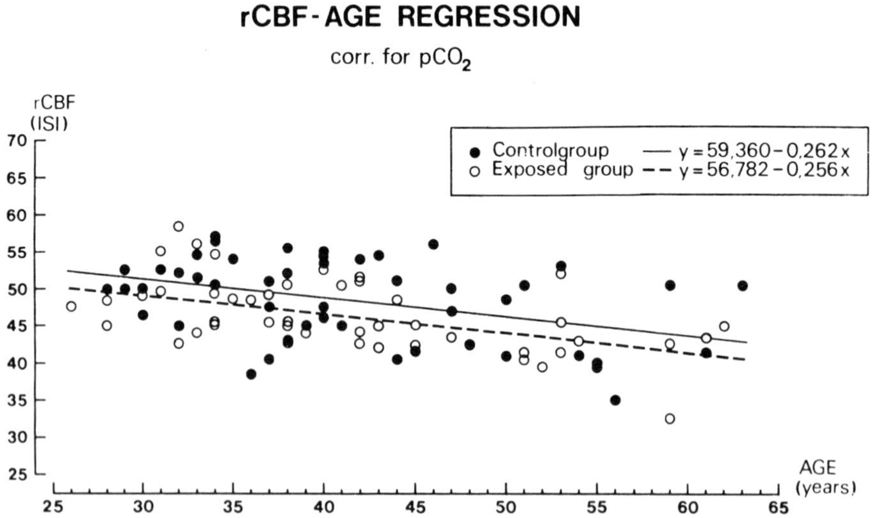

Figure 1. The relationship between rCBF, (ISI) at rest and age for the exposed and the controls. rCBF-values are corrected for pCO_2-variations.

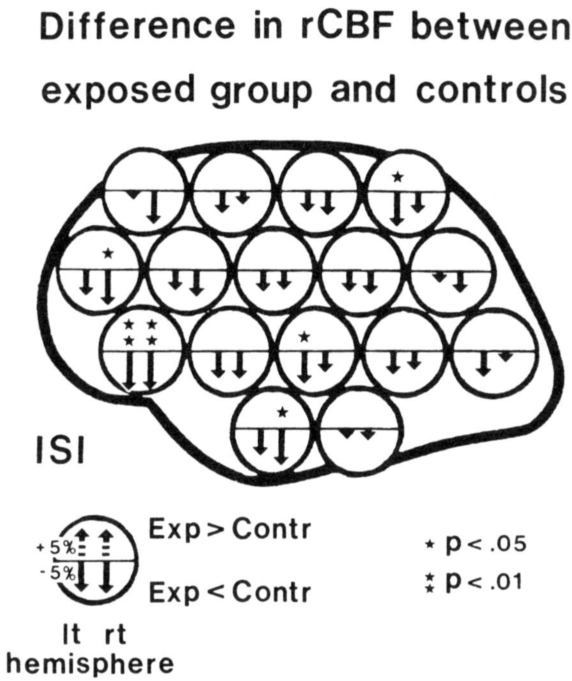

Figure 2. Regional CBF differences between exposed and controls.

Group	n	ISI:Rest	ISI:Test	Diff	Score
Total exposed	50	46.5 (5.5)	47.1 (5.9)	0.6	6.6 (9.3)
control	50	48.5 (5.6)	47.4 (6.8)	-1.1	10.2 (12.0)
Low dose exposed	17	49.4 (4.6)	48.7 (5.8)	-0.7	10.3 (10.7)
control	17	50.2 (5.7)	50.5 (6.2)	-0.3	12.4 (14.1)
Med dose exposed	18	47.4 (5.9)	46.2 (5.0)	-1.2	7.3 (10.2)
control	18	49.5 (7.2)	47.4 (6.8)	-2.1	12.9 (12.1)
High dose exposed	15	41.8 (4.3)	44.4 (5.3)*	2.6*	1.9 (3.2)
control	15	44.8 (8.2)	44.4 (6.8)	-0.4	3.9 (6.4)

* = $p < .05$ (t-test)

Table 2. Mean flow levels (ISI) during rest and during activation by a mental test (word pair learning) for each dose group and their matched controls. Test performance (number of correct answers) is shown to the right.

from rest to test in the total exposed group, while the controls show a slight decrease. Arterial pCO_2 values were essentially unchanged. Dividing the material in dose groups showed a significant ($p < .05$) increase in mean flow for the highest exposed group only. Also the difference in activation response between high-dose exposed and controls was significant ($p < .05$).

Table 2 also shows the mean scores in the word-pair test for the groups. The exposed group had lower scores than the controls at all levels of exposure. However, the differences were not statistically significant.

An analysis of the regional changes during activation was also made. The highest exposed group showed significant increases in parietal, parieto-occipital and pre-motor areas of both hemispheres. The regions activated agree well with earlier findings in normal subjects performing a similar type of mental task (Maximilian et al, 1978). The amplitude of the regional changes was significantly smaller in lower exposed subjects as well as in the controls. The regional changes were thus in accordance with the mean flow changes.

DISCUSSION

Although the overlap between the group exposed to organic solvents and the controls was large, some significant differences were established in resting CBF as well as in the response to mental activation. Since there were no differences between the groups regarding important background variables like age, educational background and alcohol consumption it seems likely that the observed differences in rCBF are due to exposure to organic solvents. The small decrease in mean flow level found in this study is in good agreement with the 6% decrease found in our pilot study (Maximilian et al, 1980). This study involved 32 subjects, who were investigated because of reports of symptoms with a likely relationship to organic solvent exposure. The present results also indicated larger differences in CBF between exposed and controls at higher levels of exposure, suggesting a dose-effect relationship.

These observations might be interpreted as reflecting an accelerated aging process in the brain in exposed workers. This has been discussed by Wilkinson et al, (1979) concerning chronic alcoholics and suggested in rCBF studies of chronic alcoholics (Berglund, 1981). Berglund (1981) also found decreased flow in fronto-temporal areas in older chronic alcoholics. It is possible that the fronto-temporal decreases in flow seen in the present exposed group reflect a similar process to that observed in chronic alcoholics. The rCBF response to mental activation was found to increase both regionally and globally with increased exposure. The high activation response was, however, associated with a low performance score in the test. This finding might reflect mechanisms compensating for a defective brain function in the exposed group. This result does not, however, confirm the finding in our pilot study, were the most exposed subjects had a lower activation response than the less exposed. This indicates a complex relationship between disturbances of brain function due to organic solvent exposure and rCBF-changes during activation.

ACKNOWLEDGEMENTS

Supported by grants from the Swedish Work Environment Fund, the Swedish Medical Research Council (contract 04969) and the Swedish Council for Social Science Research.

REFERENCES

Axelson, O., Hane, M. and Hogstedt, O. (1976). Läkartidningen, 73, 317-325.

Berglund, M. (1981). Alcoholism: Clinical and Experimental Research, Vol 5, No 2, 295-303.

Elofsson, S.A., Gamberale, F., Hindmarsh, T., Iregren A., Isaksson, A., Johansson, I., Knave, B., Lydahl, E., Mindus, P., Persson, H.E., Philipson, B., Steby, M., Struwe, G., Söderman, E., Wennberg, A. and Widén, L. (1979) Läkartidningen, 76, 4127-4148.

Gamberale, F., Lister, H. and Anshelm-Olson, B. (1975) Arbete och Hälsa, 8.

Gregersen, F., Mikkelsen, S., Klausen, H., Dössing, M., Nielsen, H. and Thygsen, P. (1978). Ugeskr. Laeg., 140, 1638-1643.

Gustafson, L. and Risberg, J. (1974). Acta Psychiat. Scand., 50, 516-538.

Gustafson, L. and Risberg, J. (1979). Acta Neurol. Scand. 60, Suppl. 72, 546-547.

Götell, P., Axelson, O. and Lindelöf, B. (1972). Scand. J. Work. Environ. Health, 9, 76-83.

Hane, M., Axelson, O., Blume, J., Hogstedt, C., Sundell, L. and Ydreborg, B. (1977). Scand. J. Work. Environ. Health, 3, 91-99.

Hänninen, H., Eskelinen, L., Husman, H. and Nurminen, M. (1976). Scand. J. Work. Environ. Health, 4, 240-255.

Juntunen, J., Eistola, P., Hupli, V. and Hernberg, S. (1979). Proceedings of the IIIrd Industrial and Environmental Neurology Congress, Prague.

Mallet, B.L. and Veall, N. (1965). Clin. Sci., 29, 179-188.

Maximilian, V.A., Prohovnik, I., Risberg, J. and Håkansson, K. (1978). Brain and Language, 6, 22-31.

Maximilian, V.A., Risberg, J., Prohovnik, I., Rehnström, S. and Haeger-Aronsen, B. (1980) in: Maximilian, V.A. (1980). Functional Changes in the Cortex During Mental Activation. Thesis, Lund University, Lund.

Obrist, W.D., Thompson, H.K., King, C.D. and Wang, H.S. (1967). Circ. Res., 20, 124-135.

Obrist, W.D., Thompson, H.K., Wang, H.S. and Wilkinson, W.E. (1975). Stroke, 6, 245-256.

Raichle, M.E., Grubb, R.L., Gado, M.H., Eichling, J.O. and Ter-Pogossian, M.T. (1976). Arch. Neurol. 8, 523-526.

Risberg, J., Ali, Z., Wilson, E.M., Wills, E.L. and Halsey, J.H., Jr. (1975). Stroke, 6, 142-148.

Risberg, J. (1980). Brain and Language. 9, 9-34.

Rosén, I., Haeger-Aronsen, B., Rehnström, S. and Welinder, H. (1978). Scand. J. Work. Environ. Health, 4, 184-194.

Seppäläinen, A.M. and Härkönen, H. (1976). Scand. J. Work. Environ. Health, 2, 140-147.

Wilkinson, A.D. and Carlen, P.L. (1979). Relationships of neuropsychological test performance to brain morphology in amnesic and non-amnesic chronic alcoholics. Second International Magnus Huss Symposium, Stockholm, Sweden.

Electroencephalographic Spectral Analysis in Chemical and Engineering Workers

R. Gilioli*, P. Genta*, L. Cotroneo* and P. Cannatelli**

*Institute of Occupational Health, Department of Preventive Medicine, University of Milan, Italy
**Department of Occupational Health, Desio Hospital, U.S.L. n. 63, Lombardia, Italy

ABSTRACT

162 chemical plant workers with long-term exposure to trichlorophenol and acute accidental exposure to 2.3.7.8 tetrachlorodibenzodioxine (TCDD) were investigated by means of conventional EEG recording and EEG frequency analysis. The study was part of a larger neurologic, neurophysiologic and psychological survey.
The results were compared with the data of a referent population consisting of 220 subjects with similar epidemiological characteristics but working in plants where there was no known exposure to neurotoxic chemicals.
The EEG frequency analysis was carried out by means of spectral Fourier analysis with the Berg Transform in three phases of 5 minutes each during the regular EEG recording.
The aims of the study were the validation of the method for application to occupational health and neuroepidemiological surveys, as well as the analytical comparison of the two groups for evaluation of the effects of exposure. The results obtained showed that EEG spectral analysis with the Berg Transform is useful in increasing the level of EEG information by enhancing sensitivity and by supplying quantitative parameters to work on for epidemiological surveys.
However, the problems concerning the validity of the method, which can be reduced with adequate study design, have yet to be totally solved.
The results are therefore encouraging but deserve further appraisal. As far as the effect of exposure on the cerebral structures is concerned, while significant differences were found between the two groups (slow activity more prevalent in the exposed), the non-availability of exposure-effect curves excludes any possibility of considering these findings as effects of exposure.

KEYWORDS

Neurotoxicology; neuroepidemiological surveys; neurophysiology; conventional EEG; EEG spectral analysis; CNS effects; health impairment; validity.

INTRODUCTION

Convincing data are reported in the literature on the usefulness and reliability of electroencephalographic (EEG) spectral analysis in anesthesiology, neuropharmacology, in monitoring procedures of the cerebral functions and in experimental toxicology (Bricolo et al, 1978; Weber et al, 1979; Weber, Echter and Couegnas, 1980; Levy et al, 1980; Dvorak et al, 1981).
However, applications of this methodology to occupational health and neuroepidemiological surveys, that is, surveillance of populations exposed to neurotoxic chemicals in the workplace,are lacking. Attempts to use EEG spectral analysis for these purposes failed mainly on account of inherent characteristics of the EEG signal, such as intraindividual and interindividual variability. This problem has not yet been totally solved. However, interindividual variability can be dealt with by normalizing the spectral values of each subject, i.e., by obtaining percent values of each frequency range as referred to the total spectral value in the sampling period. As to intraindividual variability, this limiting factor can be reduced by adopting rigorous standard conditions and by utilizing sufficiently long sampling periods with adequate study design.
Since one of the groups of subjects of our study was exposed to a pesticide and to a chemical of the dioxines group, available literature on the effects of exposure to these substances on the EEG is reported. Only a few studies have been made on EEG findings in acute human intoxication from pesticides. These studies (Hoogendam, 1962; Kazantzis, MacLaughlin and Prior, 1970; Prior, 1972) report a generalized bilaterally synchronous high-voltage theta activity. In acute intoxication from mixtures of organophosphorous and chlorinated pesticides, an increase in theta activity was also reported (Mellerio, 1964). Very little is known about the acute or long-term effects of exposure to polychlorinated compounds.
According to the paper by Jirasek et al (1974), 27% of the subjects exposed to 2.3.7.8 tetrachlorodibenzodioxine (TCDD) showed abnormalities in their EEG recordings.

AIMS AND SUBJECTS

This study was part of a more extensive neurologic, neurophysiologic and psychological survey of the workers involved in the Seveso accident of July 10, 1977.
At the same time, the study is also part of a programme to validate the methodology of spectral analysis in neuroepidemiological investigations. Therefore, the purpose of the investigation was two-fold: 1 - the study of the effects, if any, of long-term exposure to trichlorophenol, a chlorinated pesticide, and of acute exposure to 2.3.7.8 tetrachlorodibenzodioxine on the central nervous system; 2 - appraisal of the sensitivity and usefulness of the method for our needs as compared to conventional EEG.
Two groups of workers were investigated. The first group was composed of 162 subjects, 150 males (92.6%) and 12 females (7.4%) with a mean age of 41 years (S.D. 12.5 y). These subjects had a long-term exposure to trichlorophenol and suffered acute exposure to TCDD on July 10, 1977.
The second group consisted of 220 workers, 194 males (88.2%) and 26 females (11.8%) with a mean age of 40 years (S.D. 9.5 y). These subjects worked in two engineering plants with no known exposure to workplace neurotoxic chemicals and were selected because their epidemiological characteristics were similar to the first group (age, sex, socioeconomic background, living and drinking habits, etc.).
The second group can be therefore regarded as a referent group. The history and the neurologic examination of all subjects, details of which will be re-

ported elsewhere, ruled out the existence of diseases of the central nervous system.

METHODS

The conventional EEG recording was always performed by the same technician, in standard conditions, with an Ote-Biomedica Neurograph 14 - Polygraph. The recordings lasted 20-25 minutes with the patient at rest and eyes closed; hyperventilation was performed for three minutes. Bipolar montages were used and the leads were combined so as to obtain transverse and antero-posterior lines.
The interpretation of the recordings was always carried out by the same examiner who was not aware which of the two groups each EEG belonged to.
For the spectral analysis, the algorithm of the Berg Transform was chosen according to Sciarretta and Erculiani (1975) and Sciarretta and Erculiani (1979), since this permits better analysis of the EEG curve in the domain of frequencies with respect to the Fast Fourier Transform (Dumermuth and Keller, 1973; Dumermuth, 1977; Matousek, 1973; Gevins, 1980; Lopes da Silva, 1982). In fact, the Berg Transform can eliminate the discrimination in favor of the slow frequencies typical of the Fast Fourier Transform.
In addition, information on the position of a phenomenon within the analyzed epoch, also with typically differentiated spectral content, can be obtained with the Berg Transform. A two-channel Ote-Biomedica Berg-Fourier analyzer was utilized, one channel for each hemisphere with fronto-parietal leads, F_3-P_3 for the left hemisphere and F_4-P_4 for the right hemisphere.
The analysis was carried out in three phases: the first and the third one with subjects at rest and eyes closed while in the second phase the subjects were at rest and eyes closed but hyperventilating. Each phase lasted five minutes for a total of 15 minutes sampling. At five minute intervals a quantitating unit gave the total power of the sampled spectra with the power in single preselected frequency ranges and the number of artifacts occurring during the sampling period.
As compared to conventional EEG recording, there was an on-line sampling, presentation of the spectra and computing per frequency ranges. In agreement with Matousek (1973) and Bickford (1977), the single frequency ranges were preselected in the intervals indicated in table 1. Each value supplied by the quantitating unit was then transformed into a value of absolute power, with account taken of the calibration parameters (Sciarretta and Erculiani, 1977; Sciarretta and Erculiani, 1978).

Table 1. Selected intervals for each frequency range.

Range	Symbol	Interval
Delta	(δ)	0.5 ÷ 4 Hz
Theta	(ϑ)	4 ÷ 7.5 Hz
Alpha 1	(α_1)	7.5 ÷ 9.5 Hz
Alpha 2	(α_2)	9.5 ÷ 12.5 Hz
Alpha tot.	(α_t)	8 ÷ 13 Hz
Beta	(β)	13 ÷ 16 Hz

In addition, since we were aware of the great interindividual variability of the absolute powers, each single value was normalized and percent values were obtained referred to the total power of the subject in the sampling interval.

RESULTS

By means of conventional EEG recording, 115 tracings (70.99%) in the first group and 185 (84.1%) in the second group were classified as normal, while 35 tracings (15.9%) in the second group and 47 (29.01%) in the first group showed abnormalities (table 2). With X_2 test analysis, a significant difference was seen at the 0.01 level between the two groups with abnormal tracings.

Table 2. Classification by means of conventional EEG recording of all the EEG tracings in the two groups.

	FIRST GROUP		SECOND GROUP	
	N.	%	N.	%
EEG tracings with abnormalities	47	29.01%	35	15.9%
EEG tracings without abnormalities	115	70.99%	185	84.1%
Total	162	100.	220	100.

A further evaluation of the type of EEG abnormalities led to the following classification (fig. 1): - 64% of the subjects in the first group and 14% of the subjects in the second group showed a slowing of the background activity due to increase in theta rhythms; - focal abnormalities were detected in 26% of the subjects in the second group and in 6% of the first group; - non specific paroxysmal activity was observed in 26% of the second group and in 17% of the first group; - spike and wave activity was detected in 34% of the subjects in the second group and in 13% of the first group. X_2 analysis showed a significant difference at the 0.001 level between the two groups with regard to the type of abnormality. In fact, there is a marked tendency to slow background activity in the first group, while paroxysmal and spike and wave abnormalities are definitely more prevalent in the second group.

An evaluation of the total power of the spectrograms in the two groups (fig. 2) shows an increase in the spectral energy during hyperventilation, in agreement with the well-known finding with conventional EEG.

The increase is more evident in the first group and the spectral energy returns to baseline values at the end of hyperventilation. Since no significant difference was found between the spectral values of the first and last phases of sampling, the analysis of single frequency ranges was carried out only for the first and second sampling intervals.

The tracings classified as normal in the second group showed spectral indices with prevalent power in the alpha frequency range: from 80% to 90% of the total energy (table 3).

The values reported in table 3 and the following represent the mean percent values of the power in each frequency range in the subjects of the group (the standard deviation values are given in brackets).

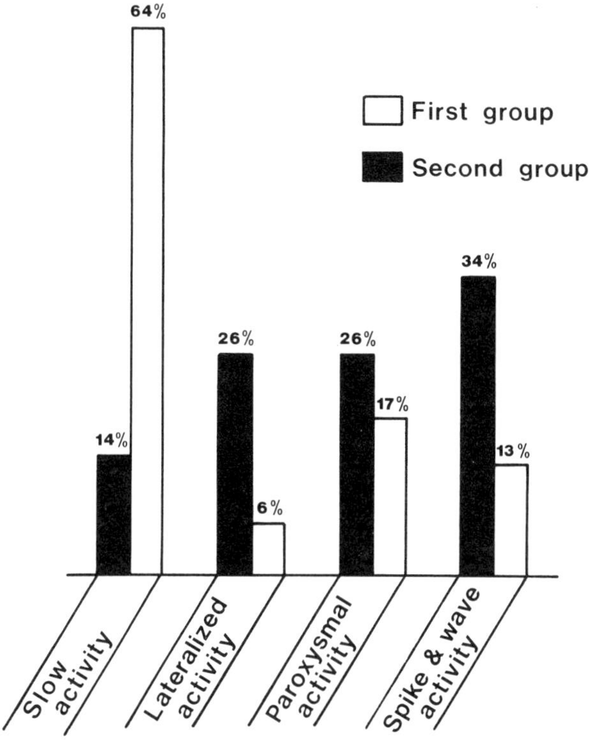

Fig. 1. Classification of the conventional EEG recordings according to the type of abnormality.

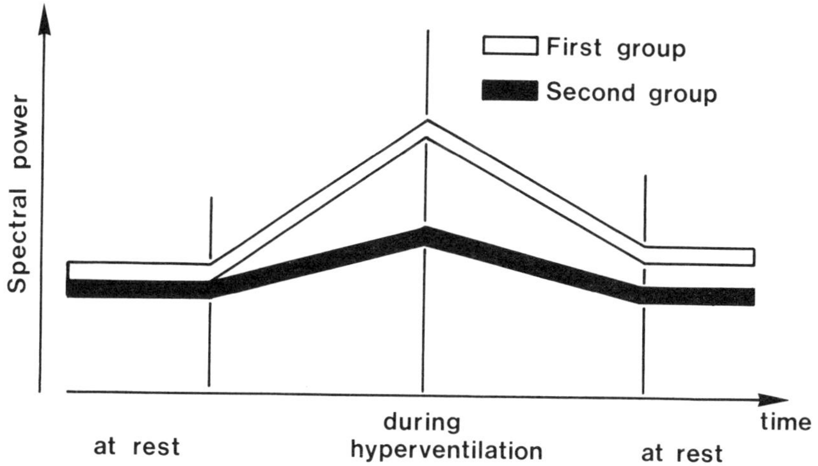

Fig. 2. Spectral power of the two groups at rest and during hyperventilation for all three phases.

Table 3. Spectral power of subjects of the second group with normal EEG tracings.

	AT REST left	AT REST right	DURING HYPERVENTILATION left	DURING HYPERVENTILATION right
Delta	5.55 (3.91)	5.49 (4.02)	5.82 (4.11)	6.02 (4.12)
Theta	7.21 (4.59)	7.62 (3.12)	7.23 (4.78)	7.11 (4.06)
Alpha 1	42.33 (13.66)	46.70 (12.53)	45.00 (12.69)	49.22 (14.55)
Alpha 2	32.64 (10.11)	37.84 (12.82)	37.21 (14.76)	41.90 (16.21)
Alpha total	81.43 (18.03)	88.07 (16.47)	86.39 (12.46)	91.18 (19.45)
Beta	4.72 (3.08)	4.91 (3.66)	5.35 (2.39)	5.84 (2.22)

Analysis of these data shows that the percentage of slow rhythms (theta, delta) is low and that there is a difference of percent power in the alpha frequency range between the two hemispheres with more power in the right hemisphere.

Since left-handedness in our subjects was not frequent, spectral analysis confirmed the evaluation by conventional EEG that the alpha rhythm in right-handed individuals is more abundant in the right hemisphere. An increase in alpha power was detected during hyperventilation.

The tracings of the second group with abnormalities yielded the following spectral values (table 4):

- a mild increase in slow frequency power
- an increase in power in fast alpha frequencies (alpha 2)
- an increase in power in the beta range.

Table 4. Spectral power of subjects of the second group with abnormal EEG tracings.

	AT REST left	AT REST right	DURING HYPERVENTILATION left	DURING HYPERVENTILATION right
Delta	10.07 (4.05)	9.40 (4.82)	8.57 (4.69)	8.00 (4.29)
Theta	9.03 (4.85)	8.77 (4.72)	8.97 (4.69)	8.60 (4.72)
Alpha 1	21.03 (8.28)	21.37 (10.37)	19.30 (12.52)	20.03 (13.75)
Alpha 2	38.87 (12.50)	40.17 (15.86)	41.93 (15.26)	43.00 (16.21)
Alpha total	65.97 (14.31)	66.93 (14.39)	66.80 (13.80)	68.18 (13.04)
Beta	9.13 (5.42)	8.87 (5.09)	14.27 (8.76)	13.95 (9.00)

With regard to the increase in alpha 2 and beta power, it is worth recalling that the spikes have a spectrum between 13 and 40 Hz; therefore, the higher prevalence of spikes observed with conventional EEG in the second group may at least partially account for the increase in alpha 2, but particularly in beta spectral power.

Hyperventilation increased only the power of beta frequencies and this may also be due to the enhancement of spike activity brought about by hyperventilation.

In the first group with abnormal tracings, the following results were obtained (table 5):

- marked increase in spectral power in the lower frequencies
- further increment of lower frequencies power during hyperventilation.

Table 5. Spectral power of subjects of the first group with abnormal EEG tracings

	AT REST		DURING HYPERVENTILATION	
	left	right	left	right
Delta	17.65 (6.50)	12.49 (6.17)	16.29 (9.49)	13.92 (8.41)
Theta	14.90 (7.46)	13.35 (4.60)	19.28 (4.29)	18.21 (5.23)
Alpha 1	22.05 (7.56)	22.15 (7.93)	23.00 (15.07)	24.87 (15.60)
Alpha 2	34.46 (8.37)	34.95 (8.15)	39.59 (14.93)	38.92 (16.25)
Alpha total	59.36 (17.90)	60.13 (17.43)	61.90 (17.18)	61.44 (14.48)
Beta	9.92 (6.41)	9.92 (6.15)	10.87 (7.26)	10.82 (7.34)

Table 6 reports the levels of statistical significance of the differences found in the spectral powers of the abnormal EEG tracings of the two groups.

Table 6. Statistical differences in spectral power between the first and second groups with abnormal EEG tracings.

	AT REST		DURING HYPERVENTILATION	
	left	right	left	right
Delta	.001	.001	.001	.001
Tetha	.001	.001	.001	.001
Alpha 1	.01	.02	.001	.001
Alpha 2	.001	.001	.01	.001
Alpha total	.001	.001	.001	.001
Beta	n.s.	n.s.	.01	.01

The high significance both at rest and during hyperventilation of the relative powers in the lower frequencies confirms the non-parametric evaluation carried out with conventional EEG.
The spectral data of the EEG tracings classified as normal with conventional reading in the first group (table 7) revealed an increase in relative power especially in the theta frequency range. This finding is statistically significant (table 8) and shows that in the first group even those subjects who were classified as normal with conventional EEG do in fact show a tendency to a greater expression of slow rhythms with respect to the second group.
Hyperventilation enhances the level of statistical significance.

Table 7. Spectral power of subjects of the first group with normal EEG tracings.

	AT REST		DURING HYPERVENTILATION	
	left	right	left	right
Delta	6.64 (3.48)	6.55 (3.64)	5.89 (3.17)	6.23 (3.09)
Theta	9.19 (4.15)	9.17 (4.25)	10.21 (4.86)	11.09 (4.32)
Alpha 1	40.27 (11.11)	49.53 (14.18)	41.18 (12.93)	48.11 (13.72)
Alpha 2	43.19 (10.17)	51.20 (12.13)	47.91 (10.03)	54.63 (12.84)
Alpha total	80.53 (13.21)	81.56 (17.53)	84.07 (12.95)	90.72 (15.55)
Beta	3.87 (2.27)	3.96 (2.59)	3.12 (2.56)	3.47 (2.69)

Table 8. Statistical differences in spectral power between the first and second groups (normal EEG tracings).

	AT REST		DURING HYPERVENTILATION	
	left	right	left	right
Delta	n.s.	n.s.	n.s.	n.s.
Theta	.05	.05	.01	.01
Alpha 1	.01	.01	.05	.05
Alpha 2	.01	.01	.001	.001
Alpha total	.01	.001	.01	.01
Beta	n.s.	n.s.	.05	.05

DISCUSSION AND CONCLUSIONS

The results obtained will be briefly analyzed in the light of the two main objectives of the study, i.e.:

- the aptitude of the data to validate the method of EEG spectral analysis in neuroepidemiological surveys
- the significance of the results in terms of effect on the central nervous system of the occupational and accidental exposures that occurred.

With regard to the first issue, the results obtained indicate that spectral analysis met our expectations as to the ability of increasing the resolving power of EEG interpretation. In fact, besides confirming and quantitating the data obtained with conventional EEG, the results show that subtle shifts of the EEG to low frequency ranges were detected with spectral analysis even when the conventionally interpreted EEG was normal. It can therefore be concluded that this automatic method of signal analysis can detect variations of the frequency ranges otherwise undetectable by the human observer, thus greatly enhancing the sensitivity of EEG information. However, before this method can be proposed for health surveillance of workers at high risk of neurotoxic chemicals, the following critical issues need further appraisal and validation:

1 - the significance in terms of health impairment of these spectral shiftings
2 - the reproducibility of the spectral indices under diverse conditions (age, sex, genetic background, circadian rhythms, fatigue, stress, etc.)
3 - the correlations of these spectral indices with other neurophysiological parameters, such as brain evoked potentials, with psychometric testing and with actual exposure (length of exposure and indicators of dose and/or effect).

As regards point 1, while the excess of theta or the presence of delta activity conventionally determined have clear correlations with brain or brain stem dysfunctions, the meaning of these subtle automatically detected variations in terms of health impairment has yet to be clearly established. However, we believe it can be anticipated that the decrease in alpha and beta activities along with decrease in theta frequencies, as observed in the exposed group, suggest a decrement of physiologic mechanisms underlying the processes of activation of brain functions, such as vigilance. In any case, the observed variations may point to a reduction in the safety range of these functions.
This interpretation is at least in part supported by the findings of the psychometric study; in fact a significant increase ($p<0.05$) was observed in the 30-40 y age group of the exposed subjects for simple visual reaction times and a significantly higher prevalence ($p<0.01$) of visual reaction time values exceeding the mean of the reference population by 1 S.D. was also found.
With regard to the second objective of the study, it must be pointed out that the results of the conventional EEG in the first group, that is, in the exposed subjects, are in agreement with the findings of other investigators (Hoogendam, 1962; Kazantzis, Mac Laughlin and Prior, 1970; Prior, 1972; Mellerio, 1974) who also found increases in theta activity but mainly in acute exposure to pesticides.
In the second group, the referents, EEG paroxysmal activity (paroxysmal + spike + wave) was found definitely more prevalent (fig. 1) with respect to the exposed. Since paroxysmal activity represents mainly a common model of response of the brain structures to insults such as perinatal damage, infectious diseases, brain tumors and head injuries, the finding of unequal

distribution in the two groups (60% vs30%) is difficult to interpret on the basis of the history and neurologic examination which ruled out neurologic defects.

With regard to the results of spectral analysis, the increment in the theta frequencies and the parallel decrese in alpha and beta activities observed in the exposed subjects has already been commented on.

These differences may suggest an effect of exposure to trichlorophenol and/ or TCDD on the brain structures. The analogy with the quoted data of other investigators seems to be in favor of an acute effect of TCDD. However, in our opinion, the data relative to differences between the two groups without demonstrable correlations with exposure and other neurophysiological parameters are insufficient to draw any such conclusion.

To sum up, our study has demonstrated the usefulness of spectral analysis in enhancing the resolving power of the EEG, thus increasing the EEG information; another important advantage is represented by precise quantitative parameters to work on for epidemiological evaluations.

However, before spectral analysis can be proposed for health surveys in subjects with high risk from neurotoxic workplace chemicals, further appraisal must be made as regards the reproducibility of the method under diverse conditions and exposure effect relationships must be established. While the differences found between the two groups studied are suggestive of an effect of exposure on cerebral functions, the non-availability of exposure-effect curves does not permit such conclusions to be drawn.

REFERENCES

Bickford, R.G. (1977). In: EEG Informatics. A didactic review of methods and applications of EEG data processing. Edit by A. Rémond, Elsevier/ NHBP.

Bricolo, A., S. Turazzi, F. Faccioli, F. Odorizzi, G. Sciarretta and P. Erculiani (1978). Electroencephal. Clin. Neurophysiol., 45, 211-225.

Dolce, G. and K. Kunkel (1975). CEAN Computerized EEG Analysis. Fisher.

Dumermuth, G. and E. Keller (1973). In: Automation of clinical EEG. Ed. by P. Kellaway and I. Petersén, Raven Press, New York.

Dumermuth, G. (1977). In: EEG Informatics. Ed. by A. Rémond, Elsevier.

Dvoràk, J., Formanek J., J. Kubàt, Jarmila Plevova, Milena Vanïckovà, M. Fires, J. Andel, T. Cipra, L. Tomasek, Zuzana Pràskovà, Eva Holoubkovà and Z. Fabiàn (1981). Activ. nerv. Sup. (Praha), 23, no. 2.

Gevins, A.S. (1980). In: Electrodiagnosis in clinical neurology. Ed. by M.J. Aminoff, Churchill Livingstone.

Holmstedt, B. (1980). Prolegomena to Seveso. Arch. of Toxicol., 44, 211-230.

Hoogendam, I. (1962) Arch. Environ. Health, 4, 86-94.

Jirasek, L., J. Kalensky, K. Kubec, J. Padzerova and E. Lukas (1974). III C.S. Derm., 49, 145-157.

Kazantzis, G., Mac Laughlin A.I.G. and P.F. Prior (1970). Brit. Journ. Ind. Med., 21, 1, 46-51.

Levy, W.J., H.M. Shapiro, G. Maruchak and E. Meathe (1980).Anesthesiology, 52, 223-236.

Lopes da Silva, F. (1982). In: Electroencephalography. Basic principles, clinical applications and related fields. Ed. by E. Niedermeyer and F. Lopes da Silva, Urban & Schwarzenberg.

Matousek, M. (1973). Handbook of enectroencephalography and clinical neurophysiology. Vol. 5,part A, Elsevier.

Mellerio, F. (1964). L'électroencéphalopathie dans le intoxications aiguës. Masson.

Prior, P.F. (1972). Electroencephal. clin. Neurophysiol., 15, 5, 915.

Sciarretta, G. and P. Erculiani (1975). Proceedings of the 2nd Symposium of the study group for EEG methodology.

Sciarretta, G. and P. Erculiani (1977). Introduction to EEG spectral analysis. Ote-Biomedica, Firenze.
Sciarretta, G. and P. Erculiani (1978). Electroencephal. clin. Neurophysiol..
Weber, B., M. Leloup, C. Baudouin and J. Couegnas (1979). Agressologie, 20, 3, 183-192.
Weber, B., E. Echter and J. Couegnas (1980). Agressologie, 21, C: 143-151.

Comprehensive Approach to the Study of Exposed Populations

Chairmen: L. Ambrosi (*Italy*) (unable to attend)
A. Grieco (*Italy*)

Experiences with Recently Developed Techniques

Chairmen: C. L. Cazzullo (*Italy*)
G. Chiappino (*Italy*)

Opening Speech

A. Grieco

Institute of Occupational Health, Department of Preventive Medicine, University of Milan, via S. Barnaba 8, 20122 Milan, Italy

The first two days of this International Symposium have been occupied by sessions of intense and profitable work held in the splendid surroundings of Villa Olmo in Como.

The papers presented by the best experts in the field, from six different countries, besides Italy of course, all showed a level of scientific rigour which I consider to be much more than satisfactory, with the use also of extremely clear and sometimes very elegant illustrative material.

The two days were industriously occupied, thanks also to the good work of the session chairmen, by papers illustrating the principles and basic methods developed by electrophysiology, neurophysiology, biochemistry, radiology, neurology, psychology and industrial toxicology, for the functional and clinical investigation of the effects due to occupational exposure to toxic agents.

We thus learnt what the contributions and limits are of clinical and neurological examination of workers, of electromyography, electrooculography, electroneurography, electroencephalography with automatic spectral analysis, evoked brain potentials, brain hematic flows, computerized axial tomography, enzymatic tests, tremometry, and of psychodiagnostic instruments that are the basis of an extremely ambitious research project.

The result was a very rich and sophisticated high standard, confirming that the possibilities of automatically recording and analyzing data have enormously increased in this field too in the last twenty years, so that we can at last test hypotheses that are in many cases older then ourselves.

The results of application of these methods in real conditions, either alone or combined will be the topic of the Second Part of the Symposium here in Milan University.

The significance and pathological definition of measurable alterations when they are within the extreme limits of normal variability or beyond, often still appear to lack a complete explanation. However, this is not the fundamental question because in the field of preventive medicine and ergono mics we are above all interested in knowing whether or not these alterations, whatever their pathological definition, are a negative effect caused by something present in the workplace in general.

On the other hand, the fact that these signs are often aspecific, the simultaneous presence in the workplace of a large number of abnormal physical and psychological stimuli that cannot easily be differentiated, and the difficulty in obtaining suitable control groups that will furnish all the possible variables, lead to not unreasonable doubts about the real possibilities of this research model succeeding in identifying one and one only cause of those effects in the working environment, especially in cases where the levels of the individual agents are not high.

Perhaps the time has come for the ever increasing number of experts involved in the field of preventive medicine and occupational health to once more query the adequateness of the monofactorial approach inherited from internal medicine, which was in turn inherited from infectious diseases, in order to solve problems which are sometimes more complicated in the working environment than in the general environment.

In the previous two days in Como all participants were fully aware of these problems, which was also clear from the conclusions of Dr. Xintaras and Dr. Suma'mur, who stressed the need to further investigate the influence of many quite important confounding factors in the working environment, one of the most important being the stress syndrome.

But it was also clear that the strongest motivation for the attempt by researchers to clearly define the cause of any alterations in biological homeostasis is the conviction that the more precise the identification of the etiology, the more convincing and practicable will be the preventive measures.

We are encouraged in this direction by the high rank of the sponsors of this Symposium, including the World Health Organization, the International Labour Office, the Commission of the European Communities, the Permanent Commission and International Association on Occupational Health, the University of Milan, and the Italian Society of Occupational Health and Industrial Hygiene, whose President, Prof. Casula, has requested me to convey to you his greetings and good wishes.

May I also extend particular thanks to the members of the Executive Committee of the Symposium, Dr. Gilioli, General Secretary, Dr. Cassitto, and Prof. Foà, for the dedicated and friendly way they have carried out their organizing tasks.

May I now declare the Second Part of the Symposium open and invite Prof. Foà to give his introductory paper, after which he will chair the morning session.

Epidemiological Issues in the Neurobehavioral Study of Occupationally Exposed Groups

P. A. Bertazzi*, G. Assennato** and S. Guercilena*

*Institute of Occupational Health "Clinica del Lavoro L. Devoto", University of Milan, Milan, Italy

**Institute of Occupational Health, University of Bari, Bari, Italy

ABSTRACT

In general, the study of an occupationally exposed population has a three-fold aim: - to ascertain which effects are associated with the exposure; to set up methods and means for the prevention of the effects; to verify the efficiency of the control measures taken. What makes such an approach "comprehensive"? A comprehensive study should ideally take into account all the factors involved in the exposure-response relationship under study. These factors relate mainly to the definition of exposure and effect and to the selection of the index and referent population. Unlike the experimental setting, the real world of a factory makes it difficult to clearly meet many of the criteria which characterize a valid and comprehensive study. Some of these difficulties are covered and discussed with reference to the specific area of neurobehavioral "risks". First, which type of study is better suited when we are dealing with multiple exposures and different outcomes? The time aspects of the exposure - outcome relationship must be carefully considered to design a proper study. In defining the exposure, different sources and methods are available: their relative merits and drawbacks are briefly discussed. The population of interest is very seldom the one currently at work: many selection factors are working, too. The choice of a control group raises problems,especially if the control subjects are not active workers. The validity of the measurement instruments employed should also be stated. Many additional factors are of interest when studying an exposure-response relationship: they may act as confounders or as modifiers of the effect, and the difference should be appreciated. A comprehensive study ought not to be primarily "significant", but "meaningful".
On the other hand a causal inference cannot only rest on statistics, but the nature and the characteristics of the exposure - outcome association should be thoroughly examined.

KEYWORDS

Epidemiologic methods; study design; occupational exposure; confounding; effect modification; bias.

A comprehensive approach to the study of an exposed population should aim at the following objectives: ascertain the effects associated with the exposure; set up methods and means to prevent these effects; test the efficiency of the control measures taken. The focal point of this approach, from an epidemiological point of view, is the study of the nature and characteristics of the exposure-effect relationship. This is often difficult in occupational settings because we are faced with existing situations that cannot be determined beforehand or manipulated as in an experimental context. So many problems arise in the stage of study design collecting information, analysis, inference.

WHICH TYPE OF STUDY?

One of the major problems derives from the fact that in an occupational setting exposures are nearly always multiple and combined and their effects numerous and different.

A starting point

It is not possible in practice to study many exposures and many effects at the same time. A choice has to be made. Figure 1 shows, schematically, two possible choices. If the study starts with a clearly defined exposure, then a number of effects can be examined. If, on the other hand, the study starts from a single effect already present, then it is possible to study its association with a number of exposures. The first model corresponds with a follow-up or cohort type investigation; the second one with a case-control or case-referent type study.

Fig. 1. It is in practice impossible to simultaneously study the multiple relationships between different exposures (E) and different outcomes(O). A clear starting point has to be chosen.

Have a hypothesis about the exposure-effect relationship

To select the most suitable type of study, the basic characteristics of the exposure-effect relationship under study must be known. If the problem is being approached for the first time, a model of the relationship must in any case be hypothesized.

First, the time interval between exposure and effect must be considered. Thus, the acute effects of a neurotoxic agent appearing at the end of a shift which may be rapidly reversible (e.g., anaesthetic changes) may be of interest; or the irreversible chronic effects as a result of long or repeated exposures (e.g., brain damage, peripheral changes). If these time relationships are not known, at least by way of an indication, a study might be carried out at the wrong time or it might not last long enough to cover the relationship under study. Many situations may arise in practice and some examples are given in Table 1. In the first example neither the exposure nor the effect can be measured at the time of the study: but something

Table 1. Possible relationships between the detectability of exposures and effects of interest and the actual time of the study

Measurability of exposure	Measurability of effect
NO	NO
YES	NOT YET
YES	YES
YES	NO MORE
NO MORE	YES
NO MORE	NOT YET

could have been wrong in the assumed relationship, in the population selected, in the time of the investigation or in the analytical methods used. This is different from stating that the study was "negative". In the second example the exposure is present and can be measured but the effects are not yet detectable. A cross-sectional study would be falsely negative: a longitudinal approach with follow-up of the population over time is essential here. In the third example a cross-sectional study is feasible, but two problems exist: first of all, the type and degree of the exposure and the nature of the effects measurable at that moment may not correspond to reality; secondly, the population present at that moment could be heavily selected. In the following example the cross-sectional study would again be falsely negative: a historical-longitudinal study of the workers' experience must be made, otherwise only "survivors" will be examined. When the effects are present but the exposure is not measurable at that moment (fifth example) a case-control (case-referent) investigation must be made and this is feasible when at least qualitative data on past exposure are recoverable from some source. In the last case the exposure occurred in the past and is not known at present, while the effects are not yet detectable. Here it is necessary to collect all existing information on past exposure and from there start a follow-up of the exposed subjects which continues prospectively.

A good example of this problem is the study which initiated systematic research on behavioral effects in occupational health (Hänninen, 1971). Since it was a cross-sectional study, it could not exactly assess whether the behavioral changes observed had appeared after the beginning of exposure or whether they were already present beforehand; nor was it possible to assess the clinical significance of the changes (reversible or irreversible) without a follow-up. But this study also tells us that a well conducted cross-sectional study can produce hypotheses that further studies will verify. For example, follow-up studies that were able to assess the temporal aspects of the exposure/effect relationship and the nature of the adverse effects have been made for mercury (Chaffin and Miller, 1974) and lead (Araki et al, 1980). There are also examples of case-referent studies to investigate the long-term effects of exposure to solvents (Axelson et al, 1976).

The toxicological basis of the exposure-effect relationship must also be known or hypothesized in advance. For instance, repeated or long-term exposures to lipophilic compounds (e.g., solvents and anaesthetics) may

result in a functional *tolerance*: in this case neurobehavioral tests could yield falsely negative results since, from an epidemiological standpoint, the "tolerance effect" will mask the actual effect under study.
As another example, lipophilic substances may act only after metabolic activation within the CNS (e.g., carbon disulphide and n-hexane) (Savoläinen, 1982): this may account for a wide variability in the response of exposed subjects (individual *susceptibility*). There are substances, e.g., lead, for which *biological interactions* with other substances play a central role in the mechanism of neurotoxicity (Silbergeld, 1982).
Finally the effects of interest must be clearly anticipated since: 1- many behavioral effects are subtle in nature; 2- different effects are associated with different exposure levels and they can mask each other (Silbergeld and Goldeberg, 1980).

DEFINE EXPOSURE

The exposure characteristics of interest are: type, level and duration.

Ideally, exposure should be measured for each subject with an indicator of dose. In practice, it is very difficult, especially in the neurobehavioral field, to have reliable indicators of the concentration of the substance in the critical organ. In general, moreover, indicators of dose only exist for a limited number of substances. For instance, lead can accumulate in the CNS without any noticeable variation in the blood lead concentration when the exposure is intermittent (Hridna et al, 1980).

A further problem is that to be informative the dose should be estimated on the basis of repeated measurements, since it is unlikely for the dose to stay constant. Which value should, then, be used? The maximum, the one closest to the investigation, a cumulative dose, or a time-weighted dose? Each choice implies a different model of interpretation. In the first case, it is assumed that the effect is associated with certain peak levels above a given threshold, and not with lower exposures, even if repeated. In the second case, it is assumed that the effect is associated only with recent exposure. In the third case, it is assumed that the effect is caused by long-term or repeated doses. In the fourth case, exposure experienced in particular periods of time is relatively more important.

Thus, dose can in general only be approximately estimated using the information sources listed in Table 2. The first type may be useful when the exposure under study is a raw material or a finished product. The second source is limited by the fact that it does not take into account inter-individual variability; also, measurements rarely exist for the past and they give no information about duration; they can, however, be used to describe the relative exposures of different occupational categories such as printers vs. foundry workers, etc. The limits of biological monitoring have been mentioned above; its advantages are that all sources and routes of exposure and interpersonal variability are considered.
Questionnaires and interviews can provide information about type and duration of the exposure, but only in a few cases will they supply semiquantitative information on the level of exposure; here, moreover, the problem of variability and selection in recall among different subjects arises. Information on duration and type of exposure may be drawn from company records when the exposure under study is defined by a given job or department in the plant; duration of exposure can help in differentiating individuals within the same occupational category.

Table 2. Sources of information on the exposure characteristics usually available in occupational health investigations

Source of information	EXPOSURE CHARACTERISTCS		
	Type	Level	Duration
Production type and processes	X		X
Environmental measurements	X	X	
Biological monitoring	X	X	
Questionnaires Interviews	X		X
Plant personnel records	X		X

POPULATION UNDER STUDY

In principle, it is fairly easy to identify the *exposed population*: however, the difference between "population at work" and "population exposed" should be kept well in mind. The population at work can be selected as compared to the exposed population as far as *time* and *place* are concerned: workers who were exposed in the past could have left the factory; workers may have moved to other departments for various reasons, one of which may be the appearance of the effects under study.

Any conclusion concerning the existence or otherwise of an association between exposure and health impairment is based on a comparison between the situation under study and a reference situation. The comparison can be made in three ways: in a group over time (e.g., before and after exposure); between sub-groups of the same population classified according to different degree and/or duration of exposure; between different groups that are comparable except for the variable of interest (e.g., exposure). The choice of a suitable *control group* is not as easy as the choice of the exposed subjects. It is advisable to select the controls from active industrial populations, preferably from within the same factory (if the exposure under study is limited to one single area), or from homogeneous industrial settings. One should be very wary in using volunteers as referents since they may turn out to be over-selected. Similar caution is required in using theoretical "normal" values, especially if they are derived by interpolation or, even more so, by extrapolation from relationships between exposure and effect observed in populations whose characteristics have not been well defined.

The size of the groups in an occupational health neurobehavioral study often has two features: it is determined beforehand and is not large. However, the size of the group under study influences the power of the study, that is, its capability of rejecting the tested hypothesis (Ho)

when it is truly false in favour of an alternative hypothesis.

Using data from studies on the neurotoxic effects of lead (Repko et al, 1978), Table 3 illustrates possible differences in mean values of median nerve MCV in exposed and non-exposed subjects; for simplicity, the standard deviation is 9 M/s in both groups; the probability "alfa" to reject the tested hypothesis when it is true is 0.05; the probability "beta" of accepting Ho when it is false is 0.10. Clearly, the sample size required for the exposed and non-exposed group varies greatly according to the value of the means difference to be tested.

Table 3. Example of sample size requirements to test differences between mean MCV in lead-exposed and non-exposed subjects (sample size calculated according to Snedecor and Cochran, 1976)

Means difference (M/sec)	Standard deviation	alfa	beta	Sample size required for each group
2	9	.05	.10	437
4	9	.05	.10	87
8	9	.05	.10	22
10	9	.05	.10	14

In some behavioral tests, sample size and study power assessment are influenced by the greater variability of the measures. Using data from a study on the effects of lead on the hand-eye coordination test (Repko et al, 1978), we obtained the results shown in Table 4.

Table 4. Example of sample size requirements to test differences between means scores of a performance test in lead-exposed and non-exposed subjects (sample size calculated according to Snedecor and Cochran, 1976)

Means difference (M/sec)	Standard deviation	alfa	beta	Sample size required for each group
4	84	.05	.10	7556
8	84	.05	.10	1880
16	84	.05	.10	472
32	84	.05	.10	118

ROLE OF OTHER FACTORS

The relationship between exposure and effect can be influenced by "extraneous" variables associated with the disease and/or exposure. They may act in two ways: confounding and effect modification. It is important that the difference be appreciated. Confounding means that the association is *distorted* by the action of other factors. Effect modification means that the association under study is *modified* when other factors are taken into account.

Confounding

A variable can be a confounder when it is a known or biologically possible risk factor of the effect (*a priori* characteristic) and it is in some way associated with exposure (characteristic in the data). Table 5 gives the hypothetical results of a neurophysiological study on exposure to solvents. Is age a confounder?

Table 5. Example for confounding. The stratum specific estimates do not differ noticeably, while the total crude estimate differs from the adjusted one (SRR) (fictitious data)

Exposure to solvents		AGE 20-30	AGE 30-40	Total
YES	MCV changes	4	56	60
	Persons	12	70	82
NO	MCV changes	16	12	28
	Persons	80	18	98
	RR	1.65	1.20	2.61
	SRR			1.35

We know *a priori* that age is a risk factor for decreasing motor conduction velocity. It can be seen that in the two age strata the ratio between frequency of alterations among exposed and non-exposed subjects (RR) do not vary ostensibly, while they differ from the crude RR. If we control for the influence of age (Standardized RR) an estimate similar to those in the strata but different from the crude one is obtained: this means that age is a confounding factor and therefore that the exposure-effect association is distorted in the crude estimate. The distortion is eliminated by controlling for the influence of age. Various methods exist such as restriction (restrict the study to subjects of comparable age), stratification (as in the example), or more complicated statistical procedures.

Effect modification

Let us look at the data in Table 6 where subjects are stratified according to alcohol consumption. There is an important difference compared to the previous table, which is the non-uniformity of the RR over the strata. When both solvent exposure and alcohol consumption are present RR is three times greater than with solvent exposure only.
This means that alcohol modifies the effect of exposure to solvents, in this case enhancing it. This is extremely important information which helps us to better understand the exposure-response relationship.

Table 6. Example of effect modification. The stratum specific estimates vary taking alcohol consumption into account (fictitious data)

Exposure to solvents		ALCOHOL Yes	No	Total
YES	MCV changes	43	8	51
	Persons	50	50	100
NO	MCV changes	13	6	19
	Persons	50	50	100
	RR	3.31	1.33	2.68

In a study, a variable may be a confounder, a modifier, both, or neither. It is advisable to give priority to effect modification: thus, if the exposure-effect association varies for different values of a third variable, this phenomenon should not be eliminated with the methods used for a confounder, but it should be described and evaluated. The confounder "confounds" the relationship under study, whereas the modifier reveals important features of it. It should be remembered that comparability *a priori* between index and referent series is only required for the variables that behave like confounders. For variables which modify the relationship, information on them must be collected and their influence assessed.

INFERENCE

The association between exposure and effect may be explained by three factors: chance, bias or causality and each of them must be taken into account and carefully discussed.

Chance

When several samples are drawn from the same population, different results (e.g., mean values) are usually obtained. Therefore, differences in mean values between, say, exposed and non-exposed people may indicate either that the two groups have been drawn from two different populations or that they are different only by chance: casually different samples from the same population. The null hypothesis (Ho = the groups are drawn from the same population) is tested and the result is usually presented by the "p value", i.e., the probability of obtaining a result like that of the study or a more extreme one, given that Ho is true. It should be emphasized that if something is statistically significant it does not necessary follow that it is important or even meaningful. If a parameter with low variability is measured it will be quite easy to obtain significant differences between group means, but these differences may not be of any biological relevance. Thus, for example, a number of studies have shown differences in motor conduction velocity between lead-exposed and non-exposed subjects, but Buchtal (1980) has shown that these differences do not reflect any particular neurotoxic mechanisms (axonal degeneration or segmental demyelination) but are only the results of slight and reversible changes in the properties of the nodal membranes. Contrarywise, differences between mean values of parameters with high variability could be biologically important even though statistically not significant.

Statistical significance is also considerably influenced by sample size. A large sample is very sensitive to small deviations from Ho, whereas a small sample is sensitive only to larger deviations. As we increase the sample size, biologically unimportant differences might turn out to be statistically significant.

Bias

A systematic error can be the actual explanation of the presence or absence of an exposure-effect association.
A *selection* bias occurs whenever the probability of entering the study depends upon the factor being studied (diseases or exposure).
For instance, the exposure may be selectively present in the cases or controls only because of the selection criteria being used (e.g., cases collected from Occupational Health Clinics). Examples of selection bias in cross-sectional studies have been discussed above.

An *information* bias is particularly common when questionnaires or interview schedules are used and is due to the different motivation for cooperating of the exposed (or cases) as compared to the non-exposed (or non-cases). Another possible source of this bias is the interviewer's awareness of which individual belongs to which group. Information bias can also arise when analytical of functional procedures are not uniformly implemented in the groups.

If *confounding* factors are present they are likely to distort the association under study as discussed above. This bias can be controlled for in the data analysis stage, while the previous two types should be taken into account in the study design.

Causality

When chance can be ruled out as an explanation of the results and the investigation has been carried out in an unbiased manner, consideration can be given to causality as an explanation of the observed association between exposure and effect.

One of the strongest points in favor of causality is the occurrence of a dose-response relationship. However, in neurobehavioral toxicology it is often difficult to measure both the dose and the effect indicative of response, as discussed above.

Furthermore, the presence of a dose-response relationship cannot be supported solely on the ground of "statistically significant" correlations. For instance, the correlation between ulnar MCV and lead in blood observed by Ashby (1979) and Repko (1978) are statistically significant, but nonetheless paradoxical: as lead in blood increases MCV increases as well.
Other factors which can corroborate the causality of an association are:

- strength of the association (magnitude of the measure of effect);
- specificity of the association (a specific outcome is associated with a specific exposure);
- consistency of the results within the study and with other studies and

investigations;
- biological and temporal plausibility of the observed association.

However, it must be emphasized that probably no single epidemiologic study in this field is capable of providing a definitive body of evidence of the causality of an exposure-effect association. If different studies are consistent, then the causality of the relationship is remarkably strengthened. As a consequence, the standardization of the techniques and procedures is a major prerequisite to be reached. This is of paramount importance in neurobehavioral toxicology, where sources of variability are so numerous that differences in the methods, although apparently irrelevant, may jeopardize the comparability of different studies.

CONCLUSIONS

Neurobehavioral studies are becoming more and more important in occupational health. This is not only due to the increasing number of substances and compounds possessing neurobehavioral effects but follows also from the necessity of using neurobehavioral techniques to detect early and subtle effects produced by relatively low concentrations of a wide range of substances.

The methodological aspects demand particular attention in this field of study. Some have been mentioned here; others have been discussed in more detail in other papers, i.e., reliability of the instruments and techniques of measurement, especially for self-reported symptoms and performance scores; sensitivity of the various tests and their specificity in relation to different exposures; influence of individual susceptibility on the appearance of effects.

Greater attention to these aspects, together with an accurate definition of the exposure, suitability of the population, and longitudinality of the observation will enable neurobehavioral toxicology to extend the already enormous progress it has made in recent years and thus make an even greater contribution towards health and safety in the workplace.

REFERENCES

Araki, S., T. Honma, S. Yanagihara and K. Ushio (1980). Int. Arch. Occup. Environ. Hlth, 46, 151-157.
Ashby, J.A.S. (1980). Br. J. Ind Med., 37, 133-140.
Axelson, O., M. Hane and D Hogstedt (1976). Scand. J. Work Environ. Hlth, 2, 14-20.
Buchtal, F. and F. Behse (1979). Br. J. Ind. Med., 36, 135-147.
Chaffin, D.B. and J. M. Miller (1974). Behavioral and neurological evaluation of workers exposed to inorganic mercury. Behavioral Toxicology. NIOSH. Washington, D.C.
Hänninen, H. (1971). Br. J. Ind. Med., 28, 374-381.
Hridna, P.D., I. Hanin and T.C. Dubas (1980). Neurochemical correlates of lead toxicity. Lead toxicity. Urban and Schwarzenberg,Baltimore, MD.
Repko, J.D., C.R. Corum, P.D. Jones and L.S. Jr. Garcia (1978). The effects of inorganic lead on behavioral and neurologic function. NIOSH. Washington, D.C.
Silbergeld, E.K. (1982). Neurochemical and ionic mechanisms of lead toxicity. Mechanisms of actions of neurotoxic substances. Raven Press, New York, N.Y.
Silbergeld, F.K. and A.M. Goldberg (1980). Problems in experimental studies of lead poisoning. Lead toxicity. Urban and Schwarzenberg, Baltimore, MD

Sävolainen, H. (1982). *Arch. Toxicol., Suppl. 5*, 71-83
Snedecor, G.W. and W.G. Cochran (1976). *Statistical Methods*. The Iowa State University.

Neurobehavioral Follow-up Study of Mercury Exposure

G. Angotzi*, D. Camerino**, F. Carboncini*, M. G. Cassitto**, F. Ceccarelli*, R. Cioni***, C. Paradiso*** and E. Sartorelli*

*Institute of Occupational Health, University of Siena, Siena, Italy
**Institute of Occupational Health, University of Milan, Milan, Italy
***Institute of Nervous and Mental Diseases, University of Siena, Siena, Italy

ABSTRACT

The effects of metallic mercury exposure were investigated in 80 workers at a mercury distillation plant, in 1978 and 1982. Besides psychological and neurophysiological tests, the workers underwent biochemical tests for minimal damage of liver and kidney.

KEYWORDS

Mercury exposure; neurotoxic effects of mercury; ocular effects of mercury.

INTRODUCTION

A group of workers occupationally exposed to low mercury levels were examined twice, in 1978 and 1982, for both the exposure indexes and toxic effects on various organs and systems. Under such working conditions, no cases of chronic mercury intoxication occur with the classical clinical picture. However, using psychodiagnostic tests consisting of psychomotor tests, measurements of cognitive functions and individual personality, different investigators demonstrated a correlation between, on the one hand, the degree of psychomotor impairment (Chaffin et al, 1978), the results of mechanic and visual memory tests, psychomotor functions, dysthymic personality involution (Forzi et al, 1976), the results of tremometry and psychomotor tests (Langolf et al, 1978), and, on the other, the levels of urinary excretion of mercury as an index of exposure. With neurophysiological investigations signs of damage to sensory and motor nervous fibers were demonstrated in workers chronically exposed to inorganic mercury (Gilioli et al, 1976; Iyer et al, 1976). Minimal kidney damage has been ascribed to mercury (Foà et al, 1976; Jesato et al, 1977; Kazantis et al, 1962). The effects of mercury, if any, on the liver of occupationally exposed people have not been established (Skerfving et al, 1972). The aim of our investigation was to study the effects of mercury in the particular exposure conditions (described below) at various body system levels.

SUBJECTS AND METHODS

Both in 1978 and 1982, the workers were divided into three groups according to jobs and exposure to mercury vapour: 1) those working full-time in the plant, 2) mechanical maintenance staff, 3) electrical maintenance staff and other workers. Until 1976, rocks containing cinnabar were the only source of mercury. After 1976 residual mud from demercurization plants of various chemical industries was also used. In the second half of 1981, workers began working again with cinnabar only. Since 1977 this mercury distillation plant has only operated for about two weeks each month and at reduced capacity, due to the economic situation. Therefore, the employees only work for limited periods (an average of between 376-737 days for each group). The turn-over of the exposed workers is presented in table 1.

Table 1. Turn-over of workers

	Workers in 1978	"Old" workers in 1982	"New" workers in 1982
Distillation plant	32	11	15
Mechanical maintenance	28	13	6
Electrical maintenance	20	9	3

"old" : people with the same job in 1978
"new" : people with other jobs in 1978.

The control group was composed of subjects not occupationally exposed to mercury living in the same area. In 1978 mercury pollution of the working environment was measured by area samples taken near the more contaminated positions.
In 1982 individual exposure was also measured with personal sampling devices. Sampling and measurement of mercury in air were carried out according to Rathie and Marcero (1976). Both in 1978 and 1982, urinary mercury concentrations were determined on one-day samples from exposed workers and controls, using an AAS method (Lindstedt, 1970). An anamnesis was made for all subjects according to a specifically planned questionnaire and all were given a clinical examination. The psychodiagnostic test battery included: Raven Progressive Matrices, one subtest of the Wechsler Intelligence Scale (block design), simple reaction times, with visual and auditory signals, choice reaction time, digit symbol, Benton visual retention test , Digit Span subtest of the Wechsler Intelligence Scale, Rey P.M.R. 1, personality questionnaire 16 P.F. by Cattel (Angotzi et al, 1980). Eye examinations consisted of a clinical examination, determination of intra-ocular pressure, and evaluation of visual fields for red and white. In 1978 the neurophysiological test battery included determination of SCV of two ulnar nerves, SCV of the left sural nerve, MCV of an ulnar nerve, EMG at the abductor muscle of the fifth finger. In view of the studies reported in the literature and our experience in 1978, in 1982 we changed the battery composition as follows: determination of SCV of right radial, medial and ulnar nerves, SCV of both sural nerves, another CV and EMG according to clinical picture and results of previous neurophysiological test. Only minor technical changes,

were made which, in our experience, do not prevent comparison of the results. The first two groups of tests started in 1982 have not yet been completed. Therefore, only the results of the neurophysiological tests are available at present. To assess liver function, serum levels of bilirubin, AST, ALT, OCT, GGT, total protein and protein pattern were determined. Kidney function was assessed by measuring urine concentrations of protein, $beta_2$-microglobulin, aminoacids.

RESULTS

Exposure Indexes. In 1978 the mercury concentrations in the working environment were between 0.098 and 0.279 mg/m^3, except for one position in the distillation plant. In 1982, the mercury concentrations of 44 area samples were between 0.002 and 0.16 mg/m^3. Mercury exposure, as determined by personal samplers, varied considerably in the distillation plant according to the job: 0.031-0.067 mg/m^3 was the lowest exposure, 0.082-0.229 mg/m^3 the highest. The mercury concentrations were between 0.008 and 0.091 mg/m^3 for mechanical maintenance staff, and between 0.02 and 0.2 for electrical maintenance staff. The ACGIH 1981 TLV is 0.05 mg/m^3. The average values of urinary mercury in exposed workers showed slight variations comparing the 1982 results with those of 1978: 109 to 108 μg/l (distillation plant workers), 62 to 84 μg/l (mechanical maintenance staff), 32 to 39 μg/l (electrical maintenance staff). In the control group, the urinary excretion of mercury was below 10 μg/l in 13 subjects, and in the other 11, between 11 and 21 μg/l. In the distillation plant, the excretion of mercury showed little variation between jobs: from 33-70 μg/l to 39 - 87 μg/l.

Three risk indexes (R.I.) were drawn up taking into account: a) the length of exposure in each group of workers with significantly different urine mercury concentrations; b) a factor reflecting these differences in the ratio of 3:2:1. The purpose of R.I. was to adequately evaluate both overall, past and recent exposure considering the different working conditions during the previous six years, and also the reversibility of some of the neurotoxic effects of mercury (Langolf et al, 1978).

Psychological parameters. In addition to age, educational level and job, the following exposure indexes were considered in the data analysis: the number of times that Hg levels exceeded 50 μg/l in three urine samples in 1 year, and urine mercury concentration in the latest samples. The age group distribution was similar in all groups of exposed workers and in the control group.

As in previous investigations, the mean of all the test variables showed significant differences between control and exposed groups (Student's t test). Furthermore, significant differences in all test variables, including Cattell's Personality Factors C, H, L, N, O, Q_4 were found between the workers who had three peak urinary mercury levels exceeding 50 μg/l during the last year and the non-exposed subjects (Student's t test). No statistical linear correlation was found between R.I. and test variables. However, significant correlations were found between the latest value of urine mercury concentration and the following test variables: Raven P.M. 38, Block design, Benton V.R.T. (correct reproductions), Benton V.R.T. (errors), Digit Symbol Simple R.T. (visual and auditory signals), Choice R.T. Table 2 shows the overall percentages of the alterations in the various psychic functions observed in the four groups. As the age of the subjects was similar in all groups, the observed differences can with all probability be attributed to environmental factors, above all with regard to cognitive functions.

Among the emotional disturbances, erethism was present either as a syndrome or as a single symptom in 42.84% of the distillation plant workers, decreasing to 17.23% in the non-exposed group.

Table 2. Percentage of alterations resulting from intelligence tests, tests for visuomotor speed, memory and personality in the three exposed groups of the mercury plant and in the non-exposed subjects.

	Cognitive functions (%)	Visual-motor functions (%)	Memory %	Emotional disturbances (%)
Distillation plant	47.61	47.61	71.42	66.66
Mechanical maintenance	16.66	44.44	66.66	43.75
Electrical maintenance	00.00	25.00	37.50	37.50
Non-exposed	00.00	24.13	17.24	34.48

Ocular examinations. The most interesting results was the narrowing of the visual field for red, (while remaining normal for white): this condition was present in 19 exposed workers (56%) and in 4 control subjects (17%). The physiopathological significance of this dysfunction in mercury exposed subjects has not yet been fully explained (Brodziek et al, 1969; Zeglio, 1958). Furthermore, minimal ocular hypertension (20.1-23.8 mmHg) was found in 5 exposed workers (3 with narrowed visual field, 2 with normal visual field).

Neurophysiological parameters. In 1978 only the ASP mean of the left sural nerve was significantly lower in the exposed subjects (Student's t test). Furthermore, no significantly lower mean of other parameters can be demonstrated either by comparing the mean of each parameter in the control group with those in each group of exposed workers, or by making the same comparison between exposed subjects with latest urine mercury concentration above 50 μg/l and the control group. No statistical linear correlation was found between R.I. and SCV. On the other hand, on the basis of clinical and neurophysiological criteria (SCV, MCV, EMG), various syndromes related to a peripheral neurological deficit were recognized (table 3).

Table 3. Neurological syndromes diagnosed in control subjects and exposed subjects in 1978.

	Control subjects	Exposed subjects
Sensory polyneuropathies	-	3
Sensory mono- and multineuropathies	5	6
Sensory motor mono- and multineuropathies	-	5
Total number of neuropathies of unknown pathogenesis	5 (19%)°	14 (27%)°
Traumatic mono- and multineuropathies	3	4
Radiculopathies	2	5
Metabolic polyneuropathies	2	2
Total number of neuropathies of known pathogenesis	7 (26%)°	11 (20%)°

° : percentage of affected subjects in the group.

It is important to point out that the 3 cases of sensory polyneuropathy and the 5 cases of sensory-motor mono- and multi-neuropathies of unknown pathogenesis were diagnosed among the exposed subjects. In 1982 comparison of the mean of the neurophysiological parameters demonstrated significant differences for SCV and ASP of the left sural nerve (Student's t test). These differences exist also when comparing the distillation plant group and the mechanical plus electrical maintenance staff with the control group. The neurological syndromes found in 1982 are illustrated in table 4.

Table 4. Neurological syndromes diagnosed in the exposed subjects in 1982.

Sensory polyneuropathies	8
Sensory multineuropathies (lower limbs)	5
Sensory mononeuropathies	2
Total number of neuropathies of unknown pathogenesis	15 (49%)°

°: percentage of affected subjects in the group

The diagnosis was made using the same clinical criteria and nearly the same neurophysiological criteria as in 1978. In fact, in 1982 SCV with more nerves was used for the diagnosis. In 1982 the number of sensory neuropathies increased. Some of these neurological dysfunctions occurred in subjects with a high consumption of alcohol (like wine) or who use vibrating tools. A comparison of the neurological examination results in 1978 with those in 1982 is given in tables 5 and 6. It is important to note the changes both in neurophysiological parameters and in neurological syndromes. As regards the former, attention must be drawn to the increase in cases with reduced SCV of the left sural nerve. All cases of neuropathies, except one, were still present in 1982 and new cases occurred among previously non-affected subjects. Only one subject, previously affected by neuropathy, did not show this syndrome in 1982.

Table 5. Evolution of neurophysiological parameters from 1978 to 1982. (N= normal; OSD = reduced by more than 1 SD; TSD = reduced by more than 2 SD).

	1978	1982
S C V		
right ulnar nerve	19 N	18 N
		1 OSD
left sural nerve	10 N	3 N
		5 OSD
		2 TSD
	1 OSD	1 N
	4 TSD	1 N
		1 OSD
		2 TSD
A S P		
right ulnar nerve	16 N	16 N
left sural nerve	10 N	2 N
		8 OSD
	6 OSD	1 N
		5 OSD

Table 6. Evolution of neurological syndromes from 1978 to 1982. (A= neurological syndromes not present; Mo= sensory mononeuropathies; Mu= sensory multineuropathies; P= sensory polyneuropathies).

	1978	1982
Neurological syndromes not present	14	9 A 1 Mo 2 Mu 2 P
Sensory mononeuropathies	1	1 Mu
Sensory multineuropathies	4	1 A 1 Mo 2 P
Sensory polyneuropathies	1	1 P

Liver and kidney tests. In 1978, bilirubin, total protein and protein pattern were always normal. Elevated ALT serum levels were found in 10 exposed subjects (17%) and in 8 controls (31%); elevated GGT in 28 exposed (48%) and in 7 control subjects (26%), elevated OCT only in 1 exposed subject (2%). In 1982, 8 exposed workers (26%) showed an increase in GGT and 3 (33%) an increase in ALT. Urinary protein concentrations slightly above normal were present in 3 exposed workers (7%) and in 1 control subject (5%). Urinary $beta_2$-microglobulin increased in 2 exposed workers only. As regards urinary aminoacid excretion, urine cystin concentration was above normal in 15 exposed (29%) and in 7 control subjects (28%), while an increase in many aminoacids was found in 8 exposed subjects (16%) and 1 control (4%).

DISCUSSION

1) In the first survey, only neurobehavioral parameters were significantly altered among the exposed workers, probably due to mercury exposure. 2) The follow-up study of neurophysiological parameters indicated a trend of sural nerve impairment. The evolution of the neuropathy cases was not homogeneous: some cases improved, others worsened. 3) At this low exposure level, we could not find widespread, minimal neurophysiological alterations among exposed workers definitely related to mercury. 4) The neurological syndromes diagnosed in some workers at this plant could have been caused by other occupational and non-occupational factors, apart from mercury, such as ethanol, arthrosis, limb trauma and vibrating tools. These factors very often affect the same subjects and it is therefore very difficult to evaluate their importance. The non-homogeneous evolution of the neuropathy cases may be influenced by the intermittent exposure both to the above factors and to mercury.

REFERENCES

Angotzi, G., M.G. Cassitto, D. Camerino, R. Cioni, E. Desideri, A. Franzinelli, R. Gori, F. Loi and E. Sartorelli (1980). Med. Lavoro, 71, 463-480.

Brodziak Krzesiekowa, K. and E. Ogielska (1969). Klin. Oczna, 39, 761-763.

Foà, V., L. Caimi, L. Amante, C. Antonini, A. Gattinoni, G. Tettamanti, A. Lombardo and A. Giuliani (1976). Arch. Occup. Environm. Hlth, 37, 115-124.

Forzi, M., M.G. Cassitto, R. Gilioli, G. Armeli and V. Foà (1976). Proceedings of 2nd International Industrial and Environmental Neurology Congress, Prague, Czechoslovakia.

Gilioli, R., C. Bulgheroni, L. Caimi, V. Foà, C. Filippini, A. Boiardi, G. Bussone, M. Quarti and R. Boeri (1976). Adverse effects of environmental chemicals and psychotropic drugs. Vol. 2, Neurophysiological and behavioral tests. Elsevier Scientific Publishing Company, Amsterdam.

Kazantis, G., K.F.R. Schiller, A.W. Asscher and R.G. Drew (1962). Quart. J. Med., 31, 403-418.

Jesato, K., M. Wakashin, Y. Wakashin and S. Toyo (1977). Ann. Intern. Med., 86, 731-737.

Jyer, K. J. Goodgol, A. Eberstein and P. Berg (1976). Arch. Neurol., 33, 788-790.

Langolf, G.D., D.B. Chaffin, R. Henderson and H.P. Whittle (1978). Amer. Ind. Hyg. Assoc. J., 39, 976-984.

Lindstedt, G. (1970). Analyst, 95, 264-271.

Rathje, A.O. and D.H. Marcero (1976). Amer. Industr. Hyg. Ass. J., 37, 311-314.

Skerfving, S. and J. Vostal (1972). Mercury in the environment. Chemical Rubber Co. Press, Cleveland.

Zeglio, P. (1958). Atti XXII Congresso Nazionale di Medicina del Lavoro, Roma, Italia.

Neurobehavioral Evaluation in Workers Following a Brief Exposure to Lead

M. L. Bleecker*,** J. Agnew**, J. P. Keogh*** and D. S. Stetson***

*The Johns Hopkins Medical Institutes, Department of Neurology, Baltimore City Hospitals, Baltimore, Maryland, USA
**The Johns Hopkins School of Hygiene and Public Health, Department of Environmental Health Sciences, Division of Occupational Medicine, Baltimore, Maryland, USA
***Baltimore City Hospitals, Division of Occupational Medicine, Baltimore, Maryland, USA

ABSTRACT

Thirteen workers with no known prior exposure to lead had an exposure history from two weeks to eight months with a median blood lead level of 72 μg/dl (37-115 μg/dl). Free erythrocyte protoporphyrin (FEP) levels, ranging from 68 to 479 μg/dl with a median of 265 μg/dl, were highly correlated with the length of exposure. A neurobehavioral battery which emphasized memory functions was administered to the lead-exposed and non-exposed control workers. Significant differences were obtained in tests of visual intelligence and visuomotor functions. A new finding was the significant decrement in performance on an auditory task for word span. These findings have never been previously described in a group with such brief lead exposure.

KEYWORDS

Lead; lead poisoning; CNS dysfunction; behavior

INTRODUCTION

In a 1975 review of chemical neurotoxins in industry and the environment, it was noted that industrial lead intoxication has become quite uncommon in the United States (Allen, 1975). This may be true when considering the classic presentation of lead intoxication, however, many recent publications have dealt with the neurotoxicity of lead at supposed safe levels of exposure (Seppäläinen, 1975). Sensitive neurophysiologic techniques have shown that subclinical neuropathies often occur in lead workers free from neurological symptoms (Catton, 1970, Seppäläinen, 1972)

Repko (1976,1978) developed a battery of neurobehavioral tests for the detection of subclinical neurological changes due to occupational exposure to inorganic lead. He found a significant negative correlation between subjective duration of an audible tone and blood lead levels. These changes in auditory function in lead exposed workers were considered by the author to be consistent with reduction in peripheral nerve conduction velocities due to lead exposure as reported by Seppäläinen and Hernberg (1972). Repko also reported decreases in psychomotor performance related to indices of lead absorption, but no effect on intellectual functions was observed. Grandjean (1978) described decreased psychomotor speed in lead exposed workers besides decreased performance in long-term memory and verbal and visuospatial abstraction.

Impaired psychological performances in workers whose highest blood lead concentration never exceeded 70 μg per 100 ml have also been demonstrated (Hanninen, 1978). The performances most affected by lead were dependent on visual intelligence and visuomotor functions (Baloh, 1979). Mantere (1975, 1981) examined a group of workers after one and two years of exposure with results suggesting exposure relationships with block design and Santa Ana tests. However, psychological test methods have virtually been used exclusively with populations experiencing long-term lead exposure.

In the present investigation we report decreased performances in visual and auditory tasks in workers with a brief exposure to lead.

DESCRIPTION OF STUDY GROUP

Thirteen male workers constituted the lead-exposed group. These individuals were exposed to inorganic lead fumes generated during short-term job assignments involving cutting or welding steel surfaces coated with lead-based paint. Worksites were generally confined spaces with poor ventilation. The group had no known prior exposure to lead, and the length of exposure ranged from two weeks to eight months with a median of 2.5 months.

Blood lead (PbB) and free erythrocyte protoporphyrin (FEP) were measured as indicators of lead uptake. Duplicate samples of the whole blood were measured for levels of lead using an aniodic stripping voltometry. FEP measurements were performed on packed red blood cells using a fluorometric assay. PbB and FEP levels at the time of examination are shown in Table 1.

Table 1. Indices of Exposure for Lead-Exposed Group

	Median	Range
Duration of exposure (months)	2.5	0.5 - 8
Blood lead (μg/dl)	72	37 - 115
FEP (μg/dl)	265	68 - 479

The 20 male workers in the control group were not currently exposed to, nor had they ever knowingly worked with inorganic lead. Controls were drawn from government workers employed as electricians. The two groups of subjects were comparable with respect to age, sex, education level, premorbid abilities, and native language (Table 2).

None of the individuals included in the study groups suffered from any neurological, psychiatric or metabolic diseases which might influence psychological test results. The alleged alcohol consumption was not extreme, with 85% of the control group and 73% of the exposed group consuming three drinks or less per day.

Table 2. Characteristics of Study Groups

	Lead-Exposed Workers (n = 13) Median (Range)	Control Workers (n = 20) Median (Range)
Age (years)	28 (20 - 60)	29.5 (24 - 64)
Education (years)	12 (8 - 12)	12 (9 - 13)
WAIS vocabulary score	38 (12 - 36)	43 (28 - 70)

Differences not significant (Wilcoxon rank sum test, 2 tailed)

TEST PROTOCOL

Both groups received a neurological and electromyographic examination and a neurobehavioral battery. The battery was composed of tasks covering a broad range with special emphasis on memory functions which are regarded to be sensitive to central nervous system disturbance. All behavioral testing was performed by the same examiner (J.A.).

Vocabulary, a subtest from the Wechsler Adult Intelligence Scale (WAIS), was used for the measurement of verbal intelligence. It requires the mobilization of acquired verbal material and is considered an indicator of pre-morbid ability.

Digit Span from WAIS measured the subjects' immediate memory for digits. As the digits must be recalled both forwards and backwards, the test demands attention and concentration.

Digit Symbol from the WAIS measured visual learning and psychomotor abilities. The subject is given a list in which symbols are associated with the digits from one to nine and is asked to enter the symbols into the blank spaces next to a list of random digits on a form. Immediate recall of the symbols was measured by requesting a list of all symbols used in Digit Symbol (free recall) and their respective digits (associative recall) immediately following that test.

Block design from the WAIS measured visual intelligence and abstraction. Subjects are required to arrange red and white blocks to correspond to a painted design. There are time limits for each problem.

Associate Learning from the Wechsler Memory Scale (WMS) was used for the measurement of verbal memory and learning. It consists of 10 pairs of words to be learned during three trials. Before each trial, the word pairs are read to the subject, and the subject must recall the second word when he hears the first one.

Visual Reproduction from the WMS was used for the measurement of visual memory. In this test the subject must draw a simple figure after a 10 second presentation.

Controlled Word Association Test is a measurement of verbal fluency. It consists of three word naming trials, using the letters F, A and S respectively.

Rey Auditory Verbal Learning Test (AVLT) measures immediate word memory span. It consists of 5 presentations of a 15 word list which are scored by number of words correctly recalled. A learning curve may be generated.

Reaction times were measured by a simple reaction timer. In the task the subject reacts to visual signals by pressing a button with the dominant index finger. The signals are presented randomly with one to ten second intervals. Eleven trials were obtained.

STATISTICAL METHODS

Subtests of the WAIS and WMS were scored in the standard manner (Wechsler, 1945; Wechsler, 1955). Raw scores were used for the analysis. Non-parametric methods were employed due to the limited number of subjects.

RESULTS

Several lead workers presented with symptoms characteristic of lead poisoning such as headache, arthralgia, and gastrointestinal disorders. The complaint most commonly expressed was irritability, noted by seventy percent of the workers or their families. None of the workers were found to be anemic or to have evidence of impaired renal function.

Examination of the indices of exposure in the lead-exposed group revealed a significant correlation between FEP and the duration of exposure $p < .001$ (Spearman rank correlation coefficient $r = .839$). The FEP level in blood at a particular time reflects the lead level at the site of erythropoiesis averaged over the preceeding 4 months. The correlation of exposure duration with PbB did not reach statistical significance. This was to be expected since PbB has an equilibrium time of no more than a few days (Hopkins, 1970) and reflects recent lead absorption. PbB fluctuates more under conditions of variable exposure than the FEP level.

Table 3 presents the various performance tests used with the median scores and ranges for both groups. The performances of the exposed group were consistently inferior to the control group, with significant differences occurring in block design, digit symbol, and Rey auditory verbal learning test (AVLT).

Lezak (1976) states that the immediate memory for digits and the number of words recalled on trial 1 of the AVLT should be within one or two points of each other. Only two exposed and two controls fell within this category, however, sixty-two percent of the exposed workers had AVLT scores less than digit span forward compared to 45% of the control. Figure 1 portrays the learning curve generated in both groups for the 5 trials of the AVLT. The most significant difference occurred when comparing the initial three trials. This was followed by improvement in performance of the exposed group resulting in equivalent fifth trials.

Table 3. Neurobehavioral Test Results

	Lead-Exposed Workers (n = 13) Median (Range)	Control Workers (n = 20) Median (Range)	p *
Digit Span	10 (8 - 14)	11 (9 - 16)	.45
Block Design	32 (28 - 34)	41 (28 - 48)	.005
Digit Symbol	48 (25 - 60)	54.5 (36 - 77)	.04
Associative Learning	14.5 (8 - 19)	16.25 (7.5 - 20)	.27
Visual Reproduction	9 (6 - 13)	11 (4 - 14)	.09
Immediate Recall- Free	8 (7 - 9)	8 (3 - 9)	.53
Immediate Recall - Association	7 (6 - 9)	8 (2 - 9)	.34
Controlled Word Association	33 (17 - 56)	38 (23 - 57)	.38
Simple Visual Reaction Time	237 (199 - 271)	223 (204 - 335)	.52
Rey Auditory Verbal Learning			
Trial 1	5 (3 - 7)	6.5 (4 - 9)	.03
Trial 2	7 (4 - 10)	8.5 (5 - 11)	.04
Trial 3	9 (6 - 12)	10 (6 - 12)	.11
Trial 4	10 (7 - 14)	11 (5 - 13)	.22
Trial 5	11 (7 - 15)	11 (6 - 14)	.42
Trials 1 - 2	13 (8 - 15)	15 (9 - 20)	.008
Trials 1 - 3	21 (14 - 26)	25.5 (15 - 32)	.009
Trials 1 - 5	42 (30 - 54)	47 (29 - 59)	.07

*Wilcoxon unpaired rank sum test, 2 - tailed

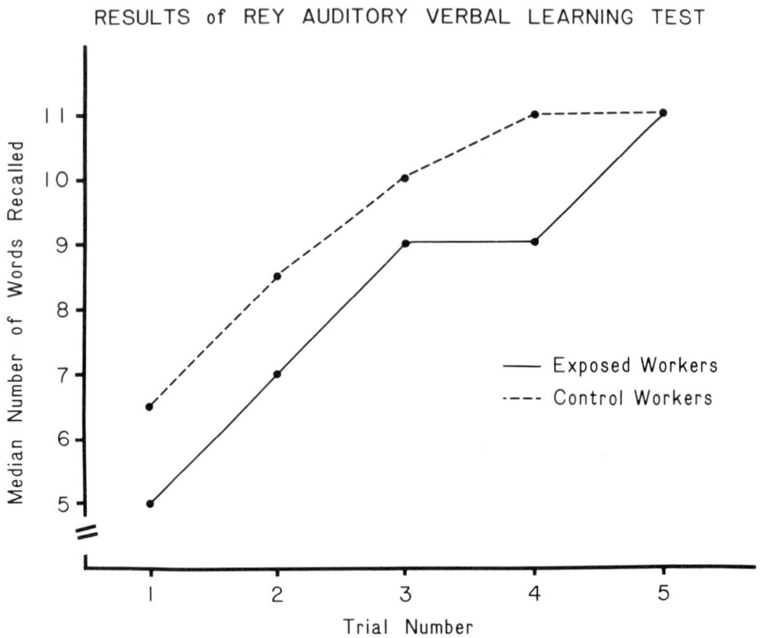

Figure 1.

The digit symbol and the block design tests which are known to be sensitive to brain damage, and the AVLT were not correlated with PbB or FEP levels. An explanation may be the nearly significant correlation found between FEP and vocabulary, implying that the more intelligent exposed workers had higher FEP levels. The scores used in examining correlations were not adjusted for any indicator of premorbid ability.

DISCUSSION

Consistent findings in prior behavioral studies of lead-exposed workers have emphasized the areas of visual intelligence and visuomotor performance as most sensitive to the effects of lead (Hänninen, 1978; Valciukas, 1978; Mantere, 1975). The present study is in complete agreement demonstrating significant decrement in performance for block design and digit symbol. Visual reproduction also appears to be influenced.

Auditory memory as measured by Rey Auditory-Verbal Learning Test (AVLT) has not previously been examined in a lead-exposed population. Lezak (1980) believes that large differences between AVLT and digit span usually favor the digit span and seem to occur in individuals with intact immediate memory and concentration who become confused by too much stimulation. These individuals tend to have difficulty with complex material or situations of any kind, but do well with simplified structured tasks. This may be a partial explanation for the poorer performance in more complex tasks by the exposed group. The learning curve generated in the AVLT also reflects this difficulty with handling new complex material. Even though the lead-exposed worker eventually achieves the performance level of the controls, there is a significantly slower acquisition of the new material.

Central nervous system dysfunction in workers with blood lead levels below 70 μg/100 ml was demonstrated by Seppäläinen (1978) using somatosensory evoked potentials (SEP's). Increased amplitudes of the N18-P22 measures were interpreted as a direct central nervous system effect by lead. Several strands of evidence suggest that the increase in N18-P22 amplitude could reflect disinhibition of central inhibitory mechanisms. Increased amplitudes of the vertex evoked potential in humans (Otto, 1978) and the superior colliculus evoked potential in rats (Dyer and Annau, 1978) have been observed following CO exposure. Dyer and Annau attributed this finding to a release of inhibition produced by increased sensitivity of inhibitory synapses to hypoxia. The effect of lead on evoked potentials may be similar. Direct correlation between alterations in SEP's and performance on neurobehavioral testing have not yet been explored.

Lilis (1976, 1977) and Hänninen (1979) found a higher prevalence of central nervous system symptoms among secondary smelter workers as compared to controls. A comparison between the test scores in individuals with these symptoms such as fatigue, nervousness, sleep disturbances, or a combination of these and asymptomatic lead-exposed workers did not give evidence for lower test scores in individuals exhibiting symptoms than in the asymptomatic workers (Valciukas, 1978, 1981). In the present study, except for irritability, few central nervous system symptoms occurred in the lead workers. This suggests the possibility that decrements in performance tests may occur earlier than the subjective central nervous system symptoms.

The groups examined were comparable, suggesting that the differences observed in block design, digit symbol and AVLT were likely the result of exposure to lead. Although no correlation of performance and indicators of lead exposure could be demonstrated, this is thought to be due to the size of the group and the positive correlation between vocabularly and exposure. Previous studies have emphasized central nervous system dysfunction as an effect of long-term lead exposure. This study demonstrates similar findings following brief exposure to lead.

ACKNOWLEDGEMENT

The authors wish to thank Dr. Julius Chisolm for referring many of the lead exposed patients and for conducting the analysis of the PbB and FEP. This work was supported by the core support for the center of Occupational and Environmental Health at The Johns Hopkins University by NIOSH and the Educational Resource Center for the Occupational Health Nursing Program at the Johns Hopkins School of Hygiene and Public Health.

REFERENCES

Baloh, R.W., G.H. Spivey, C.P. Brown, D. Morgan, D.S. Campion, B.L. Browdy, J.L. Valentine, H.C. Gonick, F.J. Massey, Jr., and B.D. Culver (1979). J. Occup. Med, 21, 490-496.

Catton, M.J., M.J.G. Harrison, P.M. Fullerton, and G. Kafantzis (1970). British Medical Journal, 2, 80-82.

Dyer, R.S. and Z. Annau (1978). Carbon monoxide and superior colliculus evoked potentials. Proceedings IV EPIC Meeting, Hendersonville, North Carolina.

Grandjean, P., E. Arnvig and J. Beckmann (1978). Scand. J. Work Environ. and Health, 4, 295-303.

Hänninen, H., P. Mantere, S. Hernberg, A.M. Seppäläinen and B. Kock (1979). Neurotoxicology, 1, 333-347.

Hänninen, H., S. Hernberg, P. Mantere, R. Vesanto and M. Jalkanen (1978). J. Occup. Med., 20, 683-689.

Hänninen, H. (1979). Test Methods for Definition of Effects of Toxic Substances on Behavior and Neuromotor Function, Neurobehavioral Toxicology 1, 157-161.

Hopkins, A. (1970). Br. J. Int. Med., 27, 130-135.

Johnson, B.L., J.R. Burg, C. Xintaras and J.L. Handke (1980). Neurotoxicology, 1, 561-581.

Lezak, M.A. (1976). Neuropsychological Assessment Oxford University Press, New York

Lilis, R., W. Blumberg, J. Eisinger, A. Fischbein, S. Diamond, H.A. Anderson and I.J. Selikoff (1977). Arch. Environ. Hlth.,32, 256-266.

Lilis, R., A. Fischbein, J. Eisinger, W.E. Blumberg, S. Diamond, H.A. Anderson, W. Rom, C. Rise, L. Sarkosi, S. Kon, I.J. Selikoff (1977). Environ. Res., 14, 255-285.

Mantere, P., H. Hänninen, S. Hernberg and L. Tenkanen (1981). Proceedings 20th Intl. Congress Occup. Hlth., Cairo, Egypt.

Mantere, P., H. Hänninen, S. Hernberg and T. Martelin (1979). Activ. Nerv. Sup. (PRAHA), 21, 292-294.

Otto, D.A., V. Benignus, J. Prah and B. Converse (1978). Paradoxical effects of carbon monoxide on vigilance performance and event-related potentials. Proceedings IV EPIC Meeting, Hendersonville, North Carolina.

Seppäläinen, A.M. and S. Hernberg (1972). Brit. J. Industr. Med., 29, 443-449.

Seppäläinen, A.M., S. Tola, S. Hernberg and B. Kock (1975). Arch. Environ. Health, 3, 180-183.

Seppäläinen, A.M., (1978). Diagnostic utility of neuroelectric measures in environmental and occupational medicine. Proceedings IV EPIC Meeting, Hendersonville, North Carolina.

Valciukas, J.A., R. Lilis, A. Fischbein, I.J. Selikoff (1981). Science,201, 465-467.

Valciukas, J., R. Lilis, J. Eisinger, W.E. Blumberg, A. Fischbein and I.J. Selikoff (1978). Int. Arch. Occup. Environ. Hlth., 41, 217-236.

Wechsler D. (1955) Wechsler Adult Intelligence Scale Manual, The Psychological Corporation, New York.

Wechsler D., C.P. Stone (1945). Wechsler Memory Scale Manual, The Psychological Corporation, New York.

Quantitative Measurement of Visual Field in Lead Exposed Workers

A. Cavalleri*, F. Trimarchi**, C. Minoia*** and G. Gallo**

*Cattedra di Medicina del Lavoro, Università di Modena, Modena, Italy
**Clinica Oculistica, Università di Pavia, Pavia, Italy
***Centro Ricerche di Fisiopatologia e Sicurezza del Lavoro, Università di Pavia, Fondazione Clinica Lavoro, Pavia, Italy

ABSTRACT

A quantitative technique for measurement of visual field in mesopic adaption was developed in order to achieve an increased sensitivity in the evaluation of early and subtle signs of neurological damage induced by neurotoxic agents. The area of the visual field obtained by light stimuli of graded efficacy is corrected according to the efficacy itself of the light stimulus and an index of visual sensitivity is calculated.
Quantitative measures of blind spot and scotomas, if present, are similarly derived. The procedure was adopted in 35 workers occupationally exposed to lead with blood lead levels 21-82 μg/100 ml and exposure duration 2-22 y, and in 35 referents matched for age, smoking and alcohol consumption. A highly significant decrease of visual sensitivity was observed for the exposed subjects. In 10 exposed workers a central scotoma was found that was not seen in any of the referents.
The performance test of critical fusion frequency did not reveal any significant difference between the two groups. The results are suggestive of subclinical optic neuropathy not detectable by traditional visual field evaluation.

KEYWORDS

Neurotoxicity, visual field, scotoma, lead.

INTRODUCTION

The visual system is a target of a number of toxic agents and overt systemic lead poisoning may cause a variety of visual and ocular disturbances (Grant, 1962). Recent behavioral and electrophysiological techniques revealed scotopic visual deficits in monkeys (Bushnell *et al*, 1977) and changes in visual evoked response in suckling rats exposed to lead (Fox *et al*, 1977). However, available data concerning chronic exposure of humans are scanty and contradictory (Milburn *et al*, 1976; D'Andrea *et al*, 1982).

This study was partially presented at the First Symposium on Epidemiology in Occupational Health, Helsinki, 10-12 June, 1981.

Our study was aimed at the development of a sensitive method to detect early and subtle signs of visual system involvement induced by lead. This being the goal we tried to measure the visual field on a quantitative basis to improve the comparison of the results and chose mesopic conditions for evaluating the visual field to increase the sensitivity of the test.
A large range of stimulus luminances is encountered by the eye in various conditions of light (Table I.).

Table I - Range of stimulus luminances encountered by the eye

TERM	LUMINANCE (log psb)	EXAMPLE
Scotopic	2.3	Threshold
	5.0	Dark night
Mesopic	5.3	Moonlight
	6.3	Reading
Photopic	7.3	Sunlight
	12.0	Sun

(from: H.L. Evans and R.H. Garman,in: Neurotoxicity of the visual system, W.H. Merigan and B. Weiss, eds, Raven Press, New York, 1980, p.136; modified).

Scotopic vision, more familiary called night vision, at the lowest luminances involves rods that lie peripheral exclusively to the center of gaze.
Photopic vision, where visual acuity is best, is dominated by cones and occurs at the highest luminances.
Mesopic vision, the range of which is less clearly determined, lies in the middle and is thought to involve both rod and cone functions.
By decreasing stimulus luminance to mesopic levels, the threshold light intensity declines and a simultaneous evaluation of cones and rods functions is possible, thus assuring greater sensitivity of the test.
In mesopic luminance conditions a quantification of the visual field was obtained by considering the visual field itself or scotomata eventually present as tridimensional solids characterized by two linear dimensions expressing the area and by a third one, arbitrarily considered the height, which measures the retinal sensitivity to the stimulus.

MATERIALS AND METHODS

The visual field examination is performed by a Bjerrum's screen lit by a source of light giving a luminance of 6.2 log psb on the screen.
Each subject examined is previously adapted to the luminance selected for 20 min. Then a campimetry is performed in both eyes with a light spot I mm in diameter. If the perimeter obtained has a dimension of less than 20° (normal value) the test is repeated with larger diameter light spots up to a maximun of 4 cm.
The same technique is adopted for evaluating scotomata, if present, and also for measuring the blind spot and its eventual enlargement.
The various isopters are plotted on a chart and the areas measured by a planimeter.
The index of visual sensitivity (IVS) is calculated as

$$V = S_I . f_I + (S_2 - S_I) . f_2 + ... + (S_{40} - S_{30}) . f_{40}$$

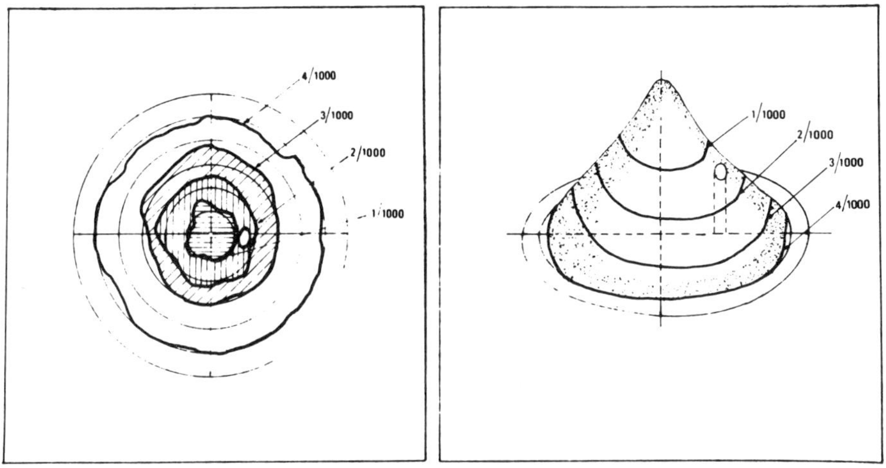

Fig.I-Visual field measured bidimensionally (left) and tridimensionally (right).The continuous lines delimits the perimeters of visual field obtained by light spots of increasing size (I/I000=I mm).The dotted line in the right hand figure represents the "height" of the visual field, defined by f (see text)

where

V is the "volume" of the tridimensional visual field,
S_I is the area obtained by the smallest spot (I mm),
$S_2..S_{40}$ is the area obtained by the spot with a diameter of 2...40 mm ,
$f_I,f_2...f_{40}$ is a coefficient by which the area obtained spot of I, 2...40 mm is multiplied and defined by

$$f_n = \frac{I}{I + 0.2\ n^2}$$

where

n is the value of the diameter of the light spot.

The function adopted was found experimentally and devised in order to decrease the contribution of values obtained with larger light spots which are clinically less significant.

The "volume" of the scotomata is similarly derived since it is defined by

$$V = (S_I - S_2).g_I + (S_2 - S_3).g_2 + ...+ (S_7 - S_8).g_7$$

where

$S_I,S_2.....S_8$ is the area of the scotoma delimited by light spot I,2.....8

and

$$g.(\ I,2......8\ )= I - f\ (I,2......8)$$

The reported test was performed on a group of male workers exposed to lead stearate in a factory producing polyvinylchloride pipes. The lead concentration in the working environment was rather homogeneous and below the threshold limit value, except in the mixing area where the limit value for lead was sometimes exceeded. Usually a rotation of duties operated and therefore it can be reasonably assumed that exposure was similar for all the exposed subjects.

The following criteria were chosen for inclusion in the study: age 20 to 60 y at the time of examination and more than 2 y of continuous exposure to lead. The 35 men who fulfilled the criteria were individually matched with the 35 referents with no history of exposure to lead, selected from metal and textile industries in the same town. Criteria for matching were formed by age (within 3 y), smoking and alcohol consumption. Diabetics were excluded and the differences between groups with respect to drug therapy were too small to have had any biasing effect.

An ophthalmoscopic examination, a test of visual acuity and a visual field examination in photopic luminance were performed in all the subjects and the normality of all tests was a prerequisite for inclusion in both groups.

After matching, the components of the groups underwent determination of lead in blood (Pb-B), lead in urine (Pb-U) and lead in plasma by atomic absorption spectrophotometry (Volosin et al, 1975; Kubasik and Volosin, 1973; Cavalleri et al, 1978); free erythrocyte porphyrin (FEP) according to the method of Chisolm and Brown (1975) and zinc protoporphyrin (ZPP) by means of a portable hematofluorometer (ESA-4000).

The visual field in mesopic luminance was then investigated according to the described method and the IVS calculated. Measures of blind spot and scotomata, if present, were similarly derived. In both groups the critical fusion frequency (CFF) was also studied for both eyes in photopic adaptation and with the use of a central light stimulus.

RESULTS

Table 2 summarizes the biochemical indicators of lead uptake and effects.

Table 2 - Indicators of lead uptake and effects in the exposed subjects and pair-matched referents

	Exposed subjects			Referents		
	Mean	SD	Range	Mean	SD	Range
Number of subjects	35			35		
Age (y)	45	14	21-59	44	11	22-59
Exposure (y)	13	6	2-22			
Lead in blood (μg/100 ml)	46	14	21-82	23	4	21-37
Lead in plasma (μg/100 ml)	1.01	0.6	0.27-2.42	0.37	0.1	0.21-0.7
Lead in urine (μg/l)	71	18	44-118	30	5	21-42
Free erythrocite porphyrin (μg/100 ml of blood)	102	44	40-180	35	13	15-67
Zinc protoporphyrin (μg/100 ml of blood)	32	18	8-95	18	11	3-48

The differences between the means of the exposed and reference groups were highly significant for all the parameters investigated. The mean level of Pb-B was 46 μg/100 ml but only 4 out of 35 subjects exceeded the value of 60 μg/100 ml.

Pb-U, FEP and ZPP levels were generally not exceedingly high, indicating a medium-to-low level of lead exposure. Mean concentrations of lead in plasma appeared significantly higher in the exposed group. The comparison of the IVS obtained as the sum of the score for both eyes showed a highly significant reduction for the visual field "volume" of the exposed workers. (Table 3).

Table 3 - Index of visual sensitivity in mesopic vision and blind spot area expressed in arbitrary units

	Exposed subjects No.	Referents No.	p-Value
	35	35	
Visual sensitivity index	1988 ± 617	2391 ± 583	0.003
Blind spot area	99 ± 13	102 ± 30	NS

NS = not significant

No significant differences were found for blind spots which in no case appeared enlarged. The CFF was the same in the exposed and in the referent groups. No significant relationship was found between IVS and the various parameters of lead absorption and effects investigated.

Only for lead in plasma values, when tested versus the IVS, a correlation coefficient of -0.226 was obtained with a p value of 0.07.

Mapping out of the visual field in mesopic luminance, blind areas inside the limits of the field were found in 10 exposed subjects, whereas none was found in the referents.

The scotomata were always central, bilateral in 8 cases, unilateral in 2. Also for scotomata, no clear relationships were demonstrated with the biochemical indicators considered. However, a relationship was found between the length of exposure and appearance of scotomata (Table 4). No scotomata were found before 5 y of exposure, while the frequency increased after more than 15 y of exposure.

DISCUSSION

The neuron that constitutes the optic nerve is highly susceptible to toxic substances and optic neuritis has long since been recognized as the most common form of visual disturbance due to lead in cases of severe poisoning (Lewin and Guillery, 1913).

With the neurophysiological tests used we were not able to localize the site of lesion exactly, but the optic nerve fibers seem the most probable target. Cones and rods, the photoreceptors, are not affected by toxic substances and poisoning by iodoacetate, which does affect photoreceptors, is of experimental importance only (Potts, 1980).

On the other hand, lesions situated in the visual cortex, as in the case of methylmercury poisoning, cause characteristic constriction of the visual field.

Usually toxic substances affecting the visual system act at the optic nerve level by inducing selective loss of central vision (methanol, carbon disulfide and tallium, for instance) or selective loss of peripheral vision (arsenicals and quinine). Our results seems to demostrate that occupational exposure to lead affects central and peripheral optic nerve fibers, since a significant reduction in visual field was found in exposed subjects in mesopic luminance when both cones and rods are activated. Further, the finding of central or bilateral scotomata in 28% of the exposed subjects indicates that a selective loss of central vision also occurs.

Table 4 - Evaluation of scotomas in the mesopic visual field of referents and exposed subjects according to the length of exposure

	Number of subjects	Blood lead (μg/100 ml) [a] Mean	SD	Number with scotoma
Referents	35	23	4	0
Subjects exposed < 5 y	6	48	18	0
Subjects exposed 5-15 y	12	46	15	3
Subjects exposed > 15 y	17	44	13	7

[a] I μg/100 ml = 0.048 μmol/l

The data obtained suggest a subclinical optic neurophathy induced by lead since visual acuity, ophthalmoscopy and visual field in photopic luminance proved normal in all the subjects included in the study.

These neurotoxic changes appear as early and subtle impairment of the visual function which seem to bear no relationship with biochemical indicators of lead exposure and effects, with the possible exception of the level of lead in plasma, but are more related to the length of occupational exposure. Further studies are certainly needed to investigate the site of action of lead in the visual system and the clinical significance of the changes found, but the above technique seems to be a promising tool in the early detection of functional impairment of the visual system and in prevention of irreversible damage from exposure to environmental and industrial pollutants.

REFERENCES

Bushnell P.H., R.E. Bowman, J.R. Allen and R.J. Marlar (1977). Science, 196, 333-335.

Cavalleri, A., C. Minoia, L. Pozzoli and A. Baruffini (1978). Brit. J. Ind. Med., 35, 21-26.

Chisolm, J.J. and D.H. Brown (1975). Clin. Chem., 21, 1669-1682.

D'Andrea, F., M. Locatelli and E. Gaffuri (1982). Med. Lavoro, 35-39.

Fox, D.A., J.P. Lewkowski and J.P. Cooper (1977). Toxicol. Appl. Pharmacol. 40, 449-461.

Kubasik, M.P. and M.T. Volosin (1973). Clin. Chem., 19, 954-958.

Lewin, L. and H. Guillery (1913). Die Wirkungen von Arzneimitteln und Giften auf das Auge. August Hirschwald, Berlin.

Milburn, H., E. Mitran and G.W. Crockford (1976). Ann. Occup. Hyg., 19, 239-249.

Potts, A.M.(1980). Duality of the optic nerve in toxicology. In: Merigan, W.H., Weiss, B., ed. Neurotoxicity of the visual system. Raven Press, New York, NY, 1980, 1-15.

Volosin, M.P., N.T. Kubasik and H.E. Sine (1975). Clin. Chem., 21, 1986-1987.

Neuropsychological Investigation on Styrene Exposed Workers

A. Mutti*, A. Mazzucchi**, G. Frigeri***, M. Falzoi*, G. Arfini* and I. Franchini*

*Institute of "Semeiotica Medica", University of Parma, Parma, Italy
**Institute of Neurology, University of Parma, Italy
***Occupational Health Service, Local Health Unit No. 5, Fidenza (PR), Italy

ABSTRACT

Higher nervous functions were investigated in 50 workers exposed to styrene. Their age averaged 40 y (SD 11.5) and their exposure 8.6 years (SD 4.5). The main styrene metabolites, mandelic acid (MA) and phenylglyoxylic acid (PGA), were measured on spot urine samples collected just before neuropsychological examination. The test battery included choice reaction time (RT), learning and memory tests, block design (BD) and vocabulary (WAIS), and visuo-spatial accuracy tests, such as digit symbol (DS) and embedded figures (EF). RT, BD, DS, and EF scores were closely related to the exposure intensity, as measured by the sum of urinary MA and PGA. RT and DS scores were also related to the duration of exposure. Age and vocabulary were eliminated as confounding factors by partial correlation procedures. After controlling for age and vocabulary, only RT remained negatively related to both the level and the duration of exposure to styrene vapours. DS score was still related to the duration of exposure, while a correlation was still present between the sum of MA and PGA, reflecting the current exposure intensity, and BD score. Using a means comparison approach, the memory tests, which did not seem to be related to the exposure intensity, were sensitive enough to show a significantly reduced performance even for very low exposure levels, e.g., lower than 25 ppm (8h TWA). These findings suggest the need for restrictive threshold limit values as health - based safety criteria.

KEYWORDS

Psychological performance; Higher nervous functions; Neurotoxins; Styrenes; Biological monitoring; Occupational exposure; Threshold limit values;

In recent years researchers' attention has been increasingly focused on the early effects of exposure to neurotoxic chemicals on the higher nervous functions. A wide number of studies has shown that exposure to organic solvents leads to changes in central nervous system functions, as measured by psychological performance tests. Sensory, perceptual and psychomotor changes have been found to be related to the inhaled concentration of organic solvents (Elofsson et al., 1980; Hane et al., 1977; Hänninen et al., 1976; Lindström, 1980).

Several field and experimental studies have shown that poor perceptual accuracy might be due to styrene exposure (Härkönen et al., 1978; Lindström et al., 1976). Nevertheless, the main effect of styrene exposure seems to be a slowing of reaction times, which have been shown to be related to exposure intensity (Stewart et al., 1968; Götell et al., 1972; Gamberale et al.,1979; Oltramare et al., 1974; Lindström et al., 1976; Härkönen et al., 1978;Cherry et al., 1980 and 1981). Despite the exposure-effect relationships, the risk of a decrement in performance capability due to long-term exposure to styrene has not yet been fully elucidated.

Moreover, the studies previously published on this subject have fallen short of a thorough analysis of the role played by confounding factors or modifiers of effect, such as intelligence and age, on such exposure-effect relationships.

The purpose of this study was to clarify, first of all, whether the relationship between exposure to styrene vapours and performance is still present when the effect due to confounding and modifier variables has been removed. Secondly, if so, whether it is possible to demonstrate the effect which may be ascribed to styrene exposure. Finally, our aim was to identify biological threshold limit values useful as health-based safety criteria.

SUBJECTS

Selection of exposed workers (No.=50) was based on information collected during individual interviews and medical examination. The workers with alcohol intake higher than 80 ml ethanol per day and those who smoked more than 20 cigarettes per day or equivalent were excluded.

Many laboratory tests enabled us to exclude the presence of chronic metabolic diseases, such as diabetes, liver and renal failure. None of the subjects suffered from epilepsy or other diagnosed neurological or psychiatric diseases which might influence neuropsychological performance.

All selected workers agreed to participate in the study. Their age averaged 40 y (SD 11.5) and their exposure time 8.6 y (SD 4.5). Sixteen subjects aged 35.1 y (SD 14.3), employed on the hospital staff, served as a control group.

EXPOSURE

The workers were exposed to styrene vapours in a factory manufacturing glass fibre silos. The main source of exposure was styrene evaporation from wide surfaces (up to 40 m^2) during manual laminating or spraying. Typical exposure ranged from 10 to 300 ppm.

The main styrene metabolites in humans, mandelic acid (MA) and phenylglyoxylic acid (PGA), were measured by high-performance liquid chromatography according to the method described by Gaetani et al, (1982). Spot urine samples were collected immediately before neuropsychological examination on the Saturday morning. The urine was adjusted to pH 2 with 1 N HCl and stored to -20°C until examination, which was always performed within a few days. The results were adjusted for creatinine measured following the procedure described by Chasson et al (1962). The urine samples of the first ten subjects were analyzed for MA and PGA by means of a capillary gaschromatographic procedure tested during a CEC-Nordic program of interlaboratory comparison of organic chemical analyses.
As part of the research project, environmental and biological monitoring was carried out simultaneously on ten workers during a whole working week. The results of this investigation (not reported here) enabled us to calculate the exposure levels corresponding to the measured concentration of styrene metabolites in the next morning's spot samples.

NEUROPSYCHOLOGICAL TEST BATTERY

Every exposed and control subject was submitted to a neuropsychological battery composed of:
Vocabulary (V), Wechsler Adult Intelligence Scale (WAIS) subtest, was performed to establish the approximate I.Q. in exposed and control subjects. In fact, it is well known that the vocabulary test is "not only" an index of schooling but also an excellent measure of "general intelligence" highly correlated with the total I.Q. (Matarazzo, 1972). The subjects were instructed to provide a definition for each word submitted by the examiner. Score: two points for a correct definition, one point for a rough definition.
Reaction Times (RT): Choice reaction times were measured using a Gerbrand's reaction time apparatus. The person was instructed to press the left lever with his left index finger when the green light appeared, to press the right lever with his right index finger when the red light appeared, to press both levers when he perceived a tone stimulus, and not to press when the orange light appeared. Score: we considered only the mean RTs obtained by the left and right hands exclusively.
Logic Memory (LM), part of the Wechsler Memory Scale. After a short passage presentation, the subject was to report the elements recalled immediately (ST) and after 30 min (LT). Score: total number of correct elements recalled immediately and after 30 min considered separately.
Verbal Memory (VM): Ten familiar (two-syllable) words were presented five times by the examiner. After each presentation (ST) the subject was to report all the words learned; after 30 min (LT) the subject was required to recall the words again. Scores: the mean of the total number reported after each of the five presentations (ST) and the number of words recalled after 30 min (LT), considered separately.
Digit Symbols (DS), WAIS subtest. In 90 sec., the subjects had to associate digits with symbols according to a reference code. Score: number of correct associations performed in 90 sec.
Block Design (BD), WAIS subtest. The subject was presented with a set of coloured cubes (four in the first six trials, nine in the last four). The task involved arranging the cubes in such a way that the top surface reproduced a design displayed by the examiner. Score: four points for each trial correctly completed in 60 sec.
Embedded Figures (EF). In 3 min, the subject had to identify as many as possible figures of the 73 present in Rey's table (Rey, 1966). Scores: one

point for each correct "complex" figure identified, half a point for each correct "single" figure identified.
Method: the test battery was performed on a Saturday morning, at least 15 hours after the last exposure to styrene vapours. The subjects were instructed to avoid consumption of alcohol and drugs during the 47 hours prior to testing.
The neuropsychological session lasted approximately one hour.

STATISTICAL PROCEDURES

Significant statistical differences between exposed and control subjects were assessed using Student's t test. Pearson's correlation coefficients were used to calculate the linear relationship of each variable with every other. Since most psychometric measures are known to be dependent on age or vocabulary, multivariate techniques were used to eliminate the effect attributable to these confounding variables.
Pearson's correlation coefficients showed some interactions between independent exposure and confounding variables. Multicollinearity was tested according to Farrar and Glauber (1967). These interactions were statistically significant and did not permit evaluation of risk of decrement in performance due to each independent (or predictor) variable. In the MANCOVA and multiple regression procedures,it can be difficult to make useful inferences because of interaction between different sources of variation. That is why the method of partial correlation was preferred to the alternative procedures of analysis to eliminate the effects caused by confounding variables.
The exposure-effect relationship between neuropsychological performance and both the level and the duration of exposure to styrene vapours was assessed by comparing the mean test scores from subgroups of workers divided according to exposure variables. Mean differences against the control group were calculated as Student's t value.
Statistical analysis was carried out on a CDC 6600 computer (Centro di Calcolo Interuniversitario dell'Italia Nord-Orientale, Casalecchio, Bologna) using programs after Nie et al (1975).

RESULTS

In Table 1, the workers and the control group are defined according to age, vocabulary, duration and intensity of exposure to styrene vapours, and performance scores. Statistically significant reductions in performance capability were found in the workers compared to the referents for all psychological tests, with the exception of DS and VM (LT) tests.
The duration as well the intensity of exposure to styrene linearly affected several performance tests. Selected RTs were closely related to the exposure intensity and, at a less significant level, to the duration of exposure.
BD test scores were related to the urinary excretion of styrene metabolites, while a relationship was found between exposure time and performance in the DS test (Fig. 1).
The first inspection of scattergrams suggests that the assumption of a linear relationship existing between the variables entered in partial correlation procedures is generally satisfied. Only RTs do not seem to increase with age before 35.

Table 1: Characterization of referent and exposed groups, according to age, vocabulary, duration and intensity of exposure to styrene vapours, and raw scores in neuropsychological tests. Mean values and SD are reported and the differences between the means are tested using Student's t values.

VARIABLE	REFERENTS		EXPOSED WORKERS	
	Mean	SD	Mean	SD
Age, y	35.1	14.3	40.0	11.5
Vocabulary	42.3	17.9	35.8	15.1
Exposure time, y	-	-	8.6	4.6
Urinary MA+PGA, μg/g creat.	-	-	413.7	271.2
RT	495.0	165.1	*623.3	168.1
LM (ST)	10.4	2.5	*8.0	3.4
LM (LT)	9.1	2.1	*6.8	3.6
VM (ST)	8.1	0.8	*6.9	1.2
VM (LT)	6.9	1.9	6.5	1.7
DS	49.5	10.5	43.6	18.6
BD	32.4	8.0	**26.7	10.1
EF	38.3	3.5	*29.1	7.5

* = $p < 0.01$ and ** = $p < 0.05$ against referents.

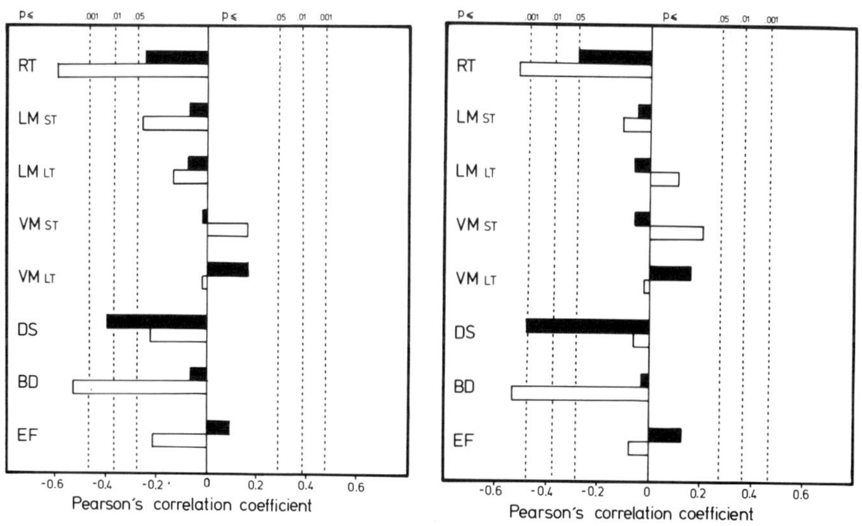

Fig. 1: Pearson's correlation coefficients and significance of the relationships between duration of exposure to styrene (dark columns) or urinary excretion of styrene metabolites (white columns) and performance in neuropsychological tests. On the left, zero order partial correlation coefficients. On the right, partial correlation coefficients after controlling for confounding variables, e.g., age and vocabulary.

Although a second-degree polynomial function gave a better description of the RT trend according to age, the linear regression was highly significant ($p<0.001$). After controlling that the assumption of linear relationship between the variables was met, partial correlation procedures were used to verify the relationships between independent exposure variables and psychological performance scores, controlling for age and vocabulary (Fig. 1, on the right).

The current exposure intensity, as measured by the urinary excretion of MA and PGA, was still closely related to RT and BD test scores. A negative relationship was also found between the duration of exposure and performance in the RT and DS tests. The partial correlation coefficients of these relationships were increased above the zero order level, after controlling for confounding factors.

In spite of the lack of relationships as measured by Pearson's correlation coefficients, short-term memory tests showed an exposure-related trend. Four subgroups of workers were identified, according to the urinary excretion of styrene metabolites immediately before testing (Table 2).

The differences between the exposed subgroups and the controls were not statistically significant for either age or vocabulary, with the exception of age, which was significantly higher for the most exposed subgroup. Exposure-effect relationships were checked by comparing the mean values for each subgroup with the referent scores.

Figure 2 shows that the short-term logic memory score, LM st, was significantly reduced for an MA and PGA sum higher than 200 mg/g creatinine. As the exposure intensity increased, no further reduction was observed in memory performance (Fig. 2a). Verbal learning, VM st, was also very sensitive to styrene exposure (Fig. 2b). The difference between exposed and control subjects was significant even in workers with an MA and PGA sum of about 100 mg/g creatinine; thus, the lack of a linear relation-

Table 2. : Characterization of four subgroups of workers according to the urinary excretion of styrene metabolites before psychological testing. Student's t test was used for comparison between each subgroup and the controls.

Group (number of subjects)		[a]MA + PGA, mg/g creatinine		Duration of exposure, y		Age, y		Vocabulary, test score	
		mean	SD	mean	SD	mean	SD	mean	SD
Controls	(16)	-	-	-	-	35.1	14.3	42.3	17.9
< 200	(14)	101	45	8.7	4.0	36.5	13.0	42.1	15.2
201-400	(9)	291	61	8.1	4.7	38.9	11.6	38.0	15.1
401-600	(14)	495	66	7.9	4.9	39.0	11.6	33.1	12.4
> 600	(13)	769	145	9.7	5.0	* 45.6	10.7	30.5	16.5

[a] Measured on Saturday morning spot samples, at least 15 hours after the last exposure to styrene vapours;

* $p<0.05$.

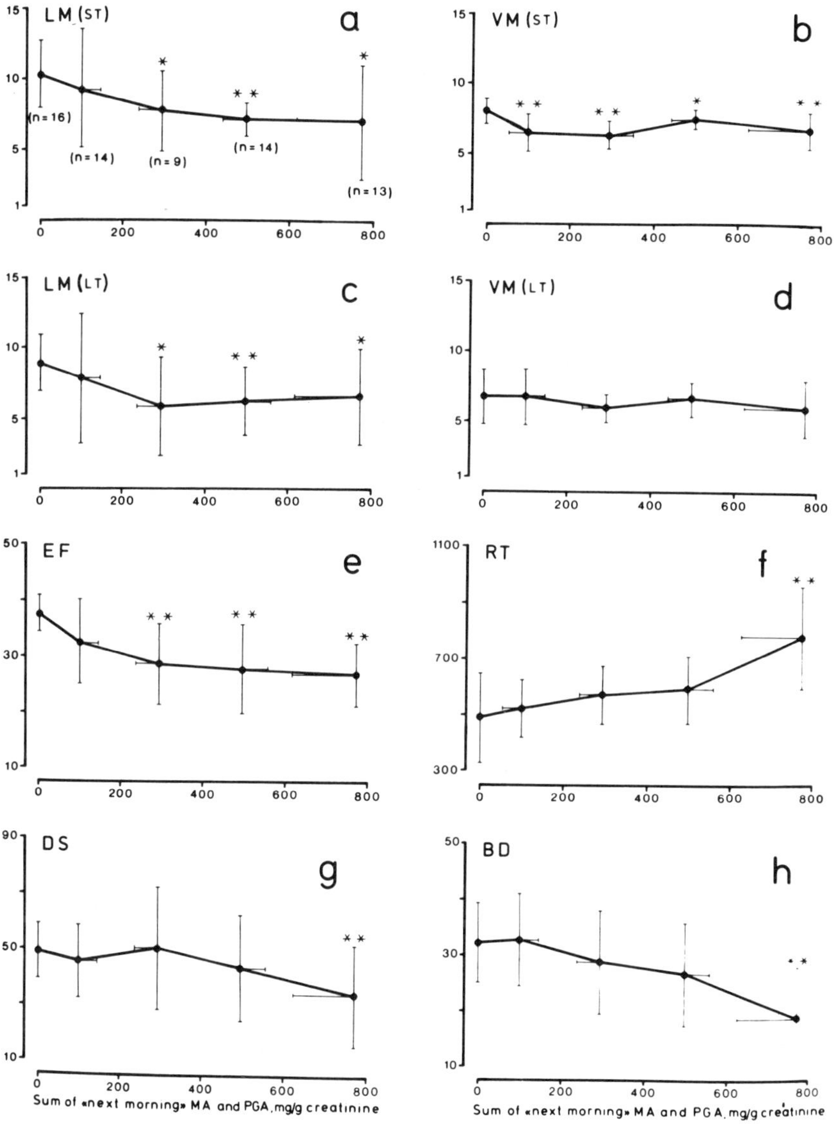

Fig. 2: Mean values and SD of neuropsychological test scores in controls and in four subgroups of workers divided according to the urinary concentration of MA and PGA in the "next morning" spot sample, as shown in Table 2. * = $p < 0.05$; ** = $p < 0.01$ against the control group (Student's t test).

ship between styrene metabolites and memory tests is explained by the fact that memory is affected early in exposed workers and no further changes occur at rising exposure levels.
When long-term memory scores of both memory tests are plotted for the aforementioned subgroups of exposed workers (Figs. 2 c and 2d), it can be seen that logic memory, LM lt, shows the same trend as short-term memory tests, while verbal learning, VM lt, does not seem to be affected in styrene workers This would mean that styrene exposure impairs learning skills but not recollection of stored material.
A trend analysis showed that only EF scores (Fig. 2e) confirm such exposure-related course as in memory tests performance. Selected RT (Fig. 2f), DS (Fig. 2g), and BD (Fig. 2h) tests were significantly affected only in workers with the highest urinary concentration of MA and PGA, i.e., higher than 600 mg/g creatinine. Workers from this subgroup were also characterized by the highest mean age and the lowest vocabulary score.
On the contrary, early changes in memory and EF tests were detectable in the subgroups of workers with age and vocabulary similar to those from the control group. The mean values of short-term verbal and logic memory were significantly reduced in workers with mean urinary concentrations of MA and PGA of 101 and 295 mg/g creatinine respectively. These biological indicators on the "next morning" spot sample correspond to 8-hour TWA styrene concentrations in the air of 25 and 60 ppm respectively.

DISCUSSION

Health-based safety criteria for occupational exposure to neurotoxic chemicals must take account of the early adverse effects on the central nervous system functions. Psychometric techniques provide sensitive and objective measures of changes in higher nervous functions.
Our group under study did not significantly differ from controls as to age and vocabulary and modifiers of effect, such as alcohol and drug consumption and preexisting diseases, were eliminated by selection procedures. However, performance in all of the performed neuropsychological tests, with the exception of long-term verbal memory recall and DS test, was significantly reduced as compared to the control group.
These results are consistent with most published psychological studies on styrene workers. We think our results agree with those of Härkönen et al. (1978) and Lindström et al. (1976). The relationship between selected RT and styrene metabolites in urine is consistent with the linear regression between simple RTs and mandelic acid in urine found by Cherry et al. (1980 and 1981). Moreover, such correlations were still present when confounding variables were eliminated by partial correlation procedures.
It can be assumed that the best exposure-related performance tests are the ones that are most sensitive to detection of the neurotoxic effects of styrene exposure. It was surprising to find that in RT and BD tests, significant changes compared to controls were present only for heavy exposure levels. It was even more surprising to find that the tests, which seemed not to be related to the exposure intensity, were sensitive enough to show a significantly reduced performance even for very low exposure levels.
Memory and EF test scores in workers with urinary MA and PGA concentrations

of 101 and 295 mg/g creatinine, respectively, were significantly lower than in a control group quite similar as regards age and vocabulary. Not only the age, but also the duration of exposure seems to be an important variable with regard to neuropsychological performance. However, the differences (not reported here) between subgroups of workers divided according to their exposure time and controls pointed to a trend of neuropsychological performance related less to the time than to the level of exposure to styrene vapours, as measured by the urinary excretion of MA and PGA. This might indicate that psychological changes are reversible enough to be independent of exposure duration.

Generally speaking, our results show greater effects of styrene exposure on higher nervous functions than reported in earlier studies by Härkönen et al. (1978), Lindström et al. (1976), and Cherry et al. (1980 and 1981). Besides consistent results, i.e., effects on visuomotor accuracy and speed, we found that also memory and visuoconstructive ability are significantly affected in styrene workers.

Several factors may account for these discrepancies: besides selection processes that make it very difficult to compare exposure-effect relationships in such selected populations during cross-sectional investigations, there are different statistical approaches that would provide complementary rather than inconsistent results. The data presented seem somewhat contradictory. The more unequivocal differences between exposed and control groups were found in those performance tests that did not appear to be related to exposure time or intensity. On the contrary, when such a relationship was found, the differences versus controls were less evident. It is obvious that a self-limiting early loss of a function or the presence in the group of very resistant subjects might lead to the absence of a relationship with the independent variable, which could further increase.

Using the comparison between the means or correlational approaches, it was possible to detect early changes in all tested neuropsychological functions, with the sole exception of long-term verbal learning. The interpretation of performance results in terms of health impairment and working capacity is difficult. Possibilities and disadvantages of neuropsychological methods have been extensively discussed by Foà et al. (1980), who in addition to the obstacles to the clinical application of experimental and field study data, also pointed out the wide variability in methodologies and results of behavioral toxicology. Nevertheless, the reliability and the validity of most tests comprising our battery have been successfully tested (Wechsler, 1955; Valciukas and Lilis, 1980). In particular, the latter Authors stated that performance tests detect neurological dysfunction earlier than overt clinical symptoms.

The unequivocal differences in performance between exposed and control groups indicate that exposure to styrene vapours causes early changes in central nervous system functions, as measured by behavioral tests. The assessment of a threshold level for neurotoxic effects greatly depends on variables which are difficult or even impossible to control. In the present study, exposure to styrene concentrations of about 25 ppm, corresponding to about 100 mg/g creatinine of MA and PGA in the "next morning" urine spot sample, seems to cause a significant impairment of short-term verbal learning. Most memory tests and the test for visual perception in masked conditions (EF) show significant changes at exposure levels of about 60 ppm and the impairment persists throughout the exposure range.

These figures strongly point to the need for very low threshold limit values as health - based safety criteria, even without considering the increasing amount of data suggesting the possible carcinogenicity or mutagenicity of styrene in humans.

ACKNOWLEDGEMENT

We would like to thank Dr P. Rustichelli and Mr P.A. Albertini for their skillful technical assistance. This study was supported in part by the Department of Social Security, Emilia-Romagna Region.

REFERENCES

Chasson, A.L., H.J. Grady, M.A. Stanley (1961). Am. J. Clin. Pathol.35,83-89

Cherry, N., H.A. Waldron, G.G. Wells, R.T. Wilkinson, H.K. Wilson, S. Jones (1980). Br. J. Ind. Med. 37, 234-240.

Cherry, N., B. Rodgers, N. Venables, H.A. Waldron, G.G. Wells (1981). Br. J. Ind. Med. 38, 346-350.

Elofsson, S.A., F. Gamberale, T. Hindmarsch, A. Iregren, A. Isaksson, I. Johnsson, B. Knave, E. Lydahl, P. Mindus, H.E. Persson, B. Philipson, M. Steby, G. Struwe, E. Söderman, A. Wennberg, L. Widen (1980). Scand. J. Work Environ. Health 6, 239-273.

Farrar, D.E., and R.R. Glauber (1967). Rev. Econ. Stat. 49, 92-107.

Foà, V., M.G. Cassitto, R. Gilioli (1980). Advances in Neurotoxicology,215-221, Pergamon Press, Oxford, UK.

Gaetani, E., C.F. Laureri, M. Vitto, M. Falzoi, A. Mutti (1982). Med. Lavoro (in press).

Gamberale, F., H.O. Lisper, B. Anshelm-Olson (1979). Adverse effects of the Environmental Chemicals and Psychotropic Drugs, 135-148, Elsevier, Amsterdam, the Netherlands.

Götell, P., O. Axelsson, B. Lindelöf (1972). Work Environ. Health 9, 76-83.

Hane, M., O. Axelsson, J. Blume, G. Hogsted, L. Sundell, B. Ydreborg (1977). Scand. J. Work Environ. Health 3, 91-99.

Hänninen, H., L. Eskelinen, K. Husman, M. Nurminen (1976). Scand. J. Work Environ. Health 4, 240-255.

Hänninen, H., M. Nurminen, M. Tolonen, T. Martelin (1978). Scand. J. Psychol 19, 163-174.

Härkönen, H., K. Lindström, A.M. Seppäläinen, S. Asp, S. Hernberg (1978). Scand. J. Work Environ. Health 4, 53-58.

Lindström, K., H. Härkönen, S. Hernberg (1976). Scand. J. Work Environ. Hlth 3, 129-139.

Lindström, K. (1980). Am. J. Ind. Med. 1, 69-84.

Matarazzo, J.D. (1972). Wechsler's Measurement and Appraisal of Adult Intelligence, 5th Ed., Williams & Wilkins Co., Baltimore, MD, USA.

Nie, N.H., C.H. Hull, J.G. Jenkins, K. Steinbrenner, D.H. Bent (1975). SPSS, Statistical Package for the Social Sciences. McGraw-Hill,New York,NY, USA.

Oltramare, M., E. Desbaumes, C. Imhoff, W. Michelis (1974). Toxicologie du styrene monomère, Ed. Médecine et Hygiène, Genève, Switzerland.

Rey, A. (1966). *Les troubles de la mémoire et leur examen psychométrique*, C. Dessart, Bruxelles, Belgique.

Stewart, R.D., H.C. Dodd, E.D. Baretta, A.W. Schaffer (1968). *Arch. Environ. Health 16*, 69-84.

Valciukas, J.A., and R. Lilis (1980). *Environ. Res. 21*, 275-297.

Wechsler, D. (1955). *Manual for the Wechsler Adult Intelligence Scale.*, the Psychological Corporation, New York, NY, USA.

Comprehensive Neurophysiological Evaluation of Lead-exposed Workers

G. Sborgia*, G. Assennato*, N. L'Abbate*, L. De Marinis*, C. Paci*, M. De Nicolo*, G. De Marinis*, N. Montrone*, E. Ferrannini**, L. Specchio**, G. Masi** and G. Olivieri**

*Institute of Occupational Health, University of Bari, Bari, Italy
**Institute of Neurology, University of Bari, Bari, Italy

ABSTRACT

A survey of the neurotoxic effects of lead in battery workers was carried out, based on the evaluation of multiple neurophysiological parameters. Thirty-one battery workers and thirty-seven controls (industrial workers from a minor iron foundry) were examined. Exposure levels in both groups were evaluated by determination of blood lead levels (PbB) and zincprotoporphyrin (ZPP). The neurophysiological tests included Visual Evoked Response-VER- (latency and amplitude in both eyes), motor conduction velocity of median, ulnar, deep peroneal and tibial nerves; sensory conduction velocity of the median nerve, recording of the smooth pursuit eye movements. Since 1979 the battery workers had undergone periodic medical surveillance, based on PbB and ZPP. Current maximal and time-weighted average PbB values were taken into account as indicators of exposure.
The survey also included a standardized questionnaire, audiometric testing and ophthalmologic and neurologic examinations.
No significant difference was detected between the two groups for the parameters under evaluation, and no consistent dose-effect relationship was found.
In our experience, no single neurophysiological parameter seems to be useful in the medical surveillance of low level lead exposed subjects working in small plants.

KEYWORDS

Lead exposure; neurophysiological tests; visual evoked response.

INTRODUCTION

Medical surveillance of lead-exposed workers should also take the potential neurotoxic effects into account. Moreover, contradictory data exist in the literature, thus making the straightforward use of neurotoxic screening tests in every-day medical surveillance impossible. As shown in Table 1,

TABLE 1. SELECTED LITERATURE ON NERVE CONDUCTION VELOCITY AND LEAD EXPOSURE

INVESTIGATORS	N. OF EXPOSED	SOURCE OF CONTROLS	EXP. VS. NON-EXP. SIGNIFICANT DIFFERENCES	EXP. VS. NON-EXP. NON-SIGNIFICANT DIFFERENCES	DOSE-EFFECT OR DOSE-RESPONSE	COMMENTS
1) CATTON (1970)	19	HOSPITAL PERSONNEL	KNEE/ANKLE MAP AMPLITUDE	LATERAL POPLITEAL MCV	NO CORRELATION WITH PbB	
2) SEPPÄLÄINEN (1975)	39	PROFESSIONAL, SKILLED, OR UNSKILLED WORKERS	ULNAR MCV MEDIAN MCV LATERAL POPLITEAL MCV, ULNAR CVSF		NEUROPHYSIOLOGICAL SCORE UNRELATED WITH DOSE MEASUREMENTS	"IT IS SUGGESTED THAT NERVE DAMAGE IS PRODUCED INDIPENDENTLY OF OTHER MANIFESTATIONS OF POISONING"
3) REPKO (1978)	85	SKILLED AND UNSKILLED WORKERS; UNEMPLOYED	MEDIAN MCV ULNAR MCV PERONEAL MCV	ULNAR SCV ULNAR CVSF	POSITIVE CORRELATION PbB VS. ULNAR MCV (?). NO CORRELATION FOR THE OTHER PARAMETERS	
4) SEPPÄLÄINEN (1979)	61	PROFESSIONAL, SKILLED, OR UNSKILLED WORKERS	POSTERIOR, TIBIAL MCV, MEDIAN SCV AND dSCV, ULNAR CVSF		MEDIAN MCV, SCV, dSCV VS., PbB MAX, ULNAR CVSF, SCV VS., PbB MAX	NON-RESPONSE LEVEL 40-49 mcg%
5) BALOH (1979)	69	INDUSTRIAL CONTROLS		ULNAR FAST AND SLOW MCV, SCV PERONEAL, SURAL SCV	CURRENT AND TWA PbB VS. ULNAR FAST AND SLOW MCV	
6) SPIVEY (1980)	55	INDUSTRIAL CONTROLS		ULNAR FAST AND SLOW MCV, SV PERONEAL MCV, SURAL SCV	CURRENT AND TWA PbB VS. ULNAR FAST MCV	FOLLOW-UP OF THE STUDY N. 5 "NERVE CONDUCTION MEASUREMENTS ARE PROBABLY NOT A GOOD EARLY INDICATOR OF LEAD EFFECT ON THE NERVOUS SYSTEM"
7) ASHBY (1980)	94	MANUAL AND OFFICE WORKERS	ULNAR MCV, MEDIAN MCV, RADIAL MCV, PERONEAL MCV	ULNAR SCV	POSITIVE CORRELATION PbB, VS ULNAR MCV (?) ULNAR MCV RELATED WITH LENGTH OF EMPLOYMENT	"IT IS ESTABLISHED THAT A SUBCLINICAL NEUROPATHY IS PRESENT IN THE LEAD EXPOSED WORKERS"
8) BUCHTAL (1980)	20	PREDICTED FROM EQUATION ON HEALTHY SUBJECTS	MEDIAN MCV MEDIAN SCV PERONEAL MCV SURAL SCV		NO CORRELATION WITH PbB	MCV AND SCV WITHIN THE NORMAL RANGE. "IT IS DOUBTFUL WHETHER THE SLIGHT SLOWING IN CONDUCTION SHOULD BE CONSIDERED TO CONSTITUTE SUBCLINICAL NEUROPATHY"
9) BORDO (1982)	62	HOSPITAL PERSONNEL	MEDIAN MCV MEDIAN SCV		NO CORRELATION WITH LENGTH OF EXPOSURE. CORRELATION WITH PbB MAX	"A DECREASE IN CONDUCTION VELOCITY OCCURS QUITE RAPIDLY AFTER EXPOSURE"

where selected literature on this subject is reported, there is considerable disagreement even about the actual occurrence of subclinical lead neuropathy: some studies showed a decrease in the peripheral nerve conduction velocity in workers with lead levels consistently below 80 μg% ml; while other studies failed to replicate such outcome (Seppäläinen, 1975, 1979; Baloh, 1979; Spivey, 1980).

Likewise, some investigators showed a clear-cut dose-effect relationship between blood lead level and neurophysiologic parameters (Ashby, 1980) while negative results were obtained by other researchers (Catton, 1970). There is no agreement even on the subclinical meaning of the alterations observed. For example, according to Buchtal (1980), the reduction in the conduction velocity is within the normal range and does not have a prognostic value since it is reversible. On the other hand, some investigators affirm that the damage occurs rather rapidly after the beginning of exposure (Bordo, 1982; Ashby, 1980), whereas, according to others,the effect is cumulative. Epidemiologic problems are probably crucial in understanding the different results: different control groups (e.g., office workers - Ashby, unemployed workers - Repko, hospital staff - Seppäläinen and Bordo, industrial controls - Baloh and Spivey), problems related to small sample size, the evaluation of multiple parameters, the use of different dose indicators (last blood lead level vs maximum blood lead level vs average blood lead level), and probable different length of exposure, different study design (cross-sectional vs prospective). It should be stressed that the only prospective study (Spivey, 1980), is essentially negative, at least for the conduction velocity.

Obviously, the biological mechanism itself, as yet unknown, could explain the contradictory results.

In spite of the limitations in the cross-sectional studies mentioned by Bertazzi (1982) in this volume, field studies are needed in every-day practice of occupational medicine. Therefore, for a correct prevention of lead toxicity, it is indispensable to obtain information by both a critical evaluation of the literature and a comprehensive survey. That is, the following questions should be addressed:

1) Do conduction velocity variations reported by many investigators in lead-exposed workers represent a subclinical neuropathy that in the natural history of neurotoxic lead pathology precedes the onset of the clinical neuropathy, or are the events completely unrelated?

2) Does a dose-effect relationship exist between dose indicators and neurophysiological parameters?

In the affirmative, it is possible to avoid the use of the neurophysiological tests in medical surveillance practice by means of biological monitoring; on the other hand, if there is no dose effect relationship, biological monitoring is not suitable and it will obviously be necessary to perform functional monitoring, based on the control of the neurophysiological parameters. In this case, owing to the impossibility of using all the neurophysiological parameters at once, the neurophysiological test which is best suited to periodic health surveillance should be used.

In view of these limited objectives, given the study design, we carried out a comprehensive survey using a group of exposed and a group of non-exposed workers.

MATERIALS AND METHODS

The group under examination consisted of 31 workers from a battery storage plant in the Apulia region. The factory was located in an area with low industrialization and therefore had a low turnover. Longitudinal measures of blood lead levels for the last four years were available for all the workers. The control group consisted of 37 volunteers from a minor iron foundry in the same area. The only difference previously known between the two groups was the longer exposure to noise in the controls.

All the exposed workers were given 1 mg of Na_2 - Ca EDTA per day for three days before the beginning of the investigation. Distribution of age, length of exposure and biological parameters in the two groups is reported in Table 2.

The obvious exponential correlation between blood lead level and ZPP (r = = .886) should be noted.

All the tests were carried out at the University Hospital of Bari. The study group and the control group were mixed up so that the workers examined did not know to which group each person belonged. During the investigation, two workers underwent the scheduled tests every day.

TABLE 2.

CHARACTERISTICS OF WORKERS EXPOSED TO LEAD (31) AND REFERENTS (37)

	EXPOSED GROUP			REFERENCE GROUP			P
	MEAN	±	SD	MEAN	±	SD	
Age (years)	42.22		8.72	41.89		6.71	N.S.
Length of employment (years)	7.84		3.14	12.78		5.19	<0.05
PbB (mcg/100 ml)	62.52		19.03	23.57		7.38	<0.001
ZPP (mcg/100 RBC)	223.71		166.14	16.86		3.27	<0.001

Biochemical tests. Blood samples were taken and the neurological test was done for each participant on the same day. The blood lead level was determined by atomic absorption spectrophotometry, and ZPP was calculated with the Buchler Haematofluorimeter.

Neurological test. A neurological test was done for each person in order both to ascertain clinically evident lesions and to detect pathologic states of a different nature that could be due to other causes. The results of the tests, which were carried out at the Neurologic Clinic, were reported on a standardized form (as used by Repko).

Nerve conduction studies. The motor nerve conduction velocity of the ulnar, median, posterior tibial and lateral peroneal nerves, the sensitive conduction velocity (proximal and distal) of the median nerve, and the distal latency and the amplitude of the potentials were determined with standardized methods.
The distal and proximal surface temperatures were measured; the values were corrected to a standard temperature of 31.5°C (by using the De Jesus equation).

Audiometric measurements. Audiometric measurements were taken at the Occupational Health Institute by an audiometrician. The test consisted of the determination of the threshold for pure tone conduction through air and bones.

Ophthalmologic tests. Each person underwent an ophthalmologic examination, with determination of natural and corrected visus, refraction test, intraocular tension, colour sense, eye movement function test and determination of the visual evoked potentials (pattern reversal). The eye tracking technique has been described elsewhere (Specchio et al, 1981).
For the visual evoked potentials, the person sits in a dark room for about 15 minutes,1 meter from the pattern stimulator. Each eye was tested separately by covering the other, after correcting any refraction defect, with the pupil dilated.
Electrodes were places as follows:

1) Recording electrodes 2.5 cm. above the occipital inion.
2) Ground electrode at the ipsilateral lobe.
3) Reference electrode at the opposite auricular lobe.

In the stimulation, the "transient" procedure was used with checkers of 21 mm., frequency of 1 Hz with slight contrast.
Signals received by the electrodes were processed by a computerized averager for about 60 times. In the analysis, the latency and the amplitude of the wave P 100 were taken into account.

Statistical Methods. Between groups, comparison of means was evaluated by Student's t-test. Frequency distribution of qualitative data was analyzed by the chi-square test. Dose-response relationship was evaluated in the exposed group by a multiple regression model where the dependent variable was the parameter under study and the independent variables were the risk factors and potentially associated findings.

RESULTS

No statistically significant difference was found between the two groups (Table 3) with regard to the variables taken from the questionnaire (eating and smoking habits, alcohol consumption, sleeping habits, occupational accident frequency, etc.).

TABLE 3.

SUMMARY OF CHI SQUARE (χ^2) ANALYSES FOR RESPONSES FROM THE PERSONAL DATA QUESTIONNAIRE

QUESTION	DF	χ^2
EATING HABITS:		
Quality of appetite	3	6.1854
SLEEPING HABITS:		
Time to go to sleep	3	1.9447
Time most alert	4	0.2065
Sleep interruption	3	0.3271
Restedness	3	0.7354
CIGARETTE SMOKING:		
Quantity per day	1	1.7341
Smoking habit	4	4.6568
DRINKING:		
Beer	4	2.5426
Wine	3	4.3043
Liquors	3	0.9658
WORK HABITS:		
Work injury	1	3.3103

The differences in clinically observed signs are reported in Fig. 1 (reflexes and reduction of muscular force). A higher frequency in reflex alteration (both increasing and decreasing) and in reduction of muscular force was observed in the exposed) even though the small number does not allow a statistical evaluation. In the exposed groups, the workers with clinical anomalies did not have mean ZPP blood lead level values significantly different from those of the normal workers. The same applies to the neurophysiological variables.

Fig. 1. PERCENTUAL DISTRIBUTION OF SELECTED CLINICAL DATA IN LEAD EXPOSED AND CONTROLS

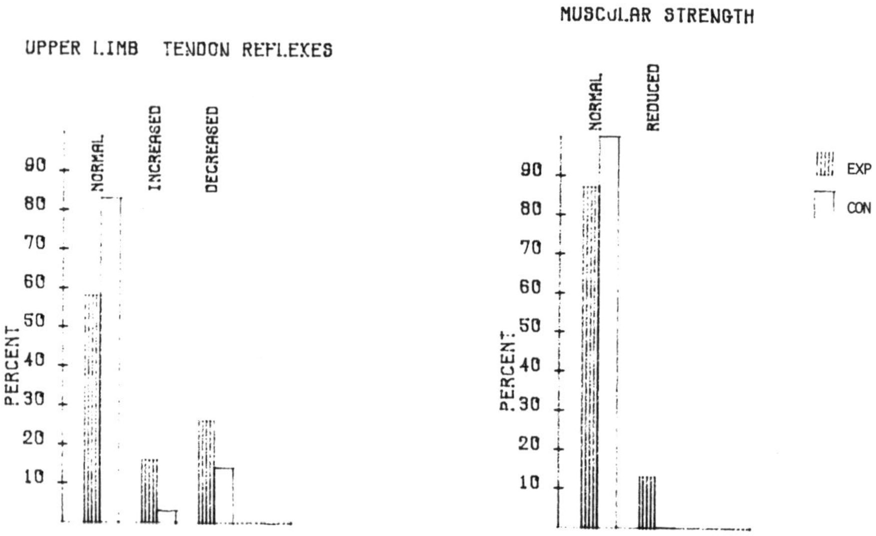

Neurophysiological parameters. The values of the neurophysiological parameters are reported in Table 4: no significant difference was observed. The data for two control workers were not collected because they had to start work again and the test was not performed. The only significant difference is observed in the lateral popliteal potential.
No dose-effect relationship was observed, except for a paradoxical positive correlation between maximal ZPP and ulnar nerve MCV.
The multiple regression model used showed a correlation between nerve MCV and age and smoking habits in the exposed, as was expected (Table 5).

Figure 2 shows the means of the audiometric values for each ear in the two groups: a clear difference is noted that must be attributed to the longer noise exposure for the controls. As for the other parameters, no correlation exists in the exposed subjects between audiometric measurements and biochemical parameters.

The data of the smooth pursuit eye movement (SPEM) were analyzed for each of the five tests. The PEV showed differences that were not statistically significant (Fig. 3). The AV revealed a significance at the first test that disappeared in the following ones, but, with the exception of test 4, the values were consistently higher for the exposed subjects (Fig. 4).

TABLE 4.

ELECTRONEUROGRAPHIC PARAMETERS OF LEAD-EXPOSED WORKERS (31) AND REFERENTS (35)

	EXPOSED GROUP		REFERENCE GROUP	
	MEAN	± SD	MEAN	± SD
MEDIAN NERVE				
FAST M.C.V. (M/S)	55.37	7.22	56.55	7.31
SLOW M.C.V. (M/S)	50.12	8.62	49.72	8.47
AMPLITUDE (MV)	15.62	4.28	14.24	5.70
DISTAL LATENCY (MS)	3.51	0.55	3.61	0.63
DISTAL S.C.V. (M/S)	63.42	10.37	62.70	9.52
PROXIMAL S.C.V. (M/S)	61.21	7.41	60.98	9.42
ULNAR NEVE				
FAST M.C.V. (M/S)	54.97	5.54	54.48	5.38
AMPLITUDE (MV)	15.90	5.17	15.73	5.38
DISTAL LATENCY	2.64	0.43	2.68	0.46
LATERAL POPLITEAL NERVE				
FAST M.C.V. (M/S)	49.28	5.18	49.98	5.57
AMPLITUDE (MV)	8.44 *	3.71	10.51	4.10
DISTAL LATENCY (MS)	3.77	0.84	3.79	0.65
TIBIAL NERVE				
FAST M.C.V. (M/S)	48.70	8.82	46.37	5.03
DISTAL LATENCY (MS)	5.17	1.08	5.37	1.02

* $p < 0.05$ (Student's t-test)

TABLE 5.

ANALYSIS OF VARIANCE FOR 11 SPECIFIED VARIABLES, ON A MULTIPLE LINEAR REGRESSION MODEL, IN THE LEAD-EXPOSED GROUP WITH ULNAR MCV AS A DEPENDENT VARIABLE

SOURCE	DF	SS	MS	F	P
TOTAL	30	922.8			
REG	11	535.7	48.7	2.4	N.S.
Age	1	143.4	143.4	7.0	< 0.05
Length of employment	1	3.6	3.6	0.2	N.S.
Current PbB	1	11.3	11.3	0.6	N.S.
Current ZPP	1	3.7	3.7	0.2	N.S.
Smoking habits	1	140.5	140.5	6.9	< 0.05
Alcohol intake	1	57.5	57.5	2.8	N.S.
Average PbB	1	17.8	17.8	0.9	N.S.
Maximal PbB	1	6.8	6.8	0.3	N.S.
PbU-EDTA	1	1.2	1.2	0.1	N.S.
Average ZPP	1	16.1	16.1	0.8	N.S.
Maximal ZPP	1	133.7	133.7	6.6	< 0.05
RESID	19	387.2	20.4		

Fig. 2. - MEAN HEARING LEVELS FOR RIGHT AND LEFT EAR OF LEAD EXPOSED AND CONTROLS

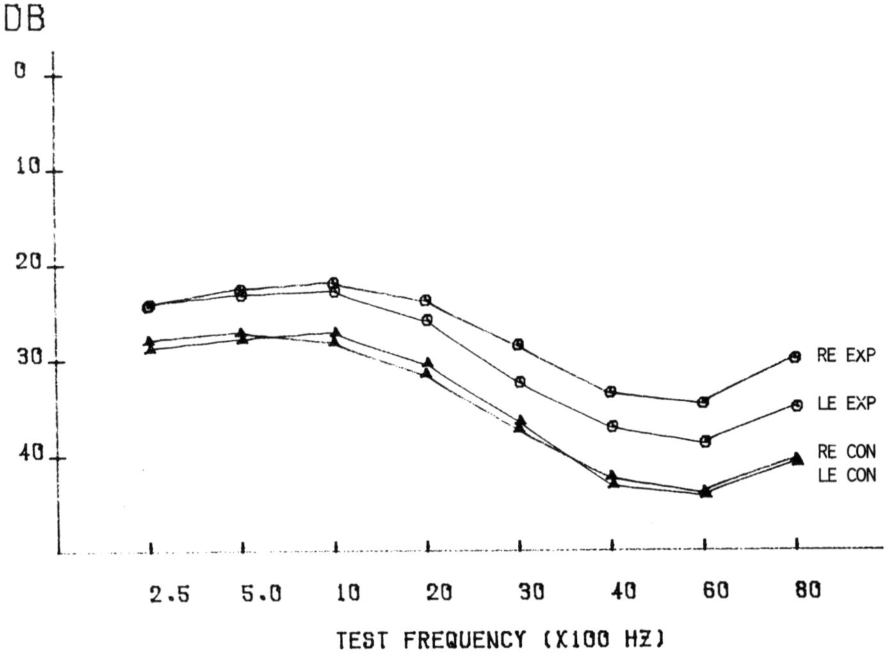

Fig. 3. MEAN VALUES (±S.D.) OF SPEM - PEV FOR EACH TRIAL IN LEAD EXPOSED AND CONTROLS

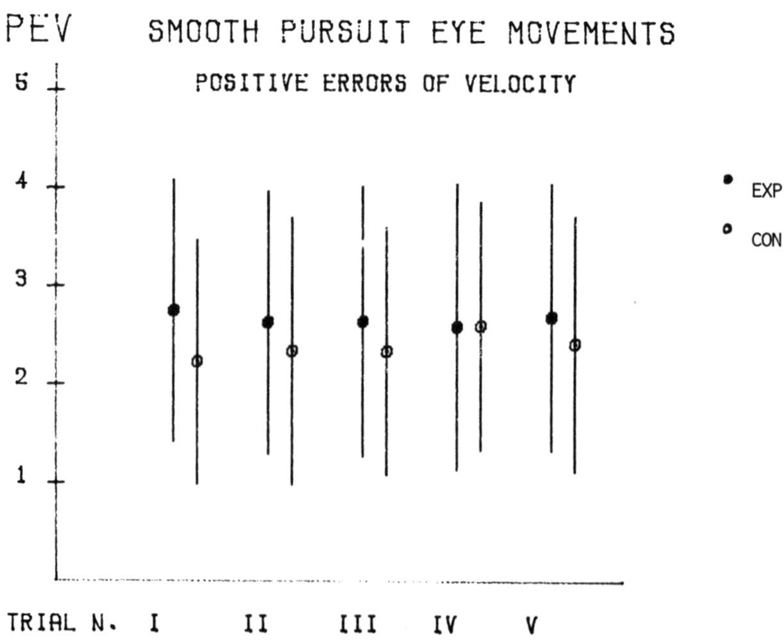

Fig.4. MEAN VALUES (±S.D.) OF SPEM - VA FOR EACH TRIAL IN LEAD EXPOSED AND CONROLS

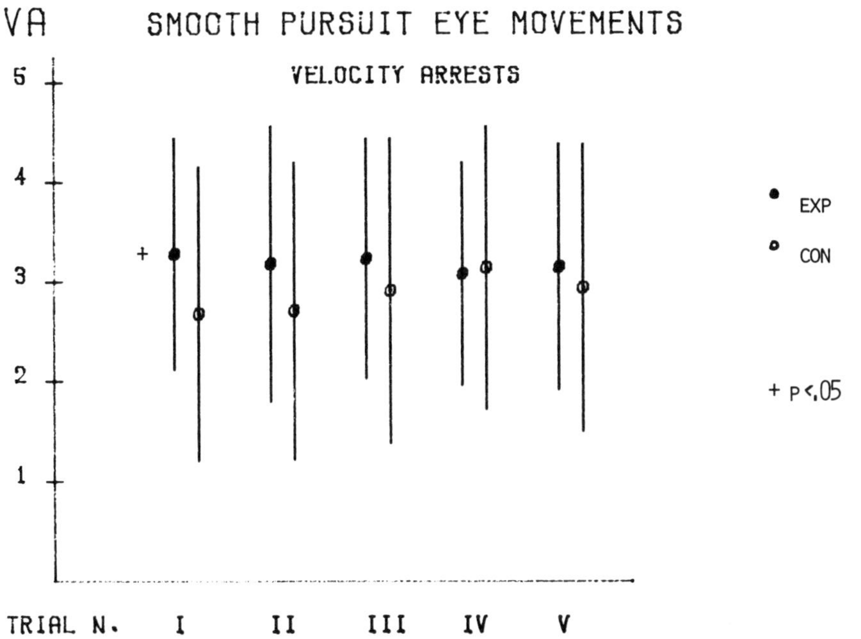

No dose-effect relationship was found between biochemical alterations and SPEM (both EP and AV).
The data relative to the ophthalmologic parameters are shown in Table 6: visual acuity was lower in the exposed subjects even though not statistically significant. Intraocular tension was normal in both groups.

TABLE 6.

OCULAR PARAMETERS AND VISUAL EVOKED RESPONSE IN LEAD-EXPOSED WORKERS (30) AND REFERENTS (37)

	EXPOSED GROUP		REFERENCE GROUP	
	MEAN ±	SD	MEAN ±	SD
Right eye visual acuity	8.42	2.72	9.32	1.94
Left eye visual acuity	7.87	3.40	8.46	2.75
Right eye tension	17.70	1.23	17.84	0.80
Right eye latency	105.60	18.80	101.22	15.84
Left eye latency	102.50	16.31	99.43	14.86
Right eye amplitude	5.36	3.49	6.49	4.36
Left eye amplitude	4.73	2.92	4.78	2.41

Measurement of P 100 latencies in the two groups are reported in Fig. 5: the differences are not statistically significant, neither for left nor right eye latency. However, the distribution of the left eye latencies of the controls shows that two values are clearly outliers: this was proved with the Dixon test for the outliers ($p < .01$); after eliminating the outliers, a statistically significant difference resulted between the left eye latencies. No dose-effect relationship was found in the exposed.

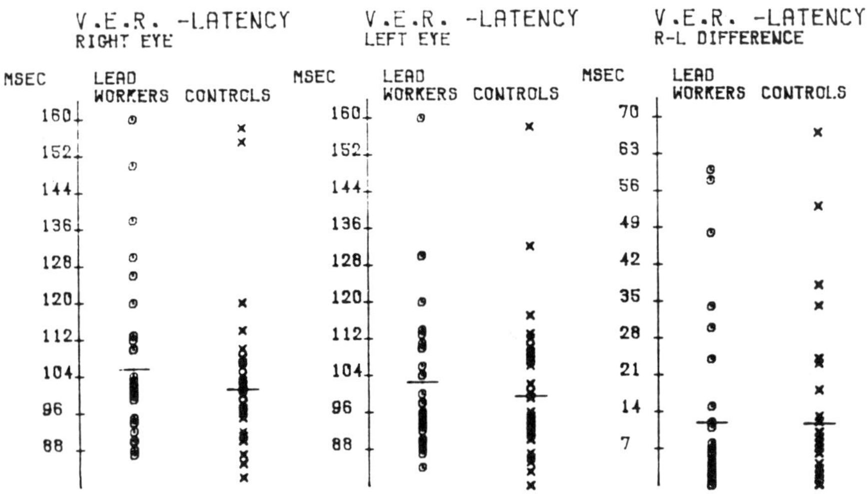

Fig. 5. VISUAL EVOKED RESPONSE (RIGHT.LEFT EYE LATENCY AND R-L LATENCY DIFFERENCE) IN LEAD WORKERS AND CONTROLS.

DISCUSSIONS AND CONCLUSIONS

The limited objectives for our cross-sectional comprehensive study aimed at improving medical surveillance of lead exposed workers: is biological monitoring able to predict the neurotoxic effects, or are functional tests needed and, in this case, which test is most suitable for these purposes? We have also dealt with the problem of critically analyzing the literature in order to evaluate the methodological or biological reasons of the contradictory results.

On the basis of our data we can exclude a dose-effect relationship between the usual dose indicators and the parameters measured; this could be due to inadequacy of dose indicators, since blood lead levels and ZPP do not indicate the dose reached in the critical organ (the nervous system), or to more complex biological characteristics: for example, in the mechanism of action, the interactions between metals could be important (e.g., the mechanism of neurotoxic action could be related to lead-calcium interactions). Hence, individual susceptibility could be very important. Moreover, the contradictory data may be explained by the characteristics of the neurotoxic mechanism: if demyelination is the pathogenetic key to lead neurotoxicity, then the pattern of remissions and exacerbations typical of demyelinating diseases (e.g., multiple sclerosis) could mask a dose-effect relationship and cause considerable disagreements between different studies.

Among the neurophysiological tests used, we can confirm that the most suitable are the visual tests. We found a certain alteration in the latencies of the P 100 wave in the exposed: this is reasonable if we assume that lead has a demyelinating action; in this case, the central fibres of the retina, that are mostly small, would be highly susceptible to demyelination.
They would be sensitive to modifications of energy metabolisms, since they have a high metabolic requirement, given the high fibre density. Given the characteristics of the VER which, according to Chiappa (1982), is extremely valuable for screening eye neuropathies, it is possible to monitor exposed subjects through periodically repeated VER measurement. In a previous study, we observed differences in eye tracking behaviour of lead-exposed subjects compared to a control group.
In this study, the differences showed a statistical significance only in the first test for the AV parameter. This could be due both to poorer cooperation of the non-exposed, many of whom had to start work again and were also probably less motivated to participate, and to a somewhat approximated signal analysis, which was done subjectively and not automatically. However, the presence of variations seems to confirm the existence of eye tracking alterations, which represent impairment of the occipito-mesencephalic pathways. The classical neurophysiological parameters do not seem to be very useful in medical surveillance of exposed subjects, at least for the exposure levels observed.
In conclusion, it is reasonable that medical surveillance in low-level lead exposure be made through the evaluation of visual parameters (VER, SPEM).
A definite evaluation of the biological meaning of such alterations cannot be drawn from this study at present, given the shortcomings of the cross-sectional study design.

ACKNOWLEDGEMENTS

We would like to acknowledge the technical support of S. Scuderi, F. Caragnano, R. Lofrumento, S. Crivellini and B. Paradiso.
We are also grateful to Prof. C. Di Benedetta for making available the facilities of the Institute of Physiology.

REFERENCES

Ashby, J.A.S. (1980). Br. J. Ind. Med., 37, 133-140.
Baloh, R.V., J.H. Spivey, C.P. Brown, D. Morgan, D.S. Campion, B.L. Browdy, J.L. Valentine, H.C. Gonick, F.J. Massey and B. D. Culver (1979). JOM, 21 (7), 490-496.
Bertazzi, P.A., G. Assennato and S. Guercilena (1982). Proceedings of the International Symposium on Neurobehavioral Methods in Occupational Health. State of the Art and Emerging Trends. Como and Milano, Italy. June 23-26, 1982.
Bordo, B.M., G. Filippini, N. Massetto, M. Musicco and R. Boeri (1982). Scand. J. Work Environ. Health, 8 (suppl. 1), 142-147.
Buchtal F. and F. Behese (1979). Br. J. Ind. Med., 36, 135-149.
Catton, M.J., M.J.G. Harrison, P.M. Fullerton and G. Kazantzis (1970). Brit. Med. J., 2, 80-82.
Chiappa, K.H. and A.H. Ropper (1982). New England Journal of Medicine, 306 (19), May 13, 1140-1150.
Repko, J.D., C.R. Corum, P.D. Jones and L.S. Garcia (1978). NIOSH Reasearch Report, DHEW (NIOSH) Pub. n. 78-128, Cincinnati, Ohio.
Seppäläinen, A.M., S. Tola, S. Hernberg and B. Koch (1975). Arch. Environ. Health, 30, 180-183
Seppäläinen, A.M., S. Hernberg and B. Koch (1979). Neurotoxicol.,1, 313-320.

Specchio, L.M., R. Bellomo, G. Pozio, F. Di Cuonzo, G. Assennato, A. Federici, G. Misciagna and F.M. Puca (1981). Clinical Toxicology, 18 (11), 1269-1276.

Spivey, G.H., R.W. Baloh, P. Brown, B.L. Browdy, D.S. Campion, J. L. Valentine, D.E. Morgan and B.D. Culver (1980). JOM, 22(9), 607-612.

Event-related Slow Brain Potential Changes in Asymptomatic Children with Secondary Exposure to Lead

D. Otto*, V. Benignus*, K. Muller**, C. Barton*** and P. Mushak****

*U.S. Environmental Protection Agency, Research Triangle Park, NC 27711, USA
**Department of Biostatistics, University of North Carolina, Chapel Hill, NC 27514, USA
***Department of Psychology, University of North Carolina, Chapel Hill, NC 27514, USA
****Department of Pathology, University of North Carolina, School of Medicine, Chapel Hill, NC 27514, USA

ABSTRACT

Slow cortical potentials were studied in young children with blood lead (PbB) levels ranging from 6 to 52 μg/dl. Slow wave (SW) voltage recorded from the scalp during sensory conditioning varied as a linear function of PbB level and age. Evidence of a persistent alteration of brain function was also found in a two-year follow-up assessment of the children.

KEYWORDS

Developmental neurotoxicology, body lead burden, slow brain potentials, sensory conditioning, CNS function.

INTRODUCTION

Children are at greater risk than adults for brain damage from lead (Pb) exposure due to increased vulnerability of the immature brain, greater absorption rate of ingested Pb, and pica behavior -- the tendency of young children to eat nonfood substances (e.g., dirt or paint chips) that contain Pb (USEPA, 1977). Although the prevalence of childhood lead poisoning has been steadily decreasing in the United States since the use of Pb in household paint was banned, inadvertent contamination of homes by workers in lead industries still poses a significant threat to young children (Dolcourt et al, 1978). Increasing evidence suggests that asymptomatic Pb exposures (i.e., blood Pb levels below 60 μg/dl) in young children are associated with subtle learning and developmental disorders (Needleman, 1980). This evidence clearly emphasizes the need for rigorous enforcement of appropriate hygienic procedures where workers handle or are exposed to Pb compounds.

Results of a preliminary study of the effects of asymptomatic Pb exposure on central nervous system function in young children will be reviewed here. These findings have been reported in more detail elsewhere (Otto, et al., 1981, and Benignus et al, 1981). Results of a two-year

year follow-up study of a subgroup of these children will also be reviewed. About half of the children in this study came from homes in which the parent(s) worked in a lead battery factory. Pb exposure in other children resulted from the ingestion of paint chips. Results indicated that slow wave (SW) voltage during sensory conditioning varied as a linear function of PbB level and age. Evidence of persistent changes in brain function was found in a subset of the children studied two years later.

METHODS

One hundred children (53 boys) aged 13 - 75 months participated in this study. Most children were from low-income Black families. All children were asymptomatic at the time of study. This study was part of a comprehensive neurobehavioral evaluation carried out at North Carolina Memorial Hospital. Pediatric and neurological examinations were administered and extensive demographic data were obtained from parent interviews.

Body Pb burden was determined from blood samples obtained from children on the day of evaluation. PbB levels were analyzed by flame atomic absorption spectrometry using the Delves-cup method (Delves 1970).

Silver-silver chloride electrodes were used to record EEG from Fpz, Cz, P3 and P4 referred to linked mastoids. The frontal electrode was used to monitor vertical eye movements. EEG was low-pass filtered at 30 Hz (3 dB down). Amplifiers were modified to yield an 8 sec time-constant. Trials with eye movements, blink, blocking, muscle or other artifacts were eliminated from averages computed off-line.

Testing was done in a small acoustic chamber. Children under 48 months usually sat in the lap of an adult. Cartoons were used to engage the attention and minimize eye/body movement artifact. In order to test very young children, a classical conditioning paradigm that required no verbal instruction or motor response was employed. A 30 msec 2 KHz tone (conditional stimulus) was paired with a 1.5 sec blackout (unconditional stimulus) of the cartoon. An electromechanical shutter attached to the movie projector was used to produce the blackout period which commenced 1.5 sec after tone onset. The tone occurred pseudorandomly every 12 - 22.5 sec.

Data from 41 children were used in the analysis. Subjects were rejected for a variety of reasons including medical, technical, or behavioral problems. At least 10 artifact-free trials were required for individual subject data to be included in the analysis. Only 13 children under 36 months of age satisfied criteria for inclusion. Slow wave (SW) amplitude was computed as the mean voltage of a 1 sec interval preceding the blackout period relative to a 350 msec pre-tone baseline. Statistical analysis was limited to the vertex in order to maximize the power of tests given a fixed α level. A stepwise multivariate procedure was used to evaluate results (see Otto et al., 1981).

RESULTS

Table I summarizes the results of the regression analysis. SW voltage varied as a linear function of PbB, but the slope of this function changed systematically with age. For children under 5 years of age, SW voltage tended to be positive at low PbB levels, but to be increasingly negative above 30 μg/dl. For children over 5 years of age, SW voltage tended to be negative at low PbB levels and less negative or even positive above 30 μg/dl. The interactive effects of age and PbB on SW voltage are illustrated in Fig. 1.

Table I. Results of Multivariate Regression Analysis

SOURCE	DF(num)	DF(den)	F	p <
Demographics	3	37	1.66	.20
Age Main Effect	3	37	.27	.85
PbB Main Effect	3	37	2.49	.08
Age x PbB Interaction	4	36	3.08	.028

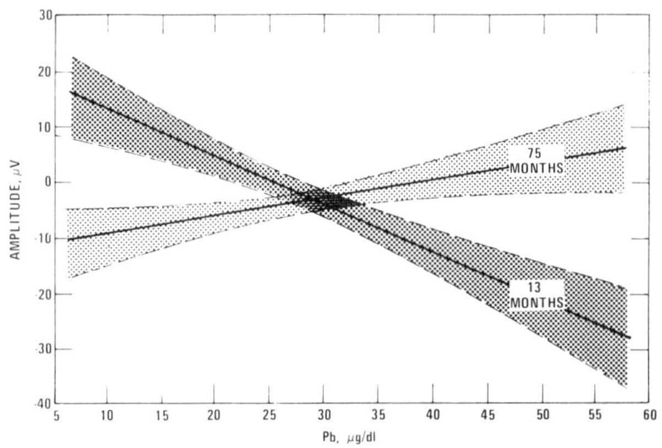

Fig. 1. Interactive effects of age and PbB levels on slow wave voltage during tone-shutter conditioning. Predicted SW amplitude and 95% confidence bounds are depicted in 13 and 75 month old children as a function of PbB. (Redrawn from Electroenceph. Clin. Neurophysiol., 1981, 52:235.)

TWO-YEAR FOLLOW-UP ASSESSMENT

Do the effects of early lead exposure in children persist over time? To answer this question with respect to slow wave voltage, a subset of the children was brought back for a two-year follow-up assessment. Several minor changes in procedures were made: more trials were recorded in the sensory conditioning paradigm to improve the signal-to-noise ratio of slow potential averages and a video cassette system was used in lieu of a movie projector to present cartoons.

Twenty-eight children (19 boys) aged 35 to 95 months participated in the follow-up study. The ranges of PbB levels were 6-55 μg/dl in the original study and 14-39 μg/dl in the follow-up study. Only three children were found to have PbBs > 30 μg/dl at follow-up. Other procedures were as described above.

TABLE II. MULTIVARIATE REGRESSION ANALYSIS OF AGE, BLOOD LEAD LEVEL (PbB1), AND SLOW WAVE VOLTAGE DURING SENSORY CONDITIONING

Source	df(num,den)	F	p<
Overall Multivariate test	6,46	3.05	0.014
All trials (1-100)			
Age	1,24	0.56	0.47
PbB1	1,24	6.97	0.015
Age X PbB1 Interaction	1,24	4.60	0.05
Early trials (1-40)			
Age	1,24	0.38	0.55
PbB1	1,24	0.09	0.76
Age X PbB1 Interaction	1,24	0.02	0.90
Late trials (41-100)			
Age	1,24	2.64	0.12
PbB1	1,24	6.23	0.02
Age X PbB1 Interaction	1,24	6.39	0.02

Table II shows the results of the multivariate analysis of slow wave voltage recorded at the vertex. SW voltage was regressed on PbB1 values obtained in the original study in order to determine if there was any residual effect of early exposure. A significant age x PbB1 interaction was observed in SW measures derived from individual averages across all artifact-free trials (1-100). The main effect of PbB1, but not of age, was also significant. The data were broken down into early (1-40) and late (41-100) trials to assess the time course of the overall effect. This breakdown revealed that the effects were present in the late trials only. Fig. 2 shows predicted SW voltage and 95% confidence bands plotted as a function of original PbB levels for the extreme ages represented in the follow-up sample.

DISCUSSION

SW voltage varied as a linear function of PbB level within an observed range of 7-52 μg/dl. The relationship of PbB and SW voltage is of particular interest since the data suggest that even very low levels of body Pb burden affect this parameter of brain function, i.e., a 'no effect' threshold does not appear to exist. These data represent the lowest PbB

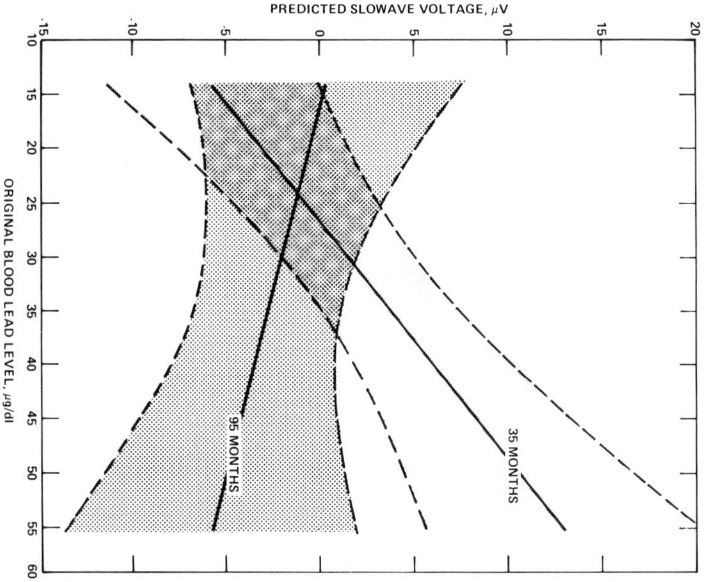

Fig. 2. Predicted SW voltage and 95% confidence bounds in 35- and 95 month old children in two-year follow-up group plotted as a function of original blood Pb level.

effect level ever observed in relation to CNS function. The only other biological function known to be affected at such low PbB levels is heme synthesis. Delta-aminolevulinic acid dehydratase activity is inhibited in children and adults at 10 μg/dl (Hernberg et al. 1970).

The slow wave observed in this study is presumed to be an index of sensory conditioning. The results, therefore, imply that Pb exposure may alter conditioning or learning in children. The effect of Pb on this electrophysiological parameter has not been studied in animal models, although impairments in operant conditioning or maze learning have been reported in several species exposed to Pb (USEPA 1977). Whether the effects are due to a general learning or specific sensory deficit has not been resolved. Fox et al. (1977) observed increased latencies in both primary and secondary components of visual evoked potentials (VEPs) of rats exposed neonatally to Pb. Results of Fox et al. suggest that both sensory and associative levels of CNS function are affected by Pb exposure since primary visual components are considered to reflect activity in specific sensory pathways, while sec-ondary visual components are considered to reflect activity in nonspecific pathways associated with attention, perception and learning.

Results of the two-year follow-up assessment suggest that the linear relationship of SW voltage and body Pb burden persists over time, despite a considerable drop in the mean PbB level of the children. These findings coupled with our original observations provide compelling support for the continuum view of Pb effects on central nervous system function. The current conception of a "safe threshold" for body Pb burden in children clearly requires a careful reappraisal. These results also reinforce the need for rigorous monitoring of worker hygiene habits to prevent the inadvertent transmission of Pb dust into the home with the consequent risk of central nervous damage to young children.

ACKNOWLEDGEMENT

Supported by NIEHS Grant #ES-01104, EPA Contract #68-02-1702, USPHS Grant #HD-03110 to the Child Development Research Institute; MCH Project 916 to the Division for Disorders of Development and Learning, and the Wake County Public Health Department. The cooperation of S. Schroeder, C. Milar, J. Dolcourt and the technical assistance of J. Nichols, J. Prah, and k. Seiple are gratefully acknowledged.

REFERENCES

Benignus, V.A., D.A. Otto, K.E. Muller and K.J. Seiple (1981). Electroenceph. Clin. Neurophysiol., 52, 240-248.

Delves, H.T. (1970). Analyst, 95, 431-438.

Dolcourt, J.L., H.J. Hamrick, L.A. O'Tuama, J. Wooten and E.L. Baker (1978). Pediatrics, 62, 563-566.

Fox, D.A., J.P. Lewkowski and G.P. Cooper (1977). Toxicol. Appl. Pharmacol., 40, 449-461.

Hernberg, S., J. Nikkanen, G. Mellen and J. Lilus (1970). Arch. Environ. Hlth, 21, 140-145.

Needleman, H.L., ed. (1980). Low Level Lead Exposure: The Clinical Implications of Current Research, Raven Press, New York.

Otto, D.A., V.A. Benignus, K.E. Muller and C.N. Barton (1981). Electroencephal. Clin. Neurophysiol., 52:229-239.

U.S. Environmental Protection Agency (1977). Air Quality Criteria for Lead, EPA-600/8-77--017. U.S. Govt. Printing Office, Washington, DC.

Relation Between Subjective Symptoms and Psychometric Test Results

P. Grandjean*, J. Beckmann** and G. Ditlev**

*Institute of Community Health, Odense University School of Medicine, Odense, Denmark
**Department of Clinical Psychology, Odense University Hospital, Odense, Denmark

ABSTRACT

Three approaches to research in human neurotoxicity are outlined: psychometric tests, symptom questionnaires, and the phenomenological approach. The advantages and shortcomings of the first two methods are illustrated by a study of lead neurotoxicity in adult men.

KEYWORDS

Lead exposure; neurotoxicity; psychometric tests; symptom questionnaire.

INTRODUCTION

Current studies of neurotoxicity in occupational health generally use neurobehavioral methods, such as psychometric tests or neurophysiological techniques. This paper will review and discuss three theoretical approaches to neurobehavioral research and will examine in detail two approaches by a concrete, empirical analysis of neurotoxic effects in lead-exposed men.

The most frequent approach to studies of human neurotoxicity attempts to combine, in a scientifically acceptable way, methods which describe bio-physicalistic phenomena and methods which describe behavior. A result of this combination is the psychometric test. The applicability and interpretation of such tests are hampered by two theoretical problems (Beckmann et al., 1981). The basic problem relates to the successful combination of methods: the validity, or "meaning", of psychometric tests is rarely critically examined, even though the underlying theoretical basis may be questionnable. Second, the scientific profile of psychometric tests is characterized by the methods from which they originated: within the objectivistic science paradigm, only objects may be examined. The basic scientific theory indicates that only the objectifying structuring of the item under study guarantees objectivity and scientific quality, because physical phenomena are public, subject-independent and may be studied by objective or intersubjective observation.

The second approach relates to the use of symptom questionnaires. The basic theoretical problem in this regard again refers to the combination of objectivistic methods with questionnaire techniques. However, this approach may not result in any deep, genuine insight into the phenomenological aspect of neurotoxicity, i.e., the individual interpretation of the dysfunctions experienced. The major shortcoming of questionnaires is that they are external and atomistic. Thus, they rather represent a refinement within the objectivistic science paradigm. Again, the phenomena are reduced to sets of physical processes or conditions, and psychology is seen from a behavioristic or physiological viewpoint. By definition, the phenomenological aspect has been eliminated. The questionnaires focus on single, selected, even isolated items, not on complex behavior mediated through individual patterns of interpretation.

Table 1. Correlations (Spearman's r_S) between neuropsychological test results and lead concentrations in blood for 67 men.

Test	r_S
Wechsler Adult Intelligence Scale:	
Information	0.22
Comprehension	0.34**
Arithmetic	0.30*
Similarities	0,55***
Digit Span	0.11
Vocabulary	0.45***
Digit Symbol	0.32**
Picture Completion	0.33**
Block Design	0.38**
Picture Arrangement	0.38**
Object Assembly	0.26*
Verbal IQ, raw score	0.40***
Verbal IQ, age adjusted	0.35**
Performance IQ, raw score	0.51***
Performance IQ, age adjusted	0.52***
Total IQ, raw score	0.50***
Total IQ, age adjusted	0.46***
Visual Gestalts:	
Learning	0.18
Reproduction	0.46***
Word Pairs:	
Learning	0.27*
Reproduction	0.18
Repetition of Sentence	0.09
Story Recall	0.20
Digit Learning	0.50***
Graphic Continuous Performance (CPG):	
I, Errors	0.14
I, Time	0.35**
II,Errors	0.14
II,Time	0.26*
Finger Tapping:	
Preferred hand	0.16
Non-preferred hand	0.34**

Two-sided probabilities of the null hypothesis:
*, p<0.05; **, p<0.01; ***, p<0.001.

The third approach appears as a consequence of the shortcomings of the methods previously discussed. The basic theoretical problem in this case consists of two questions. First, can phenomenological methods be developed in a scientifically acceptable way? Such methods would have to be based on a different interpretation of objectivity than that of the objectivistic science paradigm. Second, how can objectivistic and phenomenological methods be combined: are they complementary, incommensurable, or what? No immediate answers are available at this time.
We have employed the first two approaches to study lead neurotoxicity in an attempt to provide an example for further discussion.

METHODS AND FINDINGS.

In a cross-sectional study, we have examined 67 men with different degrees of occupational lead exposure. The study population was identified in co-operation with shop stewards and safety committees at six facilities with various levels of lead exposure: two battery plants, an automobile radiator repair shop, a lead rolling mill, a non-ferrous foundry and an oil mill. The medium age was 32 years, with a range from 18 to 50 years. The duration of occupational lead exposure in the five first-mentioned facilities varied from one month to 25 years (median, 2 years). No excess lead exposure occurred in the oil mill. The overall median blood lead concentration was 38 ug/100 ml (range, 10-99 ug/100 ml). The blood lead levels correlated closely with the concentrations of lead in hair and of zinc protoporphyrin in blood.

Table 2. Correlations (Spearman's r_S) between questionnaire answers about symptom severity and increasing lead concentrations in blood for 67 men.

Question	r_S
1. How do you get along with your immediate family?	-0.01
2. How is your psychological state of health?	0.06
3. How would you describe your mood?	0.02
4. Do you easily lose your temper?	-0.04
5. How is you self-control when something happens which you have reason to become angry about?	-0.03
6. Do you tire easily?	0.02
7. Do you have trouble sleeping?	0.08
8. How is your concentration?	0.10
9. How is your memory?	0.19
10. Do you often repeat yourself?	0.16
11. How is your ability to find your way around?	0.04
12. How well do you remember new things compared to something which happened long ago (e.g. your childhood)?	0.32**
13. Have your alcohol habits changed?	-0.23(*)
14. Do you get more easily affected by alcohol now than before?	0.20(*)
15. Are you satisfied with your life?	0.13

Two-sided probabilities of the null hypotesis:
(*), $p<0.1$; **, $p<0.01$.

The psychometric test battery consisted of a complete Wechsler Adult Intelligence Scale and seven neuropsychological tests, as described previously (Grandjean et al., 1978). The symptom questionnaire included fifteen questions relating to early symptoms of organic brain damage. Each question was scored on a scale from 0 (excellent) to 9 (severe). Both symptoms and test performance were assessed by a psychologist who was unaware of the lead exposure levels and durations.

The majority of psychometric tests correlated significantly with the exposure tests, with the best predictors of psychological dysfunction being blood lead and zinc protoporphyrin (Table 1). These results indicate that increased lead exposures cause rather generalized dysfunctions, including difficulties in transferring verbal and visual material from short-term to long-term memory; deficient verbal and visuospatial abstraction and retrieval; and decreased psychomotor speed (Grandjean et al., 1978).

In contrast to the statistically significant relationship between lead exposure and psychometric results, answers to the symptom questionnaire were not closely related to the lead exposure (Table 2). Only one question correlated significantly with the blood lead level: "How well do you remember new things compared to something which happened long ago (e.g. your childhood)?" Thus, the dysfunctions documented by the psychometric results were without apparent significance with regard to relevant symptoms, as determined by the questionnaire method.

To further examine this question, two different statistical approaches have been used to study the relationship between symptoms and psychometric test results. One method employed is the multiple regression analysis, even though the distribution of observations was non-Gaussian. The other method employed, the Automatic Interaction Detector (AID), identifies what dichotomous split on which psychometric

Table 3. Psychometric tests as predictors for questionnaire answers, determined by Automatic Interaction Detector.

Question (as in Table 2)	Best predictor (tests as in Table 1)
1. Family interaction	GCP II Time *
2. Psychological well-being	GCP II Errors *
3. Mood	Performance IQ
4. Temper	GCP II Errors *
5. Self-control	Information
6. Fatigue	GCP II Errors *
7. Sleep problems	GCP II Time
8. Concentration	Digit Learning
9. Memory	Picture Arrangement *
10. Repetitivenes	Verbal IQ
11. Orientation	GCP I Time *
12. Memory for events	Digit Learning
13. Change of alcohol habits	GCP II Errors *
14. Alcohol susceptibility	Visual Gestalts: Learning *
15. Satisfaction with life	Word Pairs: Reproduction

*, Best predictor, also, by multiple regression analysis.

test will give a maximum improvement in the ability to predict answers of the symptom question (Sonquist et al., 1974). As indicated in Table 3, the two methods agreed in eight of fifteen cases. The Graphic Continuous Performance test was found to be the best predictor for seven questions. The AID may in addition be used to obtain a family of binary splits for each of the symptom questions. Three of the most relevant questions were selected for detailed study: sleep problems, concentration and memory, symptoms frequently related to lead neurotoxicity. The binary split analyses are shown in Figs 1-3. According to the general impression of the modalities which are measured by the psychometric tests identified as significant predictors by the AID, the relations to the three symptoms examined appear to be relevant. Performance in the tests identified on Figs 1-3 may certainly be decreased by insomnia, lack of concentration or memory impairment. Thus, despite the fact that subjective symptoms do not appear to be closely related to lead exposure, as indicated by the blood lead concentration, relevant symptoms are slightly, but significantly related to the results obtained by psychometric tests.

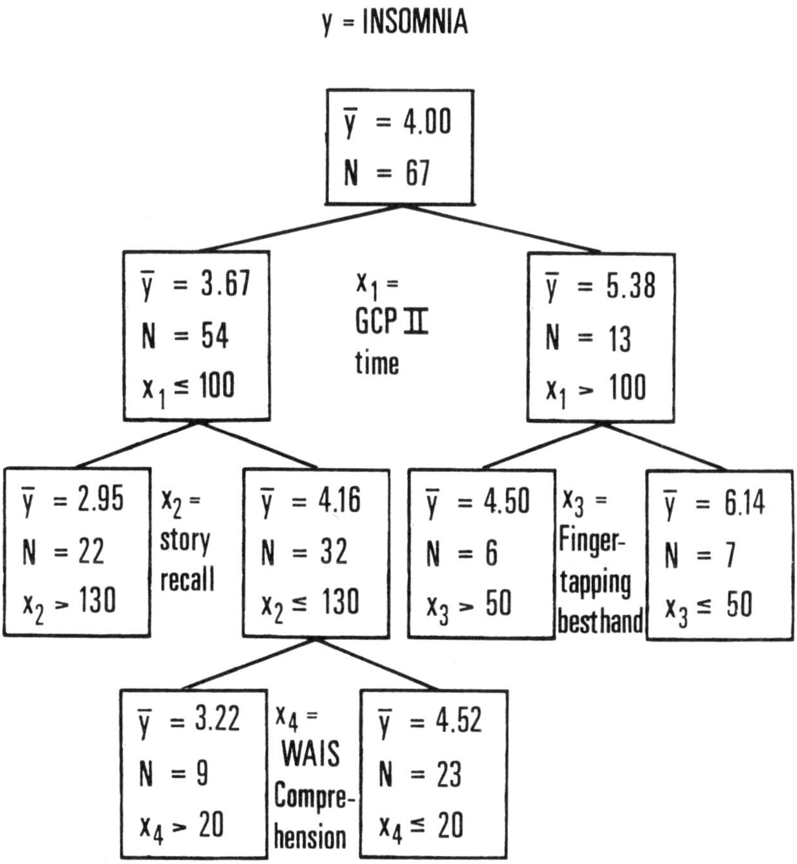

Fig, 1. Binary split analysis by Automatic Interaction Detector for answer to symptom question on sleep problems.

DISCUSSION AND CONCLUSIONS

Several studies have shown that increased lead exposures are related to impaired performance in psychometric tests (Grandjean et al., 1978; Hänninen et al., 1978; Valciukas et al., 1978). On the other hand, symptoms do not appear to be as closely related to lead exposures below the levels causing frank intoxication (Table 2). This lack of relationship could be due to several factors. In a chronic exposure situation, the individual perception of insidious changes may be a rather insensitive measure. A

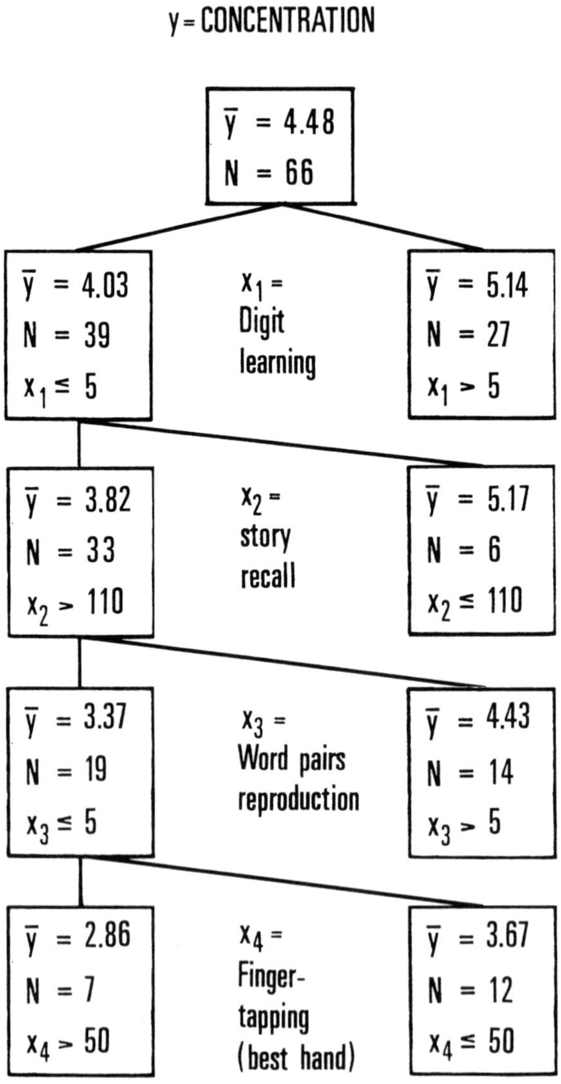

Fig. 2. Binary split analysis by Automatic Interaction Detector for answer to symptom question on concentration.

psychological adaptation to slight toxicity may develop, and defense mechanisms (repression, denial, rationalization, projection) may prevent perceived symptoms from being expressed. Also, the cognitive dysfunctions suggested by the psychometric test results may prevent symptoms from being perceived, remembered or expressed. Thus, one conclusion from this study is that psychometric tests may provide an early indicator of lead neurotoxicity. Symptoms, on the other hand, as assessed by the questionnaire used in this study, are subject to much variability which could dwarf any early effects of lead.

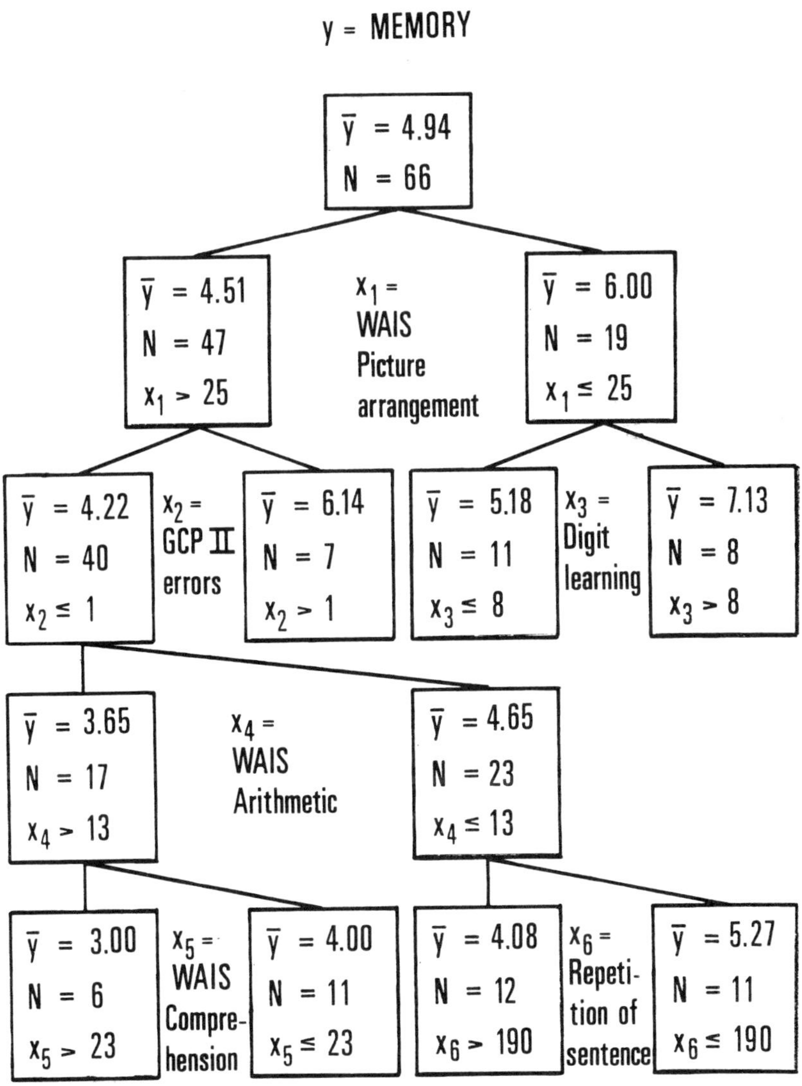

Fig. 3. Binary split analysis by Automatic Interaction Detector for answer to symptom question on memory.

The significant relationship between test results and symptom scores is of considerable interest. As discussed above, psychometric tests have generally been developed on little, if any, theoretical basis. The "meaning" of a test result is a matter of professional interpretation or a correlation with psychiatric diagnoses (Theilgaard and Beckmann, 1971). The fact that dysfunction in psychometric tests is significantly related to scores of relevant symptom questions adds considerably to the belief that such tests are valid, though objectivistic, indicators of neurotoxicity. Both symptoms and signs have, however, been assessed as single, selected and scattered items by standardized methods. Neither set of parameters offers any in-depth understanding of neurotoxicity from a phenomenological point of view. In the absence of any methods providing this approach to neurotoxicological studies, the methods currently available need to be critically examined. Further methodological studies in this area should be encouraged. A better understanding of the phenomenological aspects of neurotoxicity and the significance and validity of objective findings may be of importance for two reasons. First, this insight may help in selecting relevant, sensitive screening methods for early detection and prevention of neurotoxicity. Second, crucial information may be provided in order to define an "adverse" effect. Given the current state of the art, these two questions badly need to be solved.

REFERENCES

Beckmann, J., G. Ditlev and M. Blichert-Toft (1981). Quality of life - theoretical considerations. Proceedings of the First EORTC Quality of Life Workshop, Amsterdam, 22 May, 1981.

Grandjean, P., E. Arnvig and J. Beckmann (1978). Scand. J. Work Environ. Health 4, 295-303.

Hänninen, H., S. Hernberg, P. Mantere, R. Vesanto and M. Jalkanen (1978). J. Occup. Med. 20, 683-689.

Sonquist, J.A., E.L. Baker and J. N. Morgan (1974). Searching for Structure, Institute for Social Research, Ann Arbor, Michigan.

Theilgaard, A., and J. Beckmann (1971). Acta Psychiatr. Scand. 47, 186-205.

Valciukas, J.A., R. Lilis, J. Eisinger, W.E. Blumberg, A. Fischbein and I.J. Selikoff (1978). Int. Arch. Occup. Environ. Health 41, 217-236.

Concluding Remarks

R. Gilioli

Institute of Occupational Health, Department of Preventive Medicine, University of Milan, Italy

Drawing the conclusions of a symposium so rich in new ideas and issues relevant to occupational health is a difficult task indeed. A vast amount of information has been presented and thoroughly examined by the most eminent experts in the field of neurobehavioral studies; critical analysis of the study design, methods and applicability has been extensive, so not much is left to say for the conclusions and I must ask the audience to excuse me if there are any repetitions.

The participants in the conference may have been left with the impression that, more than stating firm points, new questions and doubts have been raised, but also that new issues and strategies have been proposed.

We all feel that neurobehavioral toxicology has reached a critical moment of its development.

Personally, I had the chance of witnessing the early days of this discipline and the pioneering efforts of many colleagues present at this symposium. Dr. Xintaras and Dr. Johnson organized the first conference to verify concepts, methods and results in Cincinnati ten years ago and this was the first important occasion to lay down strategies to develop the new area. Over the last ten years, considerable progress in neurologic, neurophysiologic, neuroradiologic and behavioral methods has been made.

However, most of these methods were confined to neurologic and psychiatric diagnoses. Therefore, we felt that the time had come to discuss and review the methods so as to assess their applicability also for the surveillance of workers exposed to neurotoxic workplace chemicals: we are well aware that the requirements the methods must meet differ in the case of clinical diagnoses as opposed to the study of people at work. This point was clearly dealt with by Dr. Xintaras from the World Health Organization, and by Dr. Suma'mur who spoke on behalf of developing countries and who was particularly worried about the complexity of some of the methods proposed in view of the limited resources available in most developing countries. It became apparent that the problems to be faced were manifold:

1. *the issue of the selective and rigorous criteria* for screening methods for application in field studies: in fact, these methods must comply with some basic requirements: - they should be apt to clearly show neurologic and/or behavioral dysfunctions; - they should be simple and easily applicable; - they should be non-invasive and with a favorable cost-benefit relationship; - they should be sensitive, specific and reproducible.

It must be immediately acknowledged that it is almost impossible to avail ourselves of a single neurobehavioral method that entirely meets these requirements. In fact, most methods proposed are highly sensitive, but lack simplicity; the specificity of the methods depends on the neurologic function they investigate. In fact, the nervous system has a pattern of response to any insult, either toxic, or vascular, traumatic or tumoral, that is directly related to the neurologic level involved.
However, there are chemicals that are active mainly or solely on the peripheral nerves and, within the nerve fiber, more on the axon that on the myelin sheath. Electromyography is specific in detecting axonal neuropathy and nerve conduction tests for myelinopathy. With regard to the problem of simplicity, the brain is such a complex organ that hardly any simple method can be devised to obtain informative data. This is why methodologies have developed to the most sensitive ones showing a high power of resolution and diagnostic penetration.
By adopting the selective criteria previously mentioned, only few methods could be accepted and even electromyography and nerve conduction tests that have been extensively validated in occupational health may not pass through the filters of all the requirements.
Most of the methods have been validated in other disciplines for diagnostic purposes. However, if we want to advance to a better understanding of the neurobehavioral changes, a progressive appraisal and introduction of the methods in occupational health,too, must be seriously considered, since it is doubtful whether we can rely only on the data from the history and the neurologic examination. The data collected through the history, especially the subjective symptoms, are sometimes intriguing being often influenced by suggestion and contaminated by other interferences. On the other hand, in order to be effective and informative, the clinical examination must be carried out by highly skilled physicians and must systematically investigate all the different neurologic functions as well as behavior; this is also quite laborious and time-consuming.
I recall the lecture given by Dr. Juntunen who thoroughly reviewed all the steps in clinical diagnosis: only the evaluation of the mental status to identify an organic brain syndrome implies the examination of 22 different items ranging from the assessment of the level of consciousness to the appraisal of behavior, memory, language, constructional abilities and higher cognitive processes.
It is impossible to review all the methods discussed now in the light of the mentioned criteria. Therefore, I will discuss some other critical issues.

2. *The problem of standardization* is of crucial importance. In the past, there has been much inconclusive debate on this topic. Most laboratories elaborate their own values and the correlations between them are insufficient. The task of standardizing neurobehavioral methods is a very difficult one. International agencies such as the WHO, the Commission of the European Communities and other scientific organizations such as the Permanent Commission and International Association on Occupational Health should promote programmes to this end, and I was very pleased to hear that for the WHO this is a priority programme.
However, standardization has limitations inherent in the parameters.
The role played by local conditions,geographical pattern, climatic conditions, nutritional status, epidemic diseases, genetic background, etc. is often unknown and should be clarified. In fact, we know that the biological response to the same toxic insult may vary in relation with many factors; for instance, carbon disulfide is reported to induce retinal micro-aneurisms in the Japanese, but not in other ethnic groups, and we heard from Dr. Seppäläinen that even nerve conduction rate variability is related to the gene-

tic background and individual height.

3. *Significance of neurobehavioral data in terms of health impairment.* One of the sessions in Milan was devoted to the comprehensive approach to the study of exposed populations. In fact, when we organized the symposium it was clear to us that a comprehensive approach is probably the only one which may yield any meaningful results.
I feel that the issue of the significance of the results should be considered in this light. Is there any possibility that single tests offer meaningful data? This possibility exists when the parameters are altered beyond doubt. This is not the most common case in monitoring procedures, where we obtain results which are often borderline or slightly altered.
What then is the meaning of a nerve conduction rate of 41 m/s, at the lower limit of normal variability? If the baseline profile is unknown, it is impossible to ascribe this finding to any change that has occurred. Moreover, it is well known that, because of its extensive compensation capacities, the nervous system has a wide range of safety for its functions.
Does a value of 41 m/s of nerve conduction rate imply a reduction of the efficiency of the system?
The answer is negative since,at this speed, the function of nerve transmission is preserved.
However, this finding can be regarded as a reduction of the range of safety of the function; this is what we might refer to as a new level of adaptation of the nervous system to changed conditions as recalled by Prof. Foà, but this adaptation to lower levels implies reduced response capacities under extreme conditions.
The same interpretation seems to be applicable to other neurophysiological and psychometric tests such as reaction times and, perhaps, also to EEG spectral analysis. In this latter instance, however, the problem is open to other interpretations since the increase in the EEG slow activity power that has been found in some studies, may account for a decreased level of arousal and subsequent alertness. As regards the interpretation of psychological tests, a low score of a single function does not necessarily imply an adverse effect.
However, when several testing performances such as the measurement of intellectual dynamism, the mnestic profile and reaction times are consistently reduced, it seems justified to consider these findings as adverse effects, especially if the subject is aware of a change in his inner experiences and shows modifications in his social adjustment.

4. *The issue of the exposure-response relationship* can be looked at in two ways: firstly, in the light of the usefulness of curve determinations for validating the methods and secondly for establishing the effects of toxic exposure.
Neurobehavioral studies utilizing dose-response curves are becoming increasingly frequent, as demonstrated also during this symposium.
The lack of these data is so important as to prevent the establishment of causality relationships, or, at least, to make it very difficult.

5. *The role played by confounding or modifying factors* is also quite well-known; confounding factors distort the exposure-response relationship while such relationship is only modified by the group of the so-called modifiers. There is a vast number of confounders or modifiers that must be taken into account and controlled during the data analysis stage. It is impossible to list all the confounders.
Alcohol consumption and the stress syndrome interfere heavily in the neurologic and psychologic changes and this must be adequately appreciated.

To sum up, critical analysis of the methods discussed revealed some inadequacies vis-à-vis an immediate utilization of the methods in field studies, but at the same time opened the way for new strategies. Clinical methods, electromyography, brain evoked potentials, EEG spectral analysis, optacon, electrooculography, regional cerebral flow, computed tomography, reaction times, neurotoxic esterase, tremometry, pupillography, computed recording of biological functions and psychological test batteries have been exaustively examined and scrutinized. For pilot studies, all these methods showed their usefulness. I believe that for the majority of these methods it is premature to propose their extensive application in neurobehavioral surveillance programmes. However, the ones that deserve special attention, besides EMG, reaction times and psychological tests which have already been accepted, are the brain evoked potentials, EEG spectral analysis, electrooculography and tremometry. Therefore, the analysis made maintains full validity, serving as basis for the proposal of future applications. The situation is rapidly evolving. Methods that were necessarily invasive only recently can now be performed with non-traumatic techniques.

The problem of making instrumentation compact and multi-functional is under study. It is likely that within the next few years we will witness a progress in the quality and use of the neurobehavioral methods.

To this end, close cooperation is necessary at an international level, and I recall again the need for coordination of efforts by WHO, CEC and the Permanent Commission.

This cooperation should cover both analysis of the methodologies and identification of new areas of high priority research, among which I recall the problems of organic solvents, standardization and confounding factors such as the stress syndrome.

It is only by the combined efforts of the coordinating Institutions and research workers that we will see a further development of our discipline for the benefit of workers' health.

Subject Index

Entries refer to first page of relevant article